basic principles of electricity

basic principles of electricity

Vester Robinson

Reston Publishing Company, Inc., Reston, Virginia

A Prentice-Hall Company

This book is dedicated to

Florence Belle

RESTON PUBLISHING COMPANY, INC.
Box 547
Reston, Virginia

10 9 8 7 6 5 4 3 2

ISBN: 0-87909-062-6

Library of Congress Catalog Card Number: 72-91115
Printed in the United States of America

preface

Electricity is the life blood of our modern civilization. All areas of technology in our materialistic society are dependent upon electricity. If all electricity could be stopped for twenty-four hours, the world would be gripped with panic. If this condition could be maintained for three weeks, the civilizations of the world would be smashed beyond any possibility of recovery.

Try to imagine life without automobiles, planes, trains, ships, artificial light, refrigeration, television, air conditioning, telephones, and radios. A world without transportation, communications, and industry would be reduced to the cannibalistic state within a few short days. This is the world without electricity.

The requirement for electric power in the United States doubles every ten years. Many modern homes are, so called, all electric. They rely on electricity for heating, lighting, cooling, cooking, communicating, and entertaining. In many places it is difficult to sharpen a pencil or boil a pot of coffee without using electricity. It is small wonder that the ever growing demand for electricians is never satisfied. In fact, the home owner or housewife of today needs to know as much about electricity as the licensed electrician needed to know twenty years ago.

Particular types of electric machines come and go, but the principles of electricity do not change. I have devoted this book to the simplification of the basic principles of electricity in order to bring them within the grasp of any person who is willing to devote a little time to serious study. I have been associated with teaching electricity and closely related subjects for a period of twenty-five years. The insight gained during that time was used to prepare this book.

The basic principles of electricity are clearly defined, explained, and illustrated. Mathematical analysis is used, but the emphasis is on clarity of explanation which will enable you to form a mental picture of the electric action.

A great many circuits are used, explained, and analyzed. Each of these circuits shows one or more practical applications of an electrical principle. In most cases, the circuits are not associated with a particular type of machine for the simple reason that they can be found in a great many different machines.

If you are a student who intends to pursue a career in electrical work, you will find this book of great value. It will move you a long way toward the realization of that career.

If you are a professional electrician, you will find this book an invaluable reference for reviewing details in electrical principles and basic circuits. It is not specifically a handbook, but it can be used as a reliable guide in numerous practical situations.

If your interest is in the direction of electronics, communications, computers, or other related areas, the knowledge of electricity is a very vital stepping stone toward your selected vocation. The contents of this book will enable you to build a solid foundation for your advanced courses.

If you are a teacher in any area of electricity or electronics, you will find this book of tremendous help in reviewing basic concepts, circuit operation, and functional analysis.

In the preparation of a book, credit is due to many people, but how do you give credit for information acquired over a twenty-five year period? I wish to express thanks to the many manufacturers of electric parts and machines for their generous supply of pictures and technical data.

I want to express a very special thanks to the two gentlemen who reviewed this manuscript:

Dean Thomas Kubala
Anne Arundel Community College
Arnold, Maryland

and

Herbert Jackson
Ontario Ministry of Colleges and Universities
Toronto, Ontario, Canada

Their careful reviews and many helpful suggestions contributed immeasurably to the quality of the book.

Vester Robinson

contents

basic principles of electricity

1

electricity

We are all familiar with the effects of electricity. Electricity has been a part of our daily lives for many years. However, until recently, few people could explain what electricity is. We knew the principles of electricity; how to produce it; how to transport it; how to use it; but not what it was. We now know that *practical electricity,* the kind we use, *is produced by small atomic particles* known as electrons. It is the movement of these particles which produce the effects of heat in the toaster and light in the lamp.

The pressure that forces these atomic particles to move, the effects they produce when they encounter opposition, and how these forces are controlled are some of the *principles* of electricity.

THE NATURE OF MATTER

Accepted atomic theory states that *all matter* is *electrical* in structure. Any object you care to name, even your own body, is largely composed of a combination of positive and negative particles of electricity. Electric current will pass through your body, over a wire, or along a stream of impure water. It can be established in some substances more readily than others, that all matter is composed of electric particles despite some basic differences in materials. The science of electricity then must begin with a study of the structure of matter.

atomic theory

Matter is defined as any substance which has mass (or weight) and occupies space. This definition should be broad enough to cover all physical objects in the universe. Wood, water, iron, and paper are some examples of matter. *Energy* is closely related to, but not to be confused with, matter. Energy does not have mass, and it does not occupy space. Heat and light are examples of energy.

The smallest particle of matter which can be recognized as an original substance was thought to be a unit called the *atom.* Recently scientists have found particles even smaller than atoms, but our theories are still based on the atom. The atom consists of a nucleus and a cloud of electrons. It is generally agreed that the electrons are small particles of electricity, which are negative in nature. These particles orbit the nucleus in much the same fashion that planets orbit a sun.

The *nucleus* is composed of protons, neutrons, and other subatomic particles. The *proton* is a relatively heavy positive particle. It has exactly the same quantity of electrical charge as the electron although its *sign* (or value) is opposite. The *proton* weighs the same as approximately 1845 electrons, and the atom contains a like number of protons and electrons. The neutron is so named because it is electrically neutral. That is: it is neither positive nor negative. We mention the neutron because it adds weight to the atom and tends to prevent movement of the protons.

We are now in the heart of the atom, the basic building block of all matter. When we examine the parts of this atom, we find minute particles with electrical charges. More specifically, we find minute particles of positive and negative electricity. The basic difference between lead and gold lies in the number of electrons and protons in the atoms which compose these metals.

The simplest atom consists of a nucleus which contains one proton, which is orbited by a single electron. This is the hydrogen atom. One of the more complex atoms is californium. This atom contains 98 protons and 98 electrons with the electrons orbiting the nucleus in seven different and distinct energy shells.

Hydrogen and californium are examples of *basic elements.* An element cannot be broken down into a simpler substance by a chemical process. It is composed of only one type of atom. There are presently 106 known elements, and several others are likely to be discovered. Of course, an atom can be split. A lot of energy is released while the atom is subdividing into its various subatomic particles. Figure 1-1 is a partial list of the 106 known elements and shows the atomic structure of each.

Atoms	Electrons	Electron shells
Hydrogen	1	1
Helium	2	1
Carbon	6	2
Sodium	11	3
Copper	29	4
Silver	47	5
Cesium	55	6
Francium	87	7

Figure 1-1. Atomic Structure of Some Basic Elements

Since there are only 106 known elements and an almost infinite number of examples of matter, there must be some method of combining atoms of elements to form other substances.

compounds and mixtures

Certain atoms will form a chemical union when they are combined with one another. Such a union results in a new substance which differs from the elements that combined to form the bond. This new substance is called a *compound.*

Hydrogen and oxygen are two gaseous elements. When they are brought together in the proper quantities (two atoms of hydrogen to one atom of oxygen), they combine into the liquid compound we know as water.

The smallest part of a compound which exhibits all the identifying characteristics of that compound is called a *molecule*. A molecule of water was described above: two atoms of hydrogen and one atom of oxygen in a chemical union. In this minute quantity (a single molecule), examination is impossible with our present laboratory instruments, but all the properties of water are present nevertheless.

Figure 1-2 is a poor representation because it is drawn on a flat surface and thus, only shows two dimensions. Actually, the electrons orbit elliptically and take many different paths. The intent here

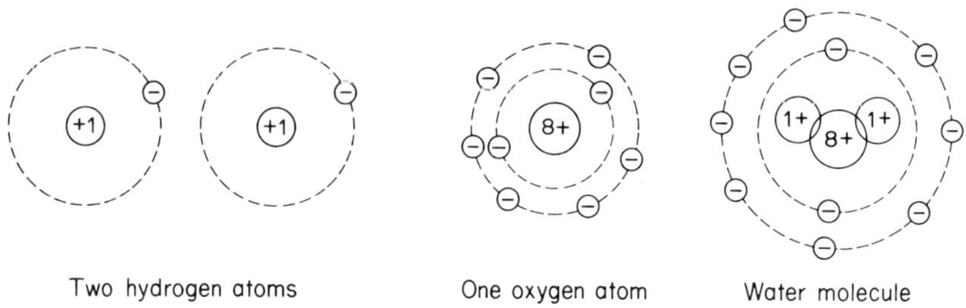

Figure 1-2. Forming a Compound

is only to illustrate the chemical bonding of two elements to form a compound. Figure 1-3 is a three-dimensional drawing that renders a slightly better representation of the orbital patterns.

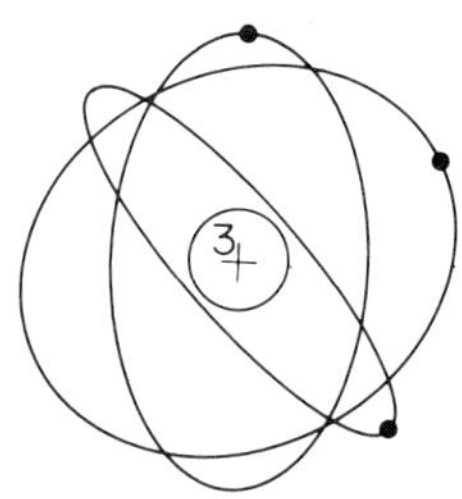

Figure 1-3. Probable Orbits Within a Lithium Atom

When two or more elements are mixed together without forming a chemical bond, the result is called a *mixture*. Oil and water will form a mixture; other elements need to be added in order to change this to a compound. When certain types of detergent are added to a oil-water mixture, a chemical combining will take place.

From the basic building blocks, we have built up to the molecular structure of matter. These molecules are very small. It has been estimated that if a drop of water could be enlarged to the size of the earth, the molecules would be about the size of baseballs. Molecules of different compounds differ only in the number and kind of atoms in their makeup.

Matter exists in *three states: solid, liquid,* and *gas.* Some types of matter may exist in any of the three states. Water is a common example. It is normally liquid; but when frozen, it is solid; and when boiled, it becomes steam (a gas). Melting iron, freezing water and breaking glass are examples of physical changes in matter. A physical change of state alters the activity of the molecules, but not the atomic structure.

STATIC ELECTRICITY

What have we proved by atomic theory? Assuming that the facts are substantially true, we have established that all matter is basically composed of two types of electricity: positive particles and negative particles. We find the negative particles to be relatively light in weight and in constant motion. These orbiting particles exhibit equal and opposite electrical characteristics to the heavier particles within the nucleus.

ions

When an atom has the same number of electrons as it has protons, it exhibits no outward electrical properties. This is because the positive and negative charges are exactly balanced. Such an atom is *electrically stable* and is said to be neutral.

Under certain influences, some atoms can be "persuaded" to part with a few of their electrons for short periods of time. They may also be influenced to accept more than their proper quota of electrons for similar short periods.

When an atom takes on an excess of electrons, it exhibits outward characteristics similar to the electron. It takes on an overall negative property. This condition is called a *negative charge,* and such a charged atom is *not* electrically stable. A *charged atom is called an ion,* and if the charge is negative, it is called a *negative ion.*

An atom which has less than its normal quota of electrons, displays a positive polarity similar to that of the proton due to the fact that it has more positive protons than it has negative electrons. This type of atom is said to assume a *positive electrical charge.* Such an atom is known as a *positive ion* while it is in this electrically unstable condition.

These charges on atoms are the simplest examples of static electricity. We stated that atoms are influenced to accept or give up electrons. When they do so, they become ions, and the process is called *ionization.*

Static electricity, as the name indicates, is *electricity at rest.* The science of electrostatics is concerned with the study of static electricity. It has useful applications, but in the field of electricity, we will be more interested in dynamic electricity. We are primarily concerned with the electricity that blows horns, rings bells, lights lamps, and drives messages to the moon. We'll get to that very soon, but it is important that we first examine some more characteristics of static electricity.

electrostatic charges

From our study of the ion we learned that an electrically unstable atom represents a charged condition. A surplus of electrons means a negative charge, while a shortage of electrons causes a positive charge. These ions are unstable because they are in an unbalanced condition. This instability will be thrown off as soon as the ion has an opportunity to neutralize itself. This is an important point, and we will return to it very soon. But first, let us think in larger terms. Instead of imagining an ion, think in terms of objects large enough to be examined with the naked eye. Positive or negative electrostatic charges can be placed on these bodies in several ways, and this larger charge will exhibit the same characteristics as would an individual ion.

One object that lends itself to such an experiment is a small pith ball suspended by a fine silk thread. When a rod of *amber* is briskly rubbed with a piece of *fur,* the rod takes on a *negative charge* because the fur gives up electrons to the rod. The negative charge can be transferred from the rod to a suspended pith ball (see Figure 1-4).

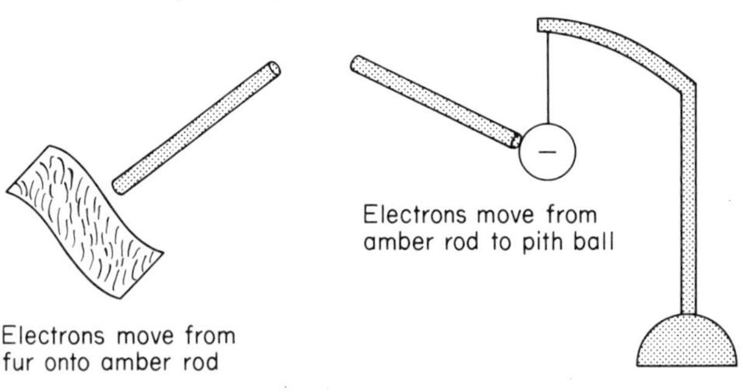

Figure 1-4. Producing and Transferring a Negative Charge

A *positive charge* can be produced by rubbing a *glass* rod with a piece of *silk*. The silk takes electrons from the glass and leaves the glass rod with a positive charge. This charge can be transferred to a suspended pith ball as shown in Figure 1-5.

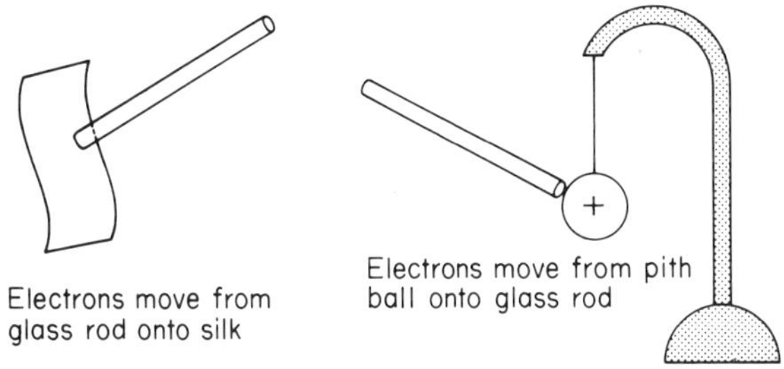

Figure 1-5. Producing and Transferring a Positive Charge

the law of charges

Now that we have examined one method of producing each polarity of charge, it is time to explore some of the characteristics of *charged bodies.* When two suspended pith balls are charged positively, we can examine their reactions to one another. When we move the two charged bodies close together, as in Figure 1-6, they will *repel* each other.

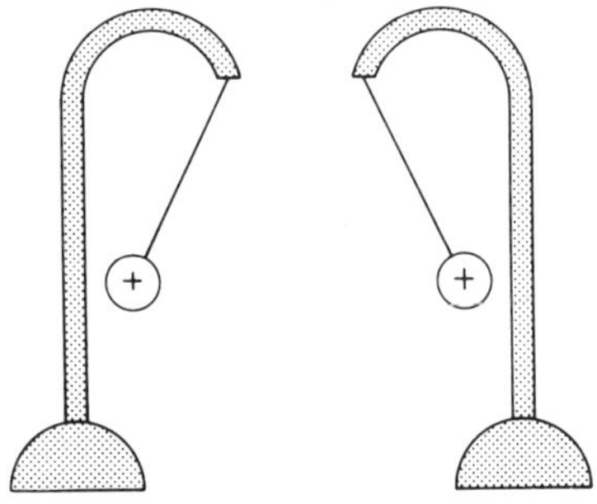

Figure 1-6. Repulsion of Two Positive Charges

Two negative charges will have the same effect on each other under the same circumstances (see Figure 1-7).

When one body is charged positively and the other negatively, they exhibit an *attraction* for each other as shown in Figure 1-8.

The above experiments, though easy to perform, illustrate very clearly the laws of charged bodies. These are: *like charges repel each other,* and *unlike charges attract each other.* These two laws

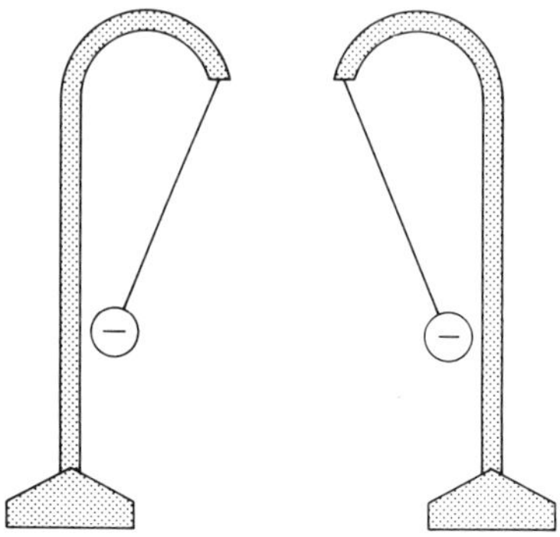

Figure 1-7. Repulsion of Two Negative Charges

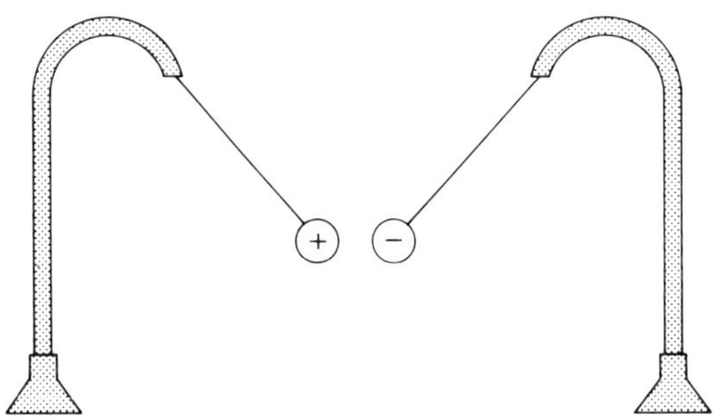

Figure 1-8. Attraction of Two Unlike Charges

may sound simple and self-evident, but they play many important roles in explaining electrical phenomena.

Another method of proving that like charges repel and unlike charges attract is by use of the *electroscope.* Its general appearance is illustrated in Figure 1-9. It is a simple device used to detect the presence of minute charges. It consists of a glass jar with a metal rod inserted through the sealed cap. The metal rod has a detector ball on one end and two gold leaves on the other. When no charge is present, the leaves hang limply as shown in the drawing. This indicates that both leaves are electrically neutral and have no influence on one another.

When an electric charge, either positive or negative, is placed near the detector ball, the leaves spring apart as in Figure 1-10. Whichever charge is placed near the ball, that same charge is transferred to both gold leaves. If both are positive, they repel; if both are negative, they repel. This *force field* which surrounds a charged body, needs to be examined a bit further.

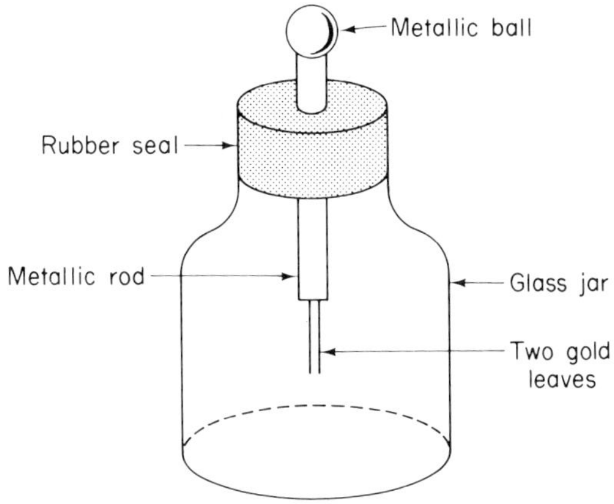

Figure 1-9. The Electroscope Registering No Charge

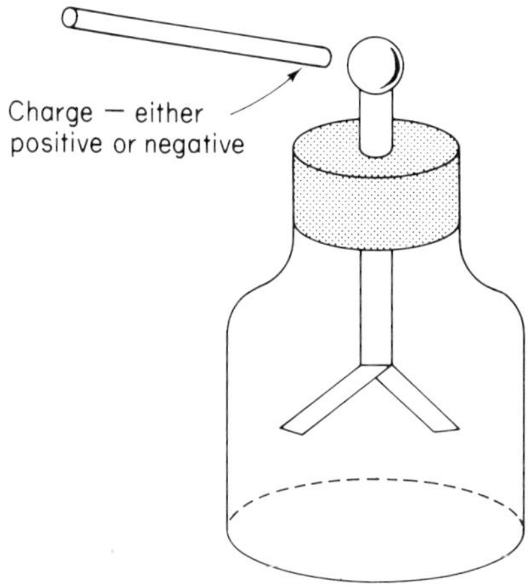

Figure 1-10. The Electroscope Registering Either a Positive or a Negative Charge

electrostatic fields

The area of invisible force that surrounds a charged body is known as an *electrostatic field.* This electrostatic force field extends outward in all directions, and the strength varies inversely with the square of the distance from the charged body. If the body is spherical, the field is round and uniform in nature. By definition, an electrostatic field is the area in the vicinity of a charge wherein its influence can be felt.

Scientists encountered some difficulty in deciding what comprised this invisible force field and in what direction the force was exerted. They have agreed that the field may be considered as being composed of a finite number of invisible *lines of force.* This agreement makes the field more tangible, but what is the direction of these lines? Again, a meeting of the minds was necessary. This time it was decided to give the lines of force the same direction that a positive charge would move if it were placed in the field. We already know that the positive charge would be attracted toward a negative charge and repelled from another positive charge. This establishes the direction of force lines to be in *toward* a negative charge and out *away from* a positive charge. This is illustrated in Figure 1-11. Again, the flat drawing only shows one plane. Imagine

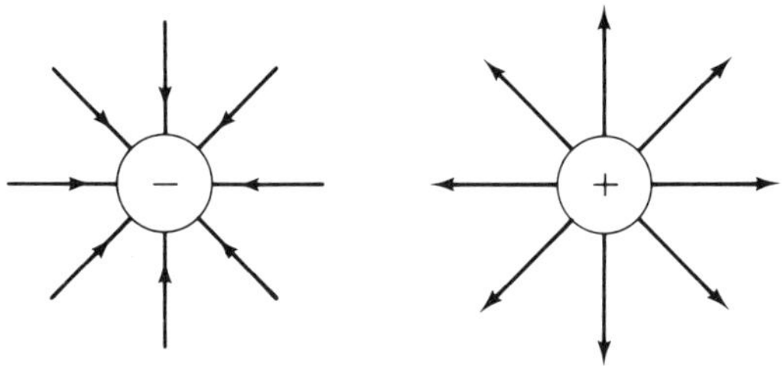

Figure 1-11. Direction of Lines of Force

the lines as pointing in from all directions toward the negative charge and out in all directions from the positive charge.

Examination of the lines of force will reveal why like charges repel and unlike charges attract. Analyze the drawing in Figure 1-12.

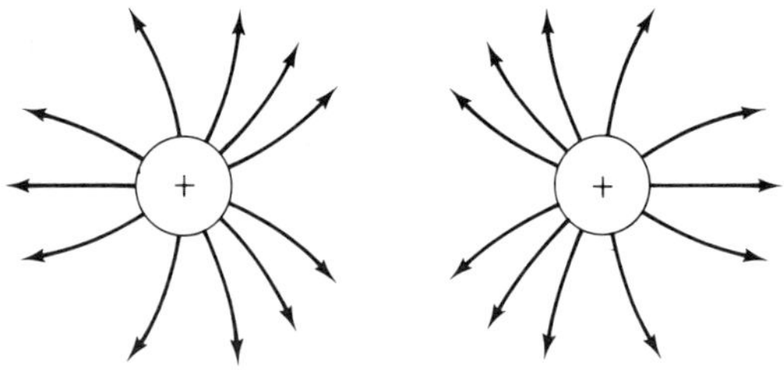

Figure 1-12. Repulsion of Like Charges

Lines of force *will not* cross. Two lines in the same direction exert a repulsive force on one another in order to stay as far apart as possible. A stress is placed on the lines between the two charges when the lines are distorted. This stress is somewhat like the stress resulting from stretching a rubber band. The collective effort of the force lines to return to their natural position exerts a mutual force of repulsion.

This conflict is not present when two opposite charges approach each other. Figure 1-13 illustrates this reaction. Notice that

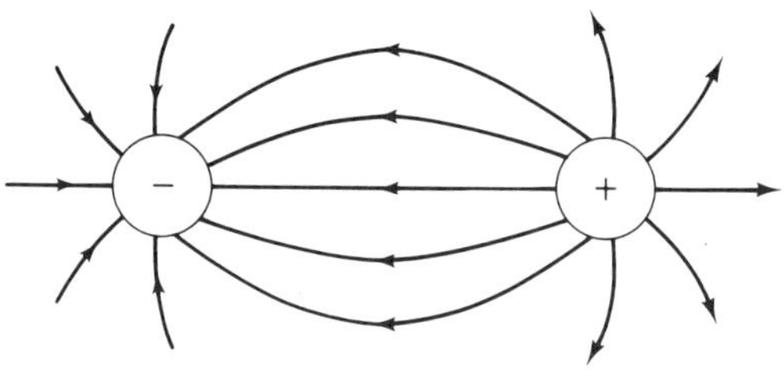

Figure 1-13. Attraction of Unlike Charges

the lines between the charges bulge away from each other. They are still distorted, but the resulting stresses are aiding one another. Between the charges, the lines combine and create a strong field of attraction which tends to pull the two bodies together.

Further study of electrostatic fields by Michael Faraday, an English scientist, established that charges distribute themselves in various patterns on a charged surface. Experiments with variously shaped objects proved that the charge concentrates at the *point of greatest curvature*—that is to say, a pointed object, when charged, will have the greatest charge at its sharpest point. Figure 1-14 shows the distribution of charges on objects of various shapes. This

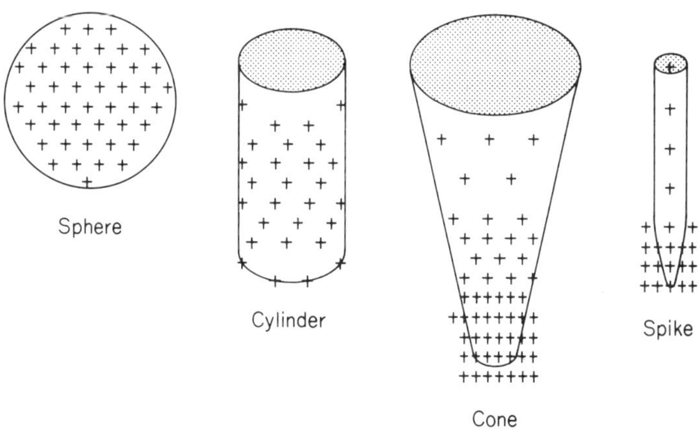

Figure 1-14. Distribution of Charges

drawing shows positive charges, but the same distribution would be present if the charges were negative.

Many people have made valuable contributions to the ever-expanding volume of facts concerning static electricity. C. A. Coulomb, a French physicist (1736–1806), worked at length with charges and the effects of the electrostatic field. It was Coulomb who proved a definite relationship between force, quantity of charge, and distance between charged bodies.

When a body has become charged, some number of electrons must have been *displaced.* The atoms in the charged body are in an electrically unstable condition. At the first opportunity they will correct this unnatural situation.

electrostatic discharge

Imagine that you have two suspended pith balls that are charged with opposite polarities. The charges are equal in magnitude, and the two balls are allowed to touch as shown in Figure 1-15.

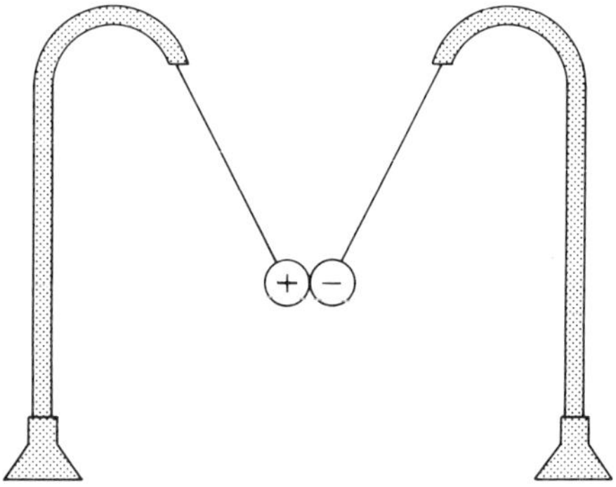

Figure 1-15. Touching Opposite Charges

What happens? The two balls cling together very briefly, then return to a neutral position as demonstrated in Figure 1-16.

Why did this happen? When the balls touched, the excessive electrons on the negatively charged body rushed onto the positively charged body. The *atoms* of both balls became electrically *balanced,* and both polarities returned to the natural (neutral) condition. This process of *neutralizing* a charge is known as *discharging.*

There are other ways to effect an electrostatic discharge. Lightning is a dramatic example. The flash of lightning occurs in an ionized air gap which allows large quantities of electrons to leap

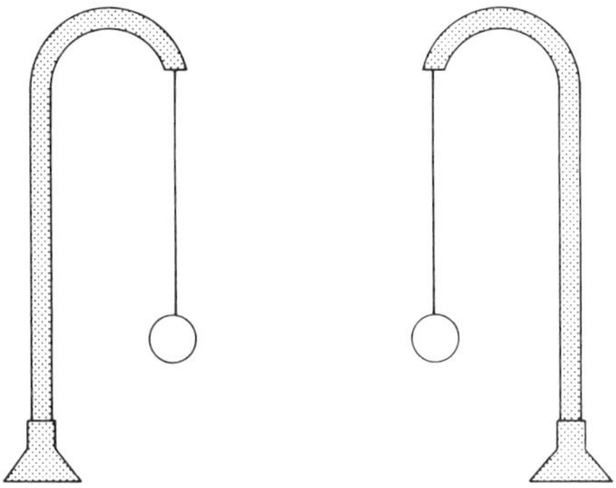

Figure 1-16. Neutralized Pith Balls

between two very large opposite charges. Another method of discharging is to provide a path for the electrons to cross from one body to the other (see Figure 1-17).

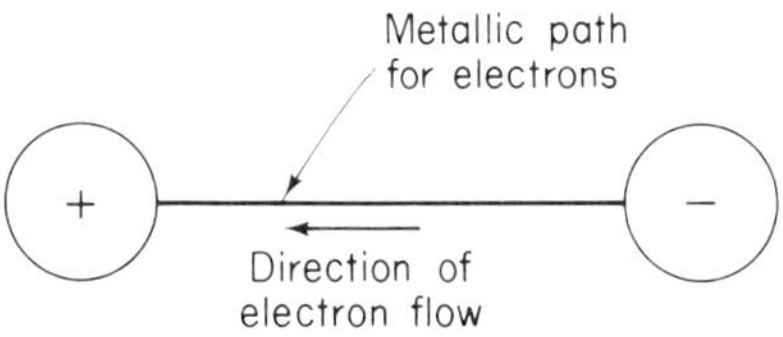

Figure 1-17. Discharge Through an Electron Path

Why do the excessive electrons rush across the path? All electrons have a negative polarity, and like charges repel each other. *Every electron repels every other electron.* This action is aided by the positive charge, which exerts a strong attraction for the negative electrons. When the two charges are equalized, the electrons cease to move along the path.

Electrostatic discharge along a path is getting very close to practical electricity. If we could find a way to increase the quantity of charge and prevent it from equalizing, we would have enough electron flow to do some work over a prolonged period of time. To sustain electron flow through the path, we must find a way to keep the polarities from returning to neutral.

DYNAMIC ELECTRICITY

As the name *dynamic electricity* indicates, this is *electricity in motion.* The name is not of great importance, but it is descriptive.

The heart of the matter is *electron movement.* The important fact is that this is *practical electricity.* Play time and abstract laboratory experiments are over. Let's harness the electron and put it to work.

voltage

In a water system, the water will flow as long as it has a path and sufficient hydraulic pressure. Pressure can be sustained by using a pump to keep moving the water back to a higher level. This principle is shown in Figure 1-18. As the water flows from Tank 1

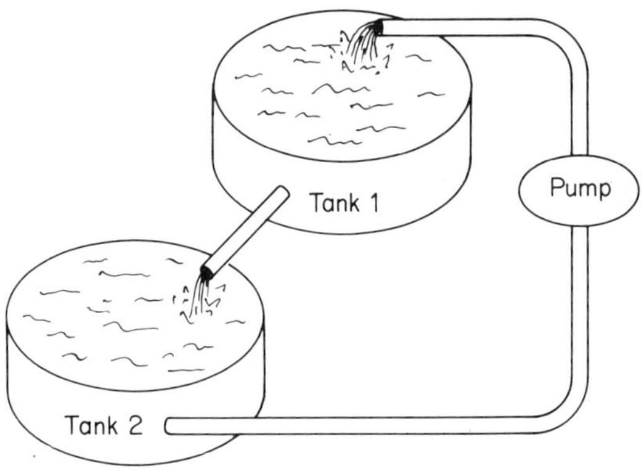

Figure 1-18. Pump Maintaining Water Pressure

to Tank 2, the pump moves the water from Tank 2 back to Tank 1. This keeps a constant pressure in Tank 1. The water will continue to flow at a steady rate as long as the pump works properly.

In an electrical system, *electrical pressure* is needed. To maintain this pressure, we need a device that will move electrons in a way similar to that in which the pump moves water. Such an electron pump is available. Probably the most familiar one is the storage battery in your automobile. This battery contains plates submerged in an electrolyte. The electrolyte is a solution of sulfuric acid and water. The plates are composed of two dissimilar metals, and these plates are charged with opposite polarities by chemical action on the plates. The parts of a storage cell, which is one unit of the storage battery, are shown in Figure 1-19.

One cell produces about two volts. The storage battery is formed by placing three cells in series for a six-volt battery and six cells in series for a twelve-volt battery. (For best results, a battery

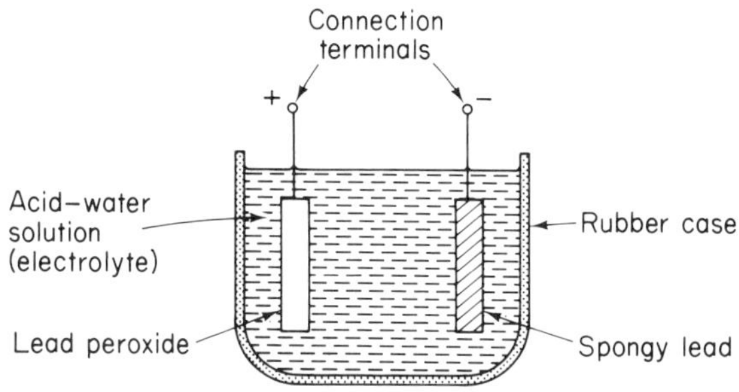

Figure 1-19. Parts of the Storage Cell

must be maintained at peak charge.) The charging process consists of forcing electrons to enter the negative terminal and exit through the positive terminal. This flow is opposite to the direction of electron flow when the battery is being used. Charging a battery is not the same process as developing an electrostatic charge. Nothing is really being changed when a battery is charged; it is actually a restorative process. The charging action restores the original condition of the electrolyte and plates. Through a process of *electrolysis,* acid is returned to the electrolyte, and harmful deposits are removed from the plates. The actual charge (difference in polarity) is a result of chemical action of the electrolyte on the dissimilar metals of the plates.

We have said that the battery will maintain pressure in an electrical system in much the same way that a pump does this in a water system. In Figure 1-19, we saw the internal, working parts of the battery. Figure 1-20 illustrates its action in a circuit.

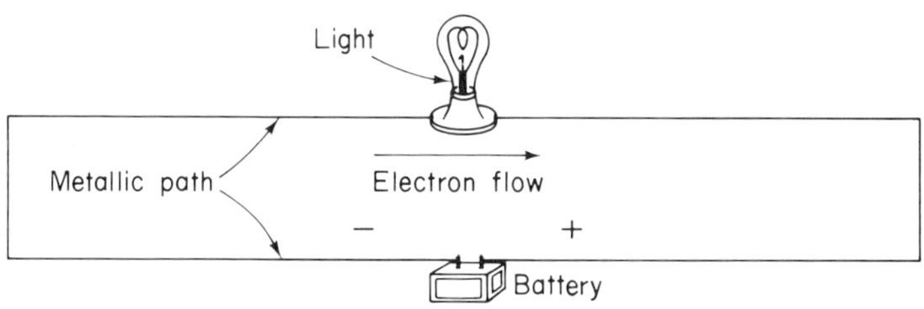

Figure 1-20. Battery Maintaining Electrical Pressure

The two terminals come directly from the plates. The *plates are charged* with opposite polarities through the *chemical action* of the electrolyte. Electrons leave the negative terminal and flow through the external path to the positive terminal. The electrolyte inside the battery moves the electrons from the positive plates back to the negative plates, and prevents the charge from equalizing.

This electrical pressure, which forces the electrons to move along the external path, is called *voltage.* Voltage is defined as the *difference in electrical potential between two points.* Our battery must have two terminals—we must have two points of reference in order to have a difference in potential. Before using the word "voltage" we spoke of electrical pressure. Even earlier, we talked of force between charged bodies, and now we speak of a difference in potential—which is correct? *All of these terms mean the same thing.* Still another term which has the same meaning as voltage is *electromotive force* (EMF). Any one of these terms can be defined as a force which causes (or tends to cause) electrons to flow. Since all these terms appear frequently in most electrical literature, it is a good idea to fix all of them in mind as soon as possible. It may seem to be unnecessary confusion at this point, but it will clear up as you gradually increase your repertoire of electric terms.

$$\text{Voltage} = \begin{cases} \text{Difference in potential} \\ \text{Electrical pressure} \\ \text{Electromotive force} \end{cases}$$

We defined voltage as a difference in potential between two points. Let's stick to that definition. The unit for measuring voltage is the volt. The symbol for voltage is E or V.

current

Current is defined as *movement of electrons.* The purist might insist that the electrons be confined to a circuit in order to qualify as current. Not so. The electrical discharge during a thunderstorm represents enough movement of electrons to qualify as current by any stretch of the imagination. We have been talking about current since we first mentioned electrostatic charges, but the word has been avoided until this time. With electrostatic charges, very few electrons moved, and movement took place over a very short period of time. Still, it represented current in very small quantities.

A. M. Ampere, the French physicist (1775–1836), worked with a charge of 6.24×10^{18} (6,240,000,000,000,000,000) displaced electrons. This quantity of charge was named in honor of C. A. Coulomb, and today, one *coulomb* means a charge of 6.24×10^{18} displaced electrons.

The unit for measuring current was, in turn, named in honor of A. M. Ampere. Thus it is called *ampere. One ampere* of current is one coulomb passing a given point in one second. The symbol for current is *I.*

resistance

Any path for current will offer some opposition to that current. This opposition is called *resistance.* The unit for measuring resistance is called the *ohm.* This unit was named for G. S. Ohm, a German physicist (1787–1854), who made many valuable contributions to the field of electrical science. The symbol for resistance is *R.*

Ohm's law

One of Ohm's major contributions was the establishment of a definite relationship between voltage, current, and resistance in a closed circuit. (A *circuit* consists of a voltage source and a complete path for current.) Ohm stated this relationship as follows:

> Current is *directly* proportional to voltage and *inversely* proportional to resistance.

As a formula, it appears like this:

$$\text{Current (in amperes)} = \frac{\text{Voltage (in volts)}}{\text{Resistance (in ohms).}}$$

In any science, it is a common, and very useful, practice to use symbols as a type of shorthand representation for units, components, and elements. In the past few paragraphs we have identified these common symbols:

$$I = \text{Current}$$

$$E = \text{Voltage}$$

$$R = \text{Resistance}$$

Restating the above formula using these symbols, we have: $I = E/R$. This formula is commonly known as *Ohm's law.* It can be transposed to solve for any of the three quantities when the other two are known. Here are the three forms: $I = E/R$, $E = IR$, and $R = E/I$.

A quick way to remember the three forms might be helpful. Place the symbols in a triangle as shown below:

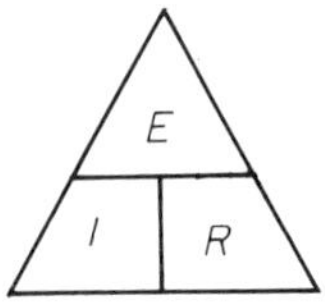

Cover the symbol that you wish to solve for. The visible portion of the triangle now indicates the proper operation to be performed on the other two. This is a bit of a crutch. If you are adept at transposing formulas, you won't need it.

Now that we know the relationship among current, voltage, and resistance, let's examine some circuit components.

circuit components

We have frequently referred to a metallic path for current. Actually, a path of any material will allow some current to exist. Some materials offer great opposition to electron movement while others oppose it very slightly. Materials that offer *slight opposition* to current are called *conductors.* The lower the opposition, the better the conductor. Materials that exert *strong opposition* are called *insulators.* The higher the opposition, the better the insulator.

Some of our best conductors are silver, copper, gold, and aluminum, in that order. Some of the best insulators are oil, mica, rubber, and glass. It should be self-evident that the best insulators are the poorest conductors, and vice versa.

Conductors are used in circuits to provide a path for current. Insulators are used to confine the current to the desired path.

The path always offers opposition to current, but if we are to do work, we must also deliberately place opposition in the current path. This opposition may be the motor of a fan, a filament in a light bulb, a heating element in a toaster, or any of literally thousands of useful items. Each one will have a certain resistance. Current being forced through this resistance does the work.

Sometimes it is necessary to reduce the magnitude of current in order to satisfy a particular requirement in a circuit. This is done by installing manufactured *resistors* in the path of current. These resistors will be discussed in detail a bit later. They come in many sizes, shapes, and ohmic values to allow great flexibility in circuit design. Figure 1-21 is the symbol for a common resistor.

Figure 1-21. Resistor Symbol

Symbols are used to represent circuit components when drawing electric circuits; a symbolized drawing of a circuit is called a *schematic.* Most people in the field of electricity are not skilled artists. The circuit symbology and schematic drawings were created for the benefit of these people. The schematic is similar to a blueprint. It shows everything that appears in the physical circuit, but it does so in a symbolical manner.

The common symbol for a battery is shown in Figure 1-22.

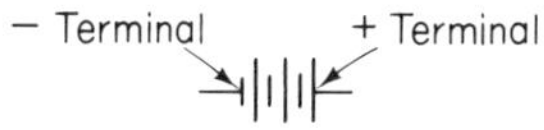

Figure 1-22. Battery Symbol

The symbol for a conductor is simply a straight line as shown in Figure 1-23. Generally, this symbol is considered to be a path

Figure 1-23. Conductor Symbol

of zero resistance. We know that this is impossible, but the difference is negligible in common circuits.

Now that we have all the necessary parts for a basic circuit, let's put them together.

basic electrical circuits

Figure 1-24 is a schematic for a basic circuit. Identify the battery, resistor, and conductor.

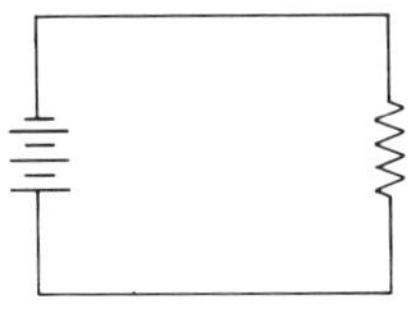

Figure 1-24. Schematic of a Basic Circuit

Let's assign values to our components and examine the circuit by use of Ohm's law. Figure 1-25 shows the direction of current and the relationship between voltage, current, and resistance in the closed circuit.

Ohm said that current is directly proportional to voltage. Let's

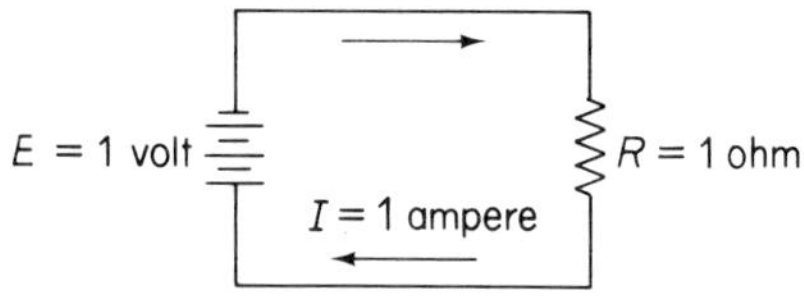

Figure 1-25. Illustration of Ohm's Law

keep the resistance at one ohm and double the voltage. This is shown in Figure 1-26.

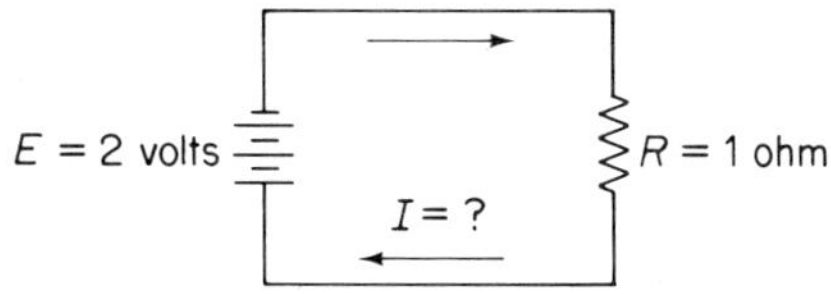

Figure 1-26. Voltage Doubled

What new value of current will exist in the circuit of Figure 1-26? The formula is: $I = E/R$.

Substitute known values: I = 2 volts/1 ohm
Divide: I = 2 amperes
Conclusion: Doubling the voltage in a given circuit causes the current to double.

In Figure 1-27 we use five times the voltage used in the previous circuit, but the resistance remains the same. Solve this circuit for current.

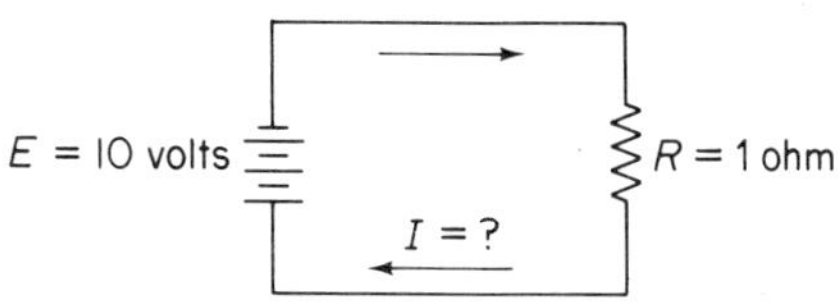

Figure 1-27. Increasing Voltage Ten Fold

Formula: $I = E/R$
Substitute known values: I = 10 volts/1 ohm
Divide: I = 10 amperes
Conclusion: Increasing voltage five-fold causes the current to increase five-fold.

In the circuit of Figure 1-28, the voltage is kept at ten volts. Let's double the resistance and solve for current.

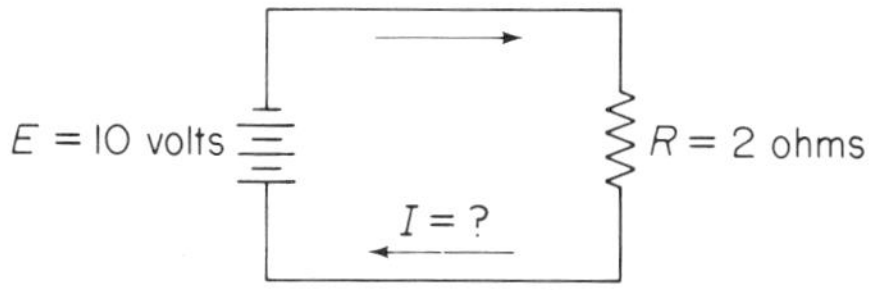

Figure 1-28. Resistance Doubled

Formula: $I = E/R$
Substitute: $I =$ 10 volts/2 ohms
Divide: $I =$ 5 amperes
Conclusion: Resistance doubles, current halves.

In Figure 1-29, the voltage is still 10 volts, but the resistance has been raised from 2 ohms to 50 ohms. What will be the new current in this circuit?

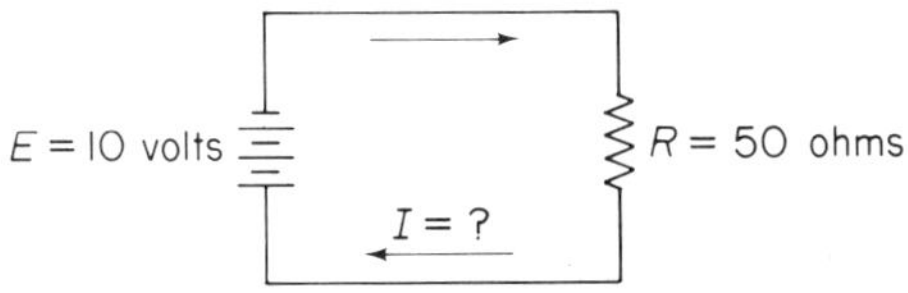

Figure 1-29. Increasing Resistance by 25 Times its Previous Value

Formula: $I = E/R$
Substitute: $I =$ 10 volts/50 ohms
Divide: $I =$ 0.2 ampere
Conclusion: Increasing resistance to 25 times its previous value reduces current to $\frac{1}{25}$ of its previous value.

Ohm's law works beautifully on paper, doesn't it? Would it work as well on practical circuits? Absolutely. A theory becomes a law only after it has been thoroughly proven. *This is a law*—it is completely reliable. Of course, you must consider the differences between the perfect circuits that are created on paper and the imperfect circuits that exist in actual practice. All practical circuits have small resistance losses. In actual practice, the measured values would be so close to those calculated here that the difference could be ignored in all cases except those measured with precision equipment.

It is time to put Ohm's law to work solving circuit problems.

solving circuit problems

In this exercise, two quantities will be given, and we'll use Ohm's law to solve for the third. How much voltage is applied to the circuit of Figure 1-30?

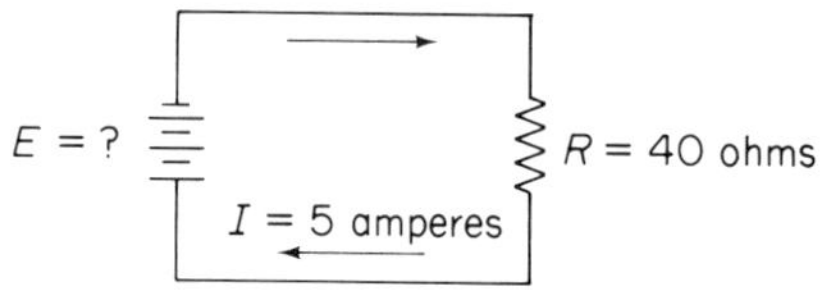

Figure 1-30. Solve for Voltage

Solution:

$$E = IR.$$
$$E = 5 \text{ amperes} \times 40 \text{ ohms}$$
$$E = 200 \text{ volts}$$

What is the value of the resistor in Figure 1-31?

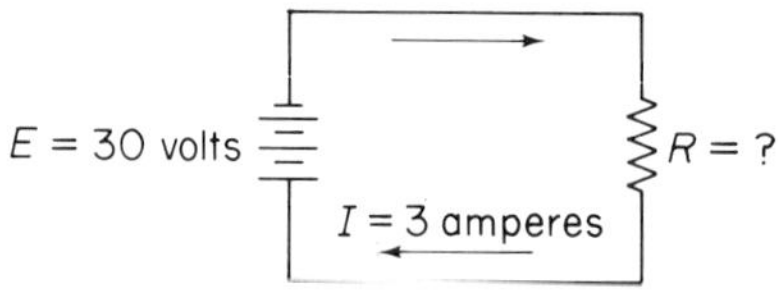

Figure 1-31. Solve for Resistance

Solution:

$$R = E/I.$$
$$R = 30 \text{ volts}/3 \text{ amperes}$$
$$R = 10 \text{ ohms}$$

What is the value of current in the circuit of Figure 1-32?

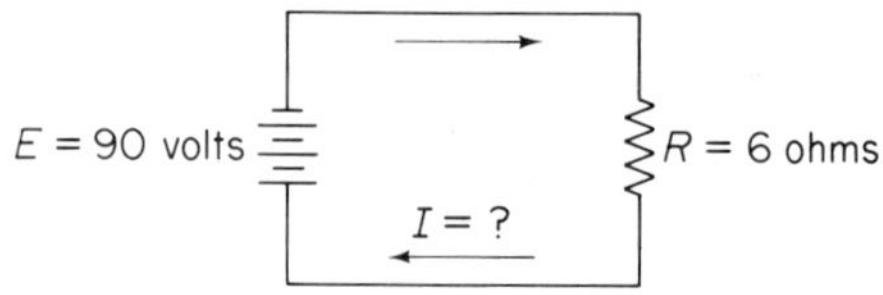

Figure 1-32. Solve for Current

Solution:

$$I = E/R.$$

$$I = 90 \text{ volts}/6 \text{ ohms}$$

$$I = 15 \text{ amperes}$$

In most cases, volts and amperes are abbreviated on schematics to reduce the writing. We shall use a capital V for volts and a capital A for amperes. The ohm has its own standard symbol for schematic uses. The symbol for ohm (or ohms) is the Greek letter omega (Ω). Figure 1-33 illustrates how these symbols will look on our schematics.

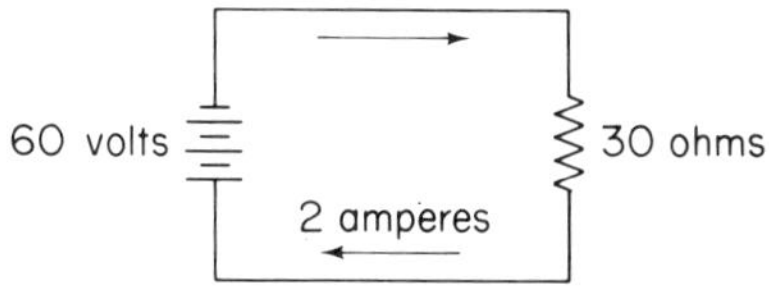

Figure 1-33. Abbreviations for Units of Measure

Solve the circuit in Figure 1-34 for the ohmic value of the resistor.

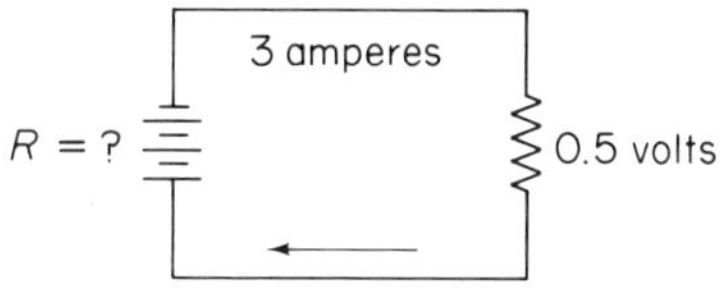

Figure 1-34. Solve for Resistance

Solution:

$$R = E/I.$$

$$R = \frac{0.5\text{V}}{3\text{A}}$$

$$R = 0.167\Omega$$

Actually, 0.5 divided by 3 is 0.1 followed by an infinite number of sixes. Here the answer is *rounded off.* This number is accurate to two decimal places with only a slight error in the third decimal place. Accuracy to two decimal places is close enough for most common problems in electricity. Thus, most of our answers will be rounded off in this manner.

The positions of the components were reversed in Figure 1-34 so that we would avoid becoming accustomed to seeing them in one particular configuration. There is no standard pattern for arrangement or position of these components.

Arrows have been used in the previous circuit diagrams to indicate the direction of current. The electrons always leave the negative terminal of the voltage source, follow the closed loop, pass through the resistor, and return to the positive terminal of the voltage source. The same value of current is present in all parts of the loop at the same instant. Imagine that current is interrupted after a fixed period of time by opening a switch. During that period of current flow, the quantity of electrons that left the negative terminal is equal to the quantity that entered the positive terminal. This same quantity of electrons could have been detected passing any given point in the circuit during that same time interval. More will be said on this subject later under the heading of "Conductance."

Solve the circuit represented in Figure 1-35 for the value of applied voltage.

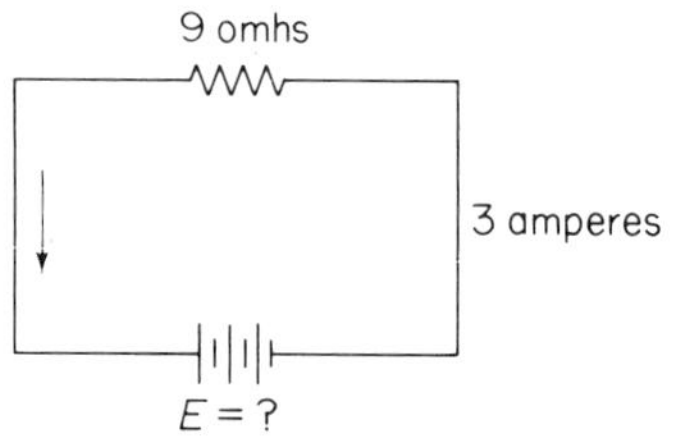

Figure 1-35. Solve for Voltage

Solution:

$$E = IR.$$

$$E = 3A \times 9\Omega$$

$$E = 27V$$

Solve the circuit in Figure 1-36 for magnitude of current.

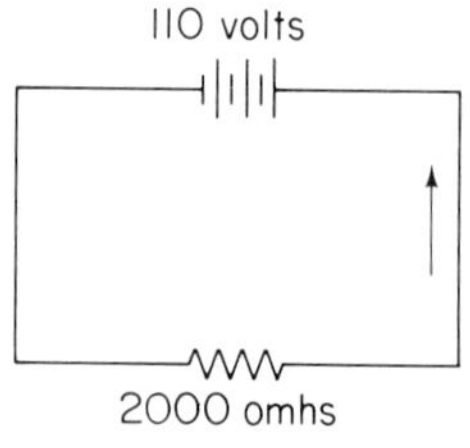

Figure 1-36. Solve for Current

Solution:

$$I = E/R.$$

$$I = \frac{110V}{2000\Omega} = 0.055A.$$

The basic principles of Ohm's law should now be clear. Numbers have been kept simple so as not to distract from the circuit function. It is important to be able to *visualize* the voltage forcing the current through all opposition and back to the source. If you can follow this process at this point, you are to be complimented on your progress. If you are unable to do so, you need to go back and review the circuit problems.

This concludes Chapter 1. The key points are covered in the following review exercises.

REVIEW EXERCISES

1. What is electricity?
2. What is the basic structure of all matter?
3. "Any material which has mass and occupies space" is a definition of ________.
4. An atom is the smallest particle of matter that can be recognized as the ________________.
5. (a) What is an electron? (b) Where does it normally exist?
6. Why do electrons move more readily than protons?
7. A substance which cannot be broken down by any chemical process is a basic ________.
8. A chemical union of two or more elements forms a ____ ____.
9. An electrically stable atom has an equal number of electrons and ________.
10. What is an ion?
11. A positive charge is a ________ of electrons.
12. State the law of charges.
13. The electroscope is used to detect the presence of a ________ and illustrates that unlike ________ repel each other.
14. The area around a charged body where its influence can be felt is an ________________.

15. State the direction of an electrostatic field.
16. Lines of force in the same direction ____________ each other.
17. State the rule for concentration of charge on a curved surface.
18. Discharging is the process of ____________ a charge.
19. Name a common device which is used to maintain electrical pressure.
20. Describe the path of current in a closed loop.
21. List four common names for the force which causes current.
22. Voltage is measured in ____________, and its symbol is ____________.
23. Current is a movement of ____________. It is measured in ____________, and its symbol is ____________.
24. A coulomb is a displacement of how many electrons?
25. Define the ampere in terms of coulombs.
26. Resistance is opposition to ____________ flow. It is measured in ____________, and its symbol is ____________.
27. State Ohm's law.
28. The formula, $I = E/R$, is a mathematical statement of ____________ law, and it demonstrates the relationship between ____________, ____________, and resistance.
29. List two good conducting materials and two good insulating materials.
30. Draw the schematic symbol for a resistor, a battery, and a conductor.
31. Draw a schematic showing a circuit with a 6-V battery connected to a 12-Ω resistor, and draw arrows on the schematic to indicate the direction of current.
32. How much voltage does the circuit in Figure P1-32 need to produce a current of 3A?

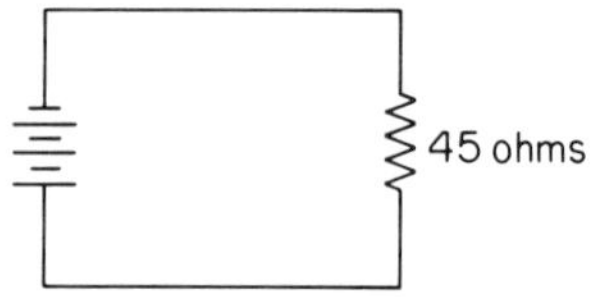

Figure P1-32

33. How much current will exist in a circuit which is composed of a 2500-Ω resistor connected to a voltage source of 50V?
34. What happens to the current in a particular circuit when the:
    (a) Resistance doubles?
    (b) Voltage doubles?
35. An automobile with a 12-V battery has a spotlight which draws 6A of current. What is the resistance of the light filaments?
36. An electric iron connected to a 110-V source draws 4.5 A. What is its resistance?
37. A light bulb has a resistance of 111 ohms. How much current will it draw when connected to a 100-volt source?
38. The filaments of a certain radio tube have a resistance of 40 ohms and require 0.3A of current for proper operation. How much voltage should be applied to these filaments?

2

circuit components

Circuit components are the items that comprise a circuit. So far, we have discussed batteries, resistors, and conductors. In this chapter, we shall examine the characteristics of these, study the function of some others, and learn a new system of expressing values.

CHARACTERISTICS OF CONDUCTORS

Why do you suppose that one material allows electrons to flow through it with relative ease while another completely blocks the current for all practical purposes? Resistivity? Well, yes. But what causes a given material to have this characteristic resistance? Basically, it is the atomic structure. The type of atoms in a material determines the number of electrons in the material and the amount of energy required to free the electrons from their atomic orbits.

free electrons

Atoms will not move when voltage is applied; but *electrons will.* This means that some electrons are torn out of their orbits and actually move through the material. These electrons will be free only a very brief time before they are captured by another atom. But it is possible for them to exist for a short while in a free state and not be associated with any atom. These free electrons make it possible to have current through a conductor.

The *resistivity* of a given material is related to the ease (or difficulty) of freeing the electrons. The best conductors have the lowest resistivity, and very little energy is required to free electrons from the atoms. On the other hand, the best insulator has the highest resistivity, and it is very difficult to free electrons from the atoms.

Figure 2-1 illustrates normal atomic action within a section of

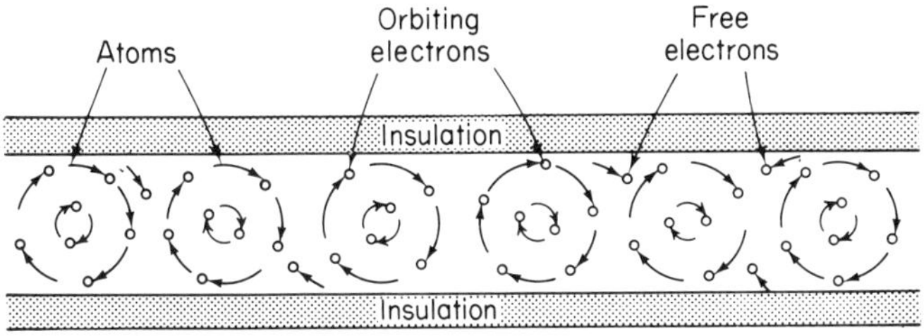

Figure 2-1. Atoms Within a Normal Conductor (no current)

a conductor. Notice that even in this state, there are a few free electrons drifting about from atom to atom. The free electrons are drifting about in random directions. This causes the overall electrical properties to be those of neutral material.

When the conductor is placed in a circuit, the electrical pressure will direct these free electrons in a definite direction as shown in Figure 2-2. Atomic activity will also have increased thus causing

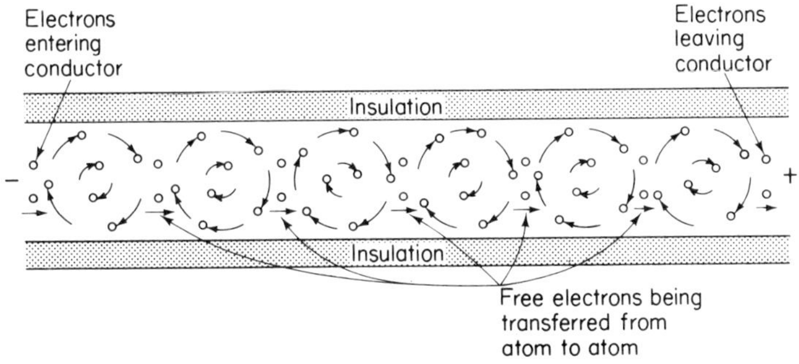

Figure 2-2. A Current Carrying Conductor

greater quantities of electrons to be freed. The negative terminal of the voltage source feeds electrons into the conductor. The electrostatic force between atoms will free other electrons and cause them to be moved from atom to atom. Movement of the free electrons is continuously away from the negative and toward the positive terminal. At the positive terminal, the electrons leave the conductor and go back into the voltage source. The number of electrons that enters the negative end of the conductor is the same as the number that leaves the positive end during a given interval of time.

The above statement does not imply that the electrons leaving a conductor are the exact same electrons that entered it. In fact, there is a great difference between the current speed and the movement of a particular electron.

speed of current

The *speed of current* is effectively the same as the *speed of light* (186,000 miles per second). The speed of movement of a particular electron is estimated at only a few miles per second. How can we account for this great difference in speed? Suppose that you have a long plastic tube that is just large enough in diameter to allow a baseball to pass through it without restriction, as represented in Figure 2-3. Push baseballs into the tube until it is com-

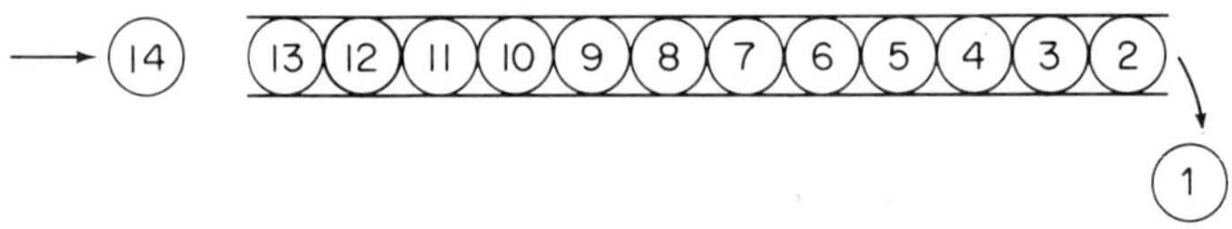

Figure 2-3. Propagation of Energy Through Baseballs in a Tube

pletely filled with balls. After the tube has been filled, a ball leaves the far end at the same instant that a ball enters the near end. As ball 13 was pushed into the tube, all balls moved one position, and ball 1 dropped out of the tube.

Regardless of the speed of the individual balls and the length of the tube, the motion is felt throughout the tube instantly. The individual baseballs may be compared to individual electrons. The speed of the propagation of energy is then comparable to the speed of current.

Figure 2-3 also illustrates the one-to-one ratio between electrons entering and leaving a conductor. Suppose that the balls leaving the tube are dropping into a machine which automatically forces them back into the near end of the tube. In this case, the machine could be compared to a battery.

conductance

The resistance of good conductors is so low that it is frequently convenient to speak in terms of ability to conduct current. While resistance is an opposition to current, *conductance* is a measure of the ability to conduct current. Conductance is the exact reciprocal of resistance. Since the ohm is the unit of measure for resistance, ohm spelled backwards (mho) is the unit of measure for conductance. When a piece of conductor material has $\frac{1}{50}$ of an ohm of resistance, it has a conductance of 50 mhos. This same pattern is followed even to the symbols. Omega (Ω) is the symbol for ohm (or ohms), and an inverted omega ($\mho$) is the symbol for mho (or mhos).

One reason for the convenience of using these reciprocal values becomes obvious when we examine the above quantities. The resistance of a conductor is usually a small fraction, which makes its reciprocal a whole number. In most calculations, the whole number is easier to handle.

Earlier, we stated that the resistance of conductors could be ignored in circuit calculations. Then why all this concern over resistance and conductance? In most *ordinary circuits,* we have a *tolerance factor* on all quantities which ranges from 5 to 20 per cent. This tolerance is wide enough to compensate for any error produced by ignoring the minute resistance of the conductor. However, one may encounter *precision circuits* that have very *close tolerances.* In a common circuit using a 110-volt source, most appliances will function on any voltage from 88 to 132 volts. This represents a 20 per cent tolerance. In satellite tracking equipment, one would probably encounter circuits that require that voltages be accurate to four decimal places. Since we are sure to encounter some precision circuits, it is worth a bit more time and space to explore some other characteristics of resistance.

factors affecting resistance

In addition to resistivity, there are *three* other factors that affect the total *resistance of a conductor*—length, cross-sectional area, and temperature.

Obviously, a long piece of wire has more resistance than a shorter section of the same wire. When both the cross-sectional area and the temperature of the two wires are equal, the comparison of length and resistance can be expressed as a ratio: $R_1/R_2 = L_1/L_2$, where: R_1 = resistance of wire 1, R_2 = resistance of wire 2, L_1 = length of wire 1, and L_2 = length of wire 2. This simply says that resistance is directly proportional to length.

Example 1: 1000 feet of No. 8 copper wire at 68 °F has a resistance of 0.628 ohm. What is the resistance of one mile of this wire?

Substitute: $0.628/R_2 = 1000/5280$.
Cross multiply: $1000R_2 = 3320$
Divide by 1000: $R_2 = 3.32$ ohms

The resistance of a specific conductor is inversely proportional to its cross sectional area. This ratio is: $R_1/R_2 = A_2/A_1$, where: R_1 and R_2 are the respective resistances in ohms, A_1 and A_2 are respective cross-sectional areas, and the length is constant.

Example 2: A square wire has a cross-sectional area of 0.1 square inch and a resistance of 0.075 ohm. What would be the resistance if the cross sectional-area were 0.02 square inch?

Substitute: $0.075/R_2 = 0.02/0.1$
Cross multiply: $0.02R_2 = 0.0075$
Divide by 0.02: $R_2 = 0.375$ ohm

Since most wires are drawn round, it is usually easier to deal with diameter than with cross-sectional area. The resistance of a round wire is inversely proportional to the square of its diameter. Thus: $R_1/R_2 = d_2^2/d_1^2$, where: d_1 and d_2 are the respective diameters.

Example 3: A round wire has a diameter of 0.25 inch and a resistance of 8 ohms. What would be the resistance if the diameter were 0.5 inch?

Substitute: $8/R_2 = 0.5^2/0.25^2$
Square diameters: $8/R_2 = 0.25/0.0625$
Cross multiply: $0.25R_2 = 0.5$
Divide by 0.25: $R_2 = 2$ ohms

The diameters of most wires are small when we deal with the inch as a unit of measurement. It is more convenient to deal with the mil as the unit. The mil is $\frac{1}{1000}$ of an inch. Thus, 94 mils = 0.094 inch.

All metals and nearly all other materials have a positive temperature coefficient. That is, the resistance is generally directly proportional to temperature. *Carbon* is one of the few materials that has a *negative temperature coefficient;* that is, its resistance is inversely proportional to temperature.

PRODUCTION OF ELECTROMOTIVE FORCE

Earlier, we discussed batteries as one means of maintaining a potential difference between two points. Actually there are several ways of producing this electromotive force.

Electricity is energy, and as we have already seen, it is a type of atomic energy. Voltage (electrical pressure or difference in potential) is frequently referred to as electrical energy. One of the fundamental laws of physics states that energy can *neither be created nor destroyed.* Then how can voltage be produced? The only means available is to *convert* some other form of energy into electrical energy. What are some of the other forms of energy? Heat is a form of energy; light is another form of energy. Others are mechanical energy and chemical energy. Which of these four forms of energy can be readily converted to electrical energy? *All four of them.*

heat energy to electrical energy

One of the simplest methods of producing electrical energy is to convert heat energy. One device employed in this operation is the *thermocouple.* The thermocouple is constructed by joining two dissimilar metals, connecting an external electrical circuit, and heating the junction of the two metals. This device has never proved very practical. In order to produce any appreciable amount of voltage, the two pieces of metal would have to be rather large, and very high temperatures would have to be used. The thermocouple is illustrated in Figure 2-4. The dissimilar metals in this example of

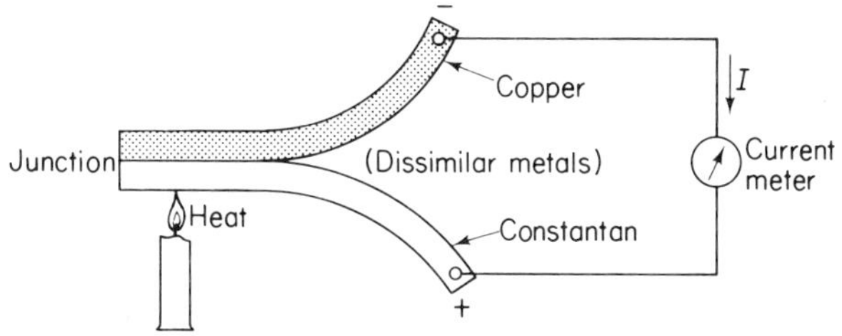

Figure 2-4. A Thermocouple

a thermocouple are copper and constantan (an alloy of copper and nickel). As the junction heats, the copper robs the constantan of electrons. This causes the copper to become negative while the constantan becomes positive. A sensitive current meter in such a circuit would detect a minute current through the conductor from the copper back to the constantan.

light energy to electrical energy

A light-sensitive surface can be used to convert light energy to electrical energy. This is known as the *photoelectric effect,* and one device used for this purpose is a photoelectric cell. The working parts of such a cell are illustrated in Figure 2-5. The small battery, added in this fashion, increases the efficiency by causing the collector to exert a definite attraction for the free electrons.

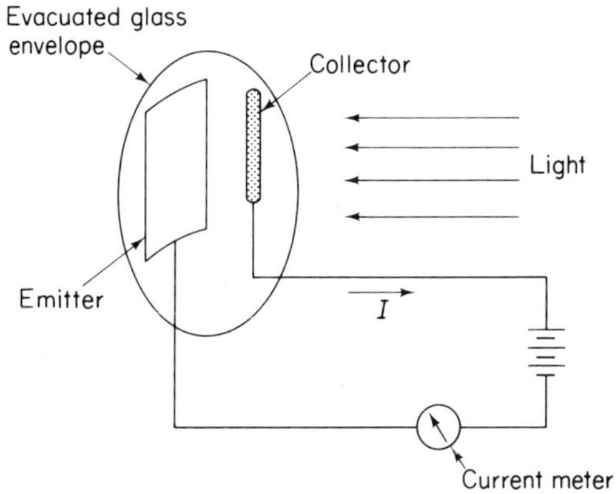

Figure 2-5. A Photoelectric Cell

As light strikes the emitter, electrons are released from the sensitive surface. These electrons are picked up by the collector and passed through the circuit back to the emitter. The sensitive surface on the emitter has been formed by successive coatings of silver, cesium, or potassium. The most common use of this device is the photographer's exposure meter.

mechanical energy to electrical energy

The simplest method of converting mechanical energy to electrical energy is to utilize the *piezoelectric effect*. When a mechanical pressure is exerted on a quartz crystal, a voltage is produced.

This phenomenon is called the piezoelectric effect. Small currents can be drawn when the crystal is connected to an external circuit. This method is used in electronic timing circuits.

A more common way of producing voltage from mechanical energy is the *generator.* The electric generator (or alternator) in the automobile is a familiar device. The automobile engine supplies the mechanical energy to turn the armature of the generator. The generator converts the mechanical energy to electrical energy. The principles of the generator will be covered in a later chapter.

chemical energy to electrical energy

Cells and batteries are devices used for converting chemical energy to electrical energy. Since these devices are encountered regularly in everyday life, they deserve a more detailed coverage than has been provided up to this point.

the dry cell

The structure of the *dry cell* is illustrated in Figure 2-6. The common flashlight cell is a good example of the dry cell in everyday use.

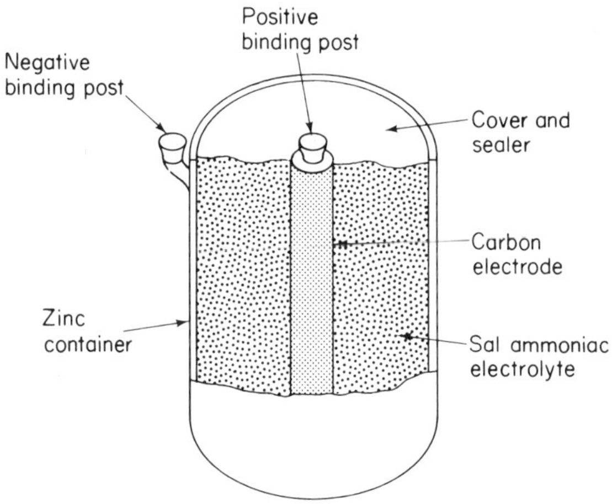

Figure 2-6. A Common Dry Cell

This simple, but useful, device consists of a zinc can filled with a diluted solution of ammonium chloride with a carbon rod running down the center of the can. The zinc container is the negative electrode (plate), and the carbon rod is the positive electrode. The diluted ammonium chloride is commonly known as *sal ammoniac* and is the active ingredient in the electrolyte. The sal ammoniac is mixed with some inert, porous material to form a paste. Binding posts provide a means of connecting the electrodes of the cell to an external circuit.

The *electrolyte moves electrons* from the carbon rod to the zinc can and maintains a difference in potential between the two binding posts. In the process, the zinc case is gradually destroyed. However, in modern cells, the case normally outlasts the electrolyte, which becomes inactive when it drys out.

An important ingredient, which does not appear in the drawing, is a depolarizing agent surrounding the carbon rod. Polarization is the process through which the cell insulates the positive electrode. The normal chemical action releases ammonia and hydrogen gases. Bubbles of these gases tend to collect on the carbon rod. The depolarizing agent is composed of carbon granules and manganese dioxide. Its function is to absorb the gases and prevent the positive electrode from becoming insulated against current.

A dry cell, composed of the materials described here, will produce a voltage of 1.5 volts. The size of the cell has no bearing on the magnitude of voltage it will produce. The size is, however, important to the amount of current that can be drawn. If a small current is needed for long periods of time, a large cell will serve better than a smaller one. The same is true when relatively large currents are required for short time intervals.

Dry cells may be externally connected in combinations that will increase either (or both) the available voltage or current capacity. If the job requires *more voltage* with small current, the cells may be connected *in series* as shown in Figure 2-7. In this combination, the voltages of the individual cells are additive to supply a larger terminal voltage. The available voltage is the sum of the individual cell voltages, but the current capacity remains that of a single cell.

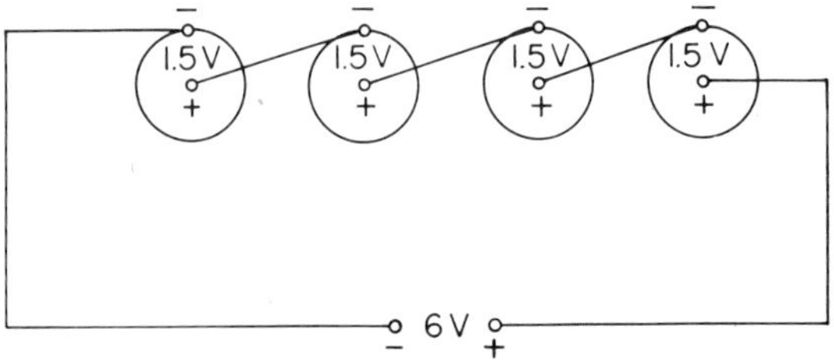

Figure 2-7. Dry Cells in Series

When the job requires *more current* but the voltage of a single cell is sufficient, the *parallel* arrangement in Figure 2-8 is used.

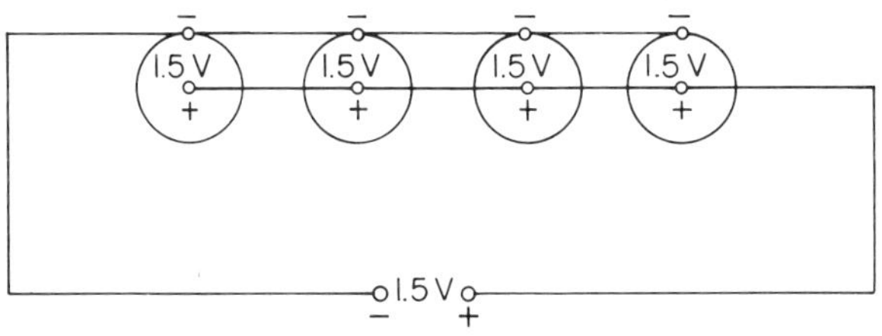

Figure 2-8. Dry Cells in Parallel

When cells are connected in this manner, the effect is the same as that of using a larger cell. The terminal voltage is the same as that provided by any single cell, but the current capacity is greater.

A *series-parallel* arrangement is shown in Figure 2-9. Cells connected in this manner will provide a *greater terminal voltage,* and at the same time, a *higher current* capacity. This particular unit will provide the terminal voltage of four cells combined with the current capacity of two cells. There are numerous such arrangements possible, but the efficiency decreases as the number of cells increases.

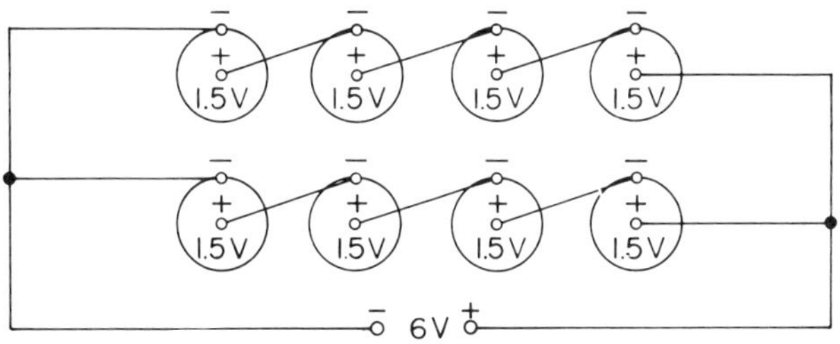

Figure 2-9. Dry Cells in Series-Parallel

the lead-acid cell and the storage battery

In a previous discussion on the storage battery, we mentioned that it is another common voltage source.

A battery is composed of cells each of which produces about one and one-half volts. The two types of plates (electrodes) were given as *spongy lead* and *lead peroxide.* The liquid electrolyte is a solution of *sulfuric acid* and water. This type of cell is known as the *lead-acid cell.* Figures 2-10 through 2-13 illustrate the chemical ac-

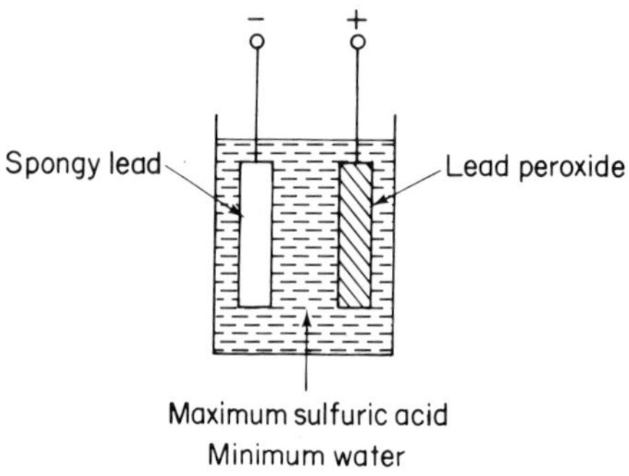

Figure 2-10. Fully Charged

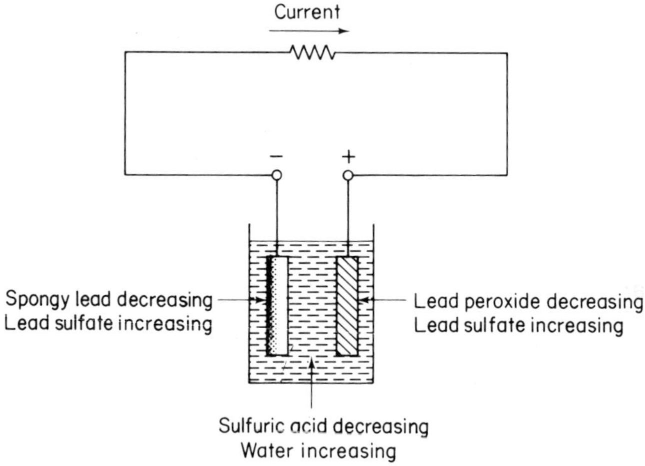

Figure 2-11. Discharging

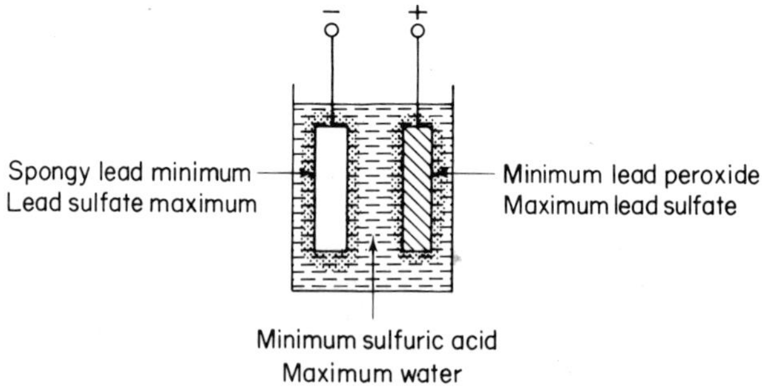

Figure 2-12. Fully Discharged

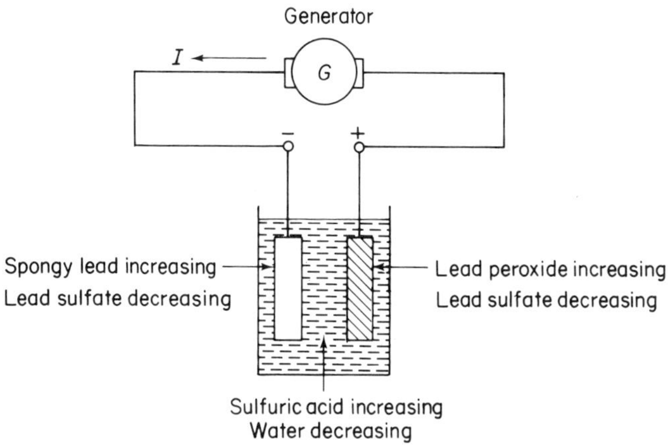

Figure 2-13. Charging

tion within this cell during each of its four possible conditions. Several of these cells connected in series provide a storage battery with the desired terminal voltage.

A useful instrument in caring for a storage battery is the *hydrometer,* which is used to determine the state of charge of the battery. It does this by measuring the *specific gravity* of the electrolyte. The general appearance of the hydrometer is illustrated in Figure 2-14. It consists of a tube, a float with scale gradations, and a rubber suction bulb for drawing the liquid electrolyte into the tube with the float. The higher the float rides in the liquid, the higher the specific gravity of the electrolyte, and the better the charge on the battery. The electrolyte from a fully charged battery will read from 1.275 to 1.300. This means that the electrolyte has maximum sulfuric acid and minimum water. A reading from a completely discharged battery will be about 1.100 which means that the electrolyte is mostly water with very little sulfuric acid.

The reading is taken on the scale where the float touches the top of the liquid. The larger numbers are at the bottom of the scale. Since acid is heavier than water, the more acid in the solution, the larger the number exposed.

Under normal conditions, there is never a need to add acid to the electrolyte. The water evaporates and must be replaced. The same quantity of acid is always there, but sometimes in a different form. The solution becomes diluted as the battery discharges, but this is corrected by the recharging process. It is best to add only

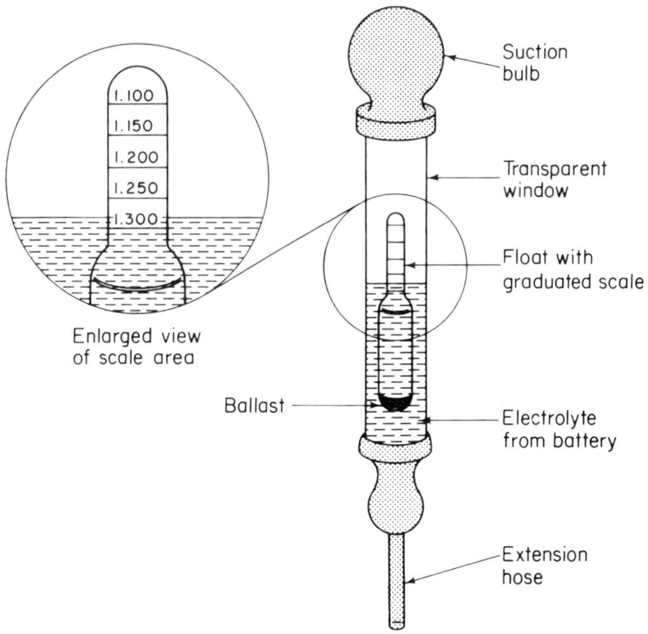

Figure 2-14. The Hydrometer

distilled water to a storage battery. While ordinary tap water will do the job, it does contain certain chemicals and minerals of its own, which may harm the battery.

internal resistance of batteries

In common equipment, battery resistance is quite properly ignored. But as in the case of conductors, the resistance is there. The electrolyte itself offers some resistance to current, and polarization causes additional internal resistance.

The actual internal resistance of a battery may be determined by the formula: $R_{\text{int}} = E_{NL} - E_L/I_L$, where: R_{int} is the internal resistance in ohms, E_{NL} is the terminal voltage with no load in volts, E_L is the terminal voltage with a load (current) in volts, and I_L is current being drawn in amperes.

A storage battery is rated by a unit of capacity called the *ampere-hour rating.* The ampere-hour rating is determined by taking the product of the current in amperes and the time in hours. Theoretically, the battery should produce satisfactorily over any combination of current and time that is within the capacity rating.

For example, a 100-ampere-hour battery should produce 1 ampere for 100 hours, 2 amperes for 50 hours, or 100 amperes for 1 hour. Actually, the capacity decreases as the current load increases. Manufacturers of batteries do not intend them to produce unreasonable amounts of current even for a short period of time. Within certain specified limits of current, the capacity rating is reasonably reliable.

The subject of resistors has been mentioned, but a great deal more needs to be said about them. This seems to be a convenient spot to look a bit deeper into this vital subject.

RESISTORS

We have said that *resistors* are manufactured in many ohmic values to allow flexibility in circuit design. They also come in several shapes and sizes.

structure

Most *resistors* that you are likely to encounter are constructed of either powdered *carbon* or high-resistance *wire.* Since carbon has a negative temperature coefficient (resistance decreases with a rise in temperature), the carbon resistor can be constructed to have a very stable resistance over a wide temperature range. A carbon resistor of a precise ohmic value, on the other hand, is very difficult to construct. A wire resistor can be easily constructed to a precise ohmic value, but its stability with temperature change is less reliable.

Carbon resistors are manufactured from ground carbon mixed with a binding agent and pressed into a mold. A proper mixture is designed so that temperature rise due to increased molecular activity will just offset the negative temperature coefficient of the carbon. All resistors heat to some extent as electrons pass through them. The carbon resistor retains a reasonably stable resistance during such normal heating.

The molding process allows the carbon resistor to be designed in a variety of sizes and shapes. The most common shape is cylindrical, with a lead wire protruding from each end of the cylinder. The physical sizes range from tiny to very large. To qualify these indefinite terms, it is not uncommon to encounter carbon resistors as small as a pencil lead or as large as your finger.

The wire resistor is most often constructed by winding a wire around an insulated cylinder. It is a simple process to determine the exact resistance of a given piece of wire. Resistors can then be constructed to rather exacting ohmic values. Figure 2-15 shows some common resistors.

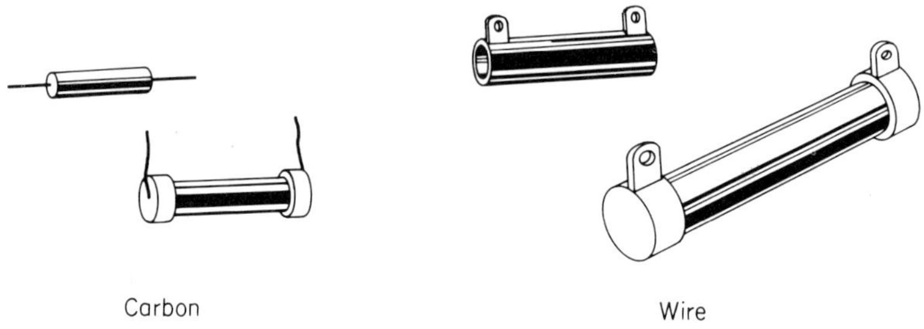

Figure 2-15. Common Resistors

In the common vernacular, *resistor size* normally refers to the *ohmic value.* Thus far, we have been careful to use either ohmic value or physical size. When speaking of physical size, we must so specify. Otherwise, ohms of resistance is assumed.

The physical size of a resistor has no bearing on its resistance. A carbon resistor with a diameter equal to that of your pencil lead may have a resistance of 1,000,000 ohms. On the other hand, a wire wound resistor the size of your wrist may have less than 50 ohms of resistance.

The physical size does, however, bear a relationship to the ability to dissipate heat. The ability to dissipate heat is expressed in a *power rating.* Power will be covered later, but a few words to clarify this point are in order. Power, in watts, is a measure of the heat dissipated. The number of watts used by a resistor is equal to the product of current squared times the resistance ($P = I^2R$). The power (or wattage) rating of a resistor represents the number of watts that that resistor can safely handle. Manufacturers rate their resistors on the basis of both power and ohms of resistance. Most circuit designers use resistors that are rated for twice the power that the circuit will require. This gives a wide margin of safety against circuit failure.

A resistor will be destroyed if it is forced to handle excessive power. The carbon resistor usually expands and separates across the middle when this happens. The wire resistor burns away a small section of the wire. In either case, current is interrupted in the circuit. Resistors are among the least expensive circuit components. For this reason, they are frequently intended to burn out in order that the more expensive circuit parts be protected from damage when excessive current is encountered.

In purchasing a 100-ohm resistor, chances are that you would be satisfied with any size from 80 to 120 ohms. This is a 20 percent *tolerance* on the specified value. Twenty percent is the normal tolerance unless a more exact value is stated. Most carbon resistors have tolerance values ranging from 5 to 20 percent. Closer tolerances may be obtained, but they usually are not needed, and they add unnecessary expense to the circuit. Tolerances as close as one-tenth of one percent may be obtained on wire resistors. Resistors are classified according to their intended use.

classification

The *three general types* of resistors are fixed, adjustable, and variable.

The *fixed resistor,* which was previously called a *basic resistor,* is used to restrict circuit current to a relatively constant value. This is done by placing a fixed resistor of specific resistance in the path of current. The resistors shown in Figure 2-15 are fixed resistors. This type of resistor appears in a schematic as this familiar symbol:  . By taking into account the value of voltage, the circuit designer can select one or more fixed resistors of a size that will provide the proper regulation of circuit current. In the normal aging process, resistance tends to increase slightly, but these resistors give years of satisfactory service with a very low failure rate.

The *adjustable resistor* is used in circuits where the amount of resistance needs to be adjusted. This is not an adjustment that can be varied with ease; rather, it is usually a one-time affair, which is made as the resistor is being installed. Figure 2-16 shows one form of an adjustable resistor. This one is wire-wound with two fixed contacts and one adjustable contact. It is sometimes called a *tapped resistor.* The resistance between the adjustable contact and either end can be altered by sliding the center ring along the length of the resistor. This sliding ring is secured by tightening the screw when the proper setting has been found.

Voltage dividers, which will be covered in a later chapter, are frequently constructed of adjustable-type resistors. These resistors are not limited to any specific number of adjustable bands. The

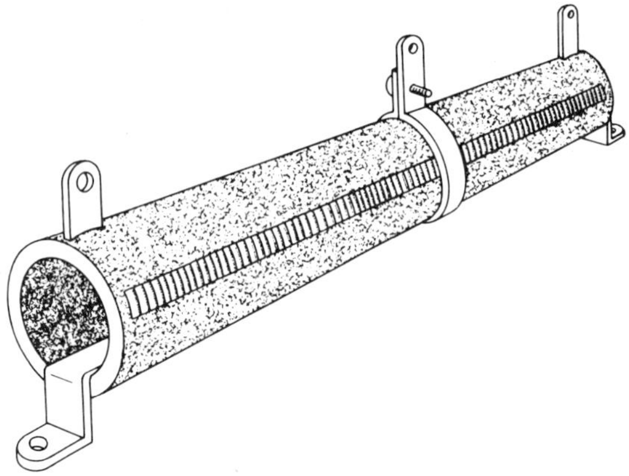

Figure 2-16. Adjustable Resistor

proper symbol for this type of resistor is:  . Each of the tapped points on the resistor represents an adjustable band where a wire may be connected.

The *variable resistor* is used for various types of controls where resistance needs to be changed frequently. The volume control on a radio is an example. These resistors are usually constructed by winding wire on a semicircular, insulating form. They may have a fixed lead on either or both ends. The variable control is provided by a wiper arm which slides along the wire resistor when a control knob is turned. Figure 2-17 illustrates the construction of two variable resistors.

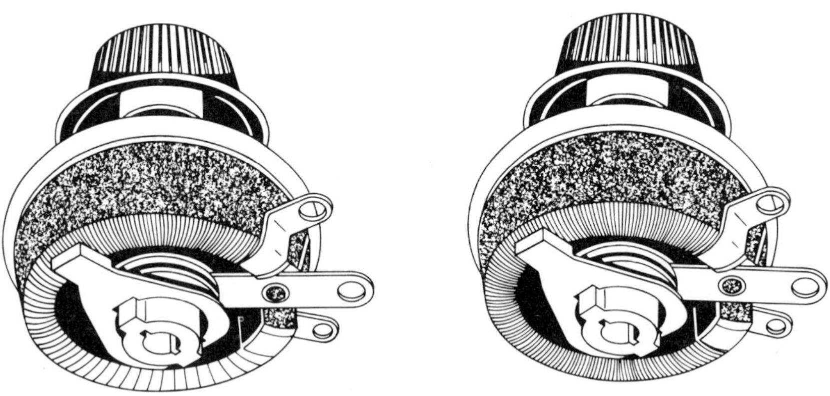

(a) (b)

Figure 2-17. Variable Resistors

Both of these pictures show wire-wound resistors with two fixed contacts and a sliding arm. The sliding arm is turned with the control knob and is electrically connected to the center contact. If all *three contacts* are connected, the variable resistor becomes a *voltage control* and is called a *potentiometer.* If one end contact is left disconnected, the resistor becomes a *current control* and is called a *rheostat.*

Notice that in Figure 2-17(a) the spacing of the windings are tapered. This provides a nonlinear control. The windings on drawing (b) are evenly spaced and provide a linear control.

Some resistors are designed specifically for use as potentiometers and others as rheostats. In this case, the potentiometer has three contacts and is intended to handle only small currents. The rheostat has only two contacts (one fixed and one variable) and is intended to handle relatively large currents. The rheostat is more ruggedly constructed than the potentiometer. It is also more expensive.

The symbols for these two resistors are:

1. Rheostat: 

2. Potentiometer: 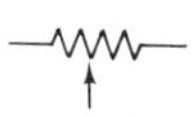

In circuits where the difference between a rheostat and a potentiometer has no bearing on the circuit, one may see this symbol: 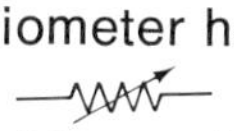 . It simply means a variable resistor, which could be either a rheostat or a potentiometer.

standard sizes

Long ago both manufacturers and users saw the impossibility of making and stocking resistors of all possible sizes. Further, it was discovered that the construction of common circuits was not an exacting science. The circuits continued to function satisfactorily with only approximate resistor sizes. Today, there are fixed resistors in 93 *standard sizes.* These are listed in Figure 2-18.

Resistors also needed a code so that their value could be identified without one's having to measure the resistance.

color codes

Fixed resistors come in a wide selection of standard sizes (as you have seen in Figure 2-18). When you need a particular size, one of these standard values will probably fall within your tolerance range. Most of these resistors have a *color code* that enables both

ohms	ohms	ohms	ohms	ohms in millions
0.47	22	1000	47,000	1
0.56	27	1200	56,000	1.2
0.68	33	1500	68,000	1.5
0.82	39	1800	82,000	1.8
1	47	2200	100,000	2.2
1.2	56	2700	120,000	2.7
1.5	68	3300	150,000	3.3
1.8	82	3900	180,000	3.9
2.2	100	4700	220,000	4.7
2.7	120	5600	270,000	5.6
3.3	150	6800	330,000	6.8
3.9	180	8200	390,000	8.2
4.7	220	10,000	470,000	10
5.6	270	12,000	560,000	12
6.8	330	15,000	680,000	15
8.2	390	18,000	820,000	18
10	470	22,000		22
12	560	27,000		
15	680	33,000		
18	820	39,000		

Figure 2-18. Standard Sizes of Fixed Resistors

merchant and customer to check the *resistance value* quickly. When analyzing circuits that are already constructed, the color code becomes very important. You may suspect that a particular resistor is measuring either too much or too little resistance. A glance at the color code on the resistor will tell you what the original value of that resistor was. In the military services, the field of electronics is barred to an individual who is color blind. One reason for this is that his defect may render him unable to read resistor color codes. Why not just stamp the number on the resistor? This is sometimes done—when the resistor is physically large enough to contain the number. However, the stamping process is highly unsatisfactory for two reasons:

1. The resistor may be installed in a position that would make reading the stamped value impossible.
2. The number expressing the ohms of resistance in many cases is entirely too long to be accommodated across the surface area of the resistor without one's having to resort to microscopic lettering.

There are *two systems* of color coding in existence at this time. They are:

1. The Band System.
2. The Body-End-Dot System.

the band system

This system of coding designates the *ohmic value,* the percentage of *tolerance,* and in some cases, the *failure rate* with a series of *colored bands* that encircle the resistor. The resistor *always* has three bands; it frequently has four bands; and occasionally, it has five bands. The latter is illustrated in Figure 2-19.

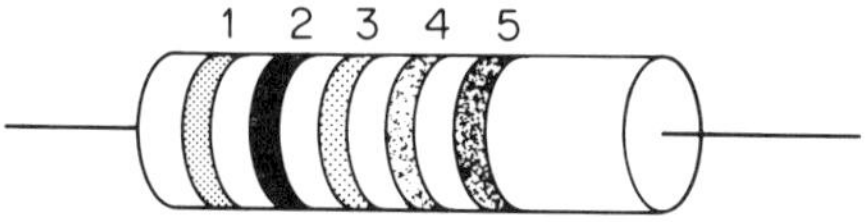

Figure 2-19. Arrangement of Color Bands

When reading this code, *start at the end which has a band nearest it.* In Figure 2-19, it happens to be the left end. The significance of each band is indicated in Figure 2-20(a) and (b).

Now comes the most difficult part: Each color signifies a specific number, and this association of color and number must be correct beyond doubt. Most serious students of electricity begin by memorizing the chart in Figure 2-20(a). The competent electrical technician associates color and number as well as most of us associate our own first and last names. He uses the code so frequently that he has no chance to forget. Figure 2-20(b) is a chart of colors that are used in this coding system. The chart also specifies meaning of each color as it appears on each of the five bands. The word "none" signifies that this color is not used on this particular band.

This is an easy chart to fix in mind. Take a little time to study it carefully before proceeding.

There are two other colors that are used, but they were omitted because they appear *only* on bands three and four. On very small precision resistors (less than 10 ohms) silver or gold may appear in the third band. When this occurs, gold means multiply the first two digits by 0.1 to get the ohmic value; silver indicates a multiplication of the first two digits by 0.01.

Gold or silver *frequently* appears in the fourth band as an additional tolerance indicator. On this band, gold indicates a tolerance of 5 percent and silver shows a tolerance of 10 percent. The use of

Color of 1st band	Color of 2nd band	Color of 3rd band	Color of 4th band	Color of 5th band
Gives the 1st digit in the ohmic value	Gives the 2nd digit in the ohmic value	Gives the number of zeros that follow the 1st two digits	Gives the percentage of tolerance on the ohmic value	Gives the percentage of failure per 1000 hours of use

Figure 2-20(a). Significance of Color Bands

Color	1st band (1st digit)	2nd band (2nd digit)	3rd band (number of zeros)	4th band (% of tolerance)	5th band (% of failure)
Black	0	0	0	None	None
Brown	1	1	1	1	1
Red	2	2	2	2	0.1
Orange	3	3	3	3	0.01
Yellow	4	4	4	4	0.001
Green	5	5	5	None	None
Blue	6	6	6	None	None
Violet	7	7	7	None	None
Gray	8	8	8	None	None
White	9	9	9	None	*
*(Used only on film type resistors. This is microelectonics and means that the resistor has a solderable connection)					

Figure 2-20(b). Relation of Colors, Bands, and Numbers

these colors is summarized in Figure 2-21. When there is *no fourth color* band, a tolerance of 20 per cent is understood.

Color	3rd band (multiplier)	4th band (% tolerance)
Gold	0.1	5
Silver	0.01	10

Figure 2-21. Use of Gold and Silver

The best way to learn the code is to use it. Let's practice reading some ohmic values.

reading the band system

What is the order of the bands on the resistor depicted in Figure 2-22?

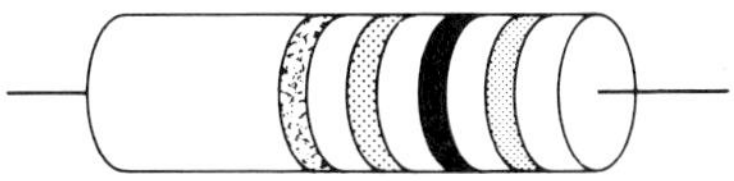

Figure 2-22. Identify Order of Bands

If you had numbered the right hand band as number one and the left as number four, you would have been correct. The first band always appears closest to an end of the resistor. The position of the resistor determines the order of reading. Since resistors are installed in every imaginable position, it is best for one not to become accustomed to seeing them in a neat, orderly fashion. Determine the order of the bands for the resistors depicted in Figure 2-23.

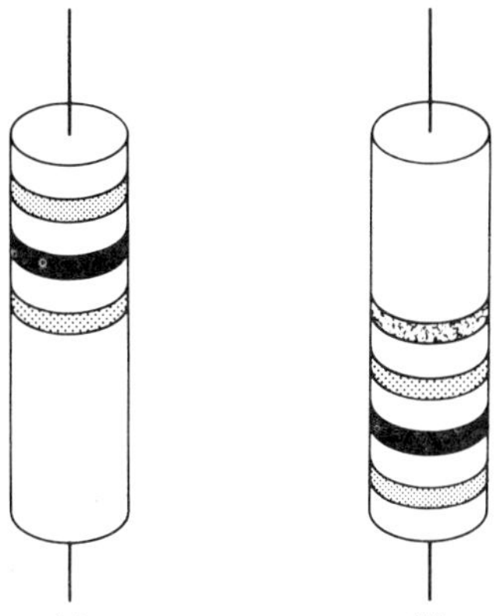

Figure 2-23. Label Order of Bands

Resistor *A* is read from top to bottom, and *B* is read from bottom to top. The colors are read according to the sample in Figure 2-24.

This resistor has an ohmic value of 4,000,000 ohms within plus or minus ten percent, and it has a predictable failure rate of one-tenth of one percent for a thousand hours of use. Any resistance value from 3,600,000 ohms to 4,400,000 ohms would fall within the ten percent tolerance range. Let's go over all of that slowly. The first band is near the left end; it is yellow. The first digit of the resistance value is 4. The second band is black which makes

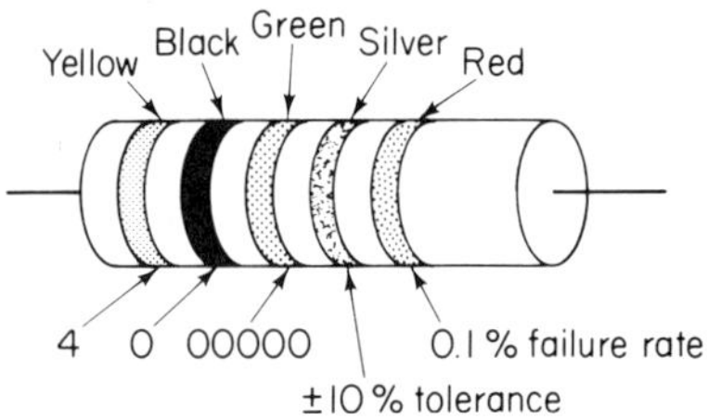

Figure 2-24. Reading a Band Coded Resistor

the second digit a 0. The fourth band is green. Green is 5; add five zeros after the first two digits. We now have 4,000,000 ohms. This is our central figure, and the tolerance tells us the percentage of this number that may be either added to or subtracted from it. The fourth band is silver which indicates ten percent tolerance. Ten percent of 4,000,000 is 400,000. 4,000,000 − 400,000 = 3,600,000, and 4,000,000 + 400,000 = 4,400,000. The fifth band is red. This gives the percentage of failures in 1000 hours of use, and red is 0.1.

Simple? Of course, but let's do a couple more for practice. What does the code in Figure 2-25 indicate?

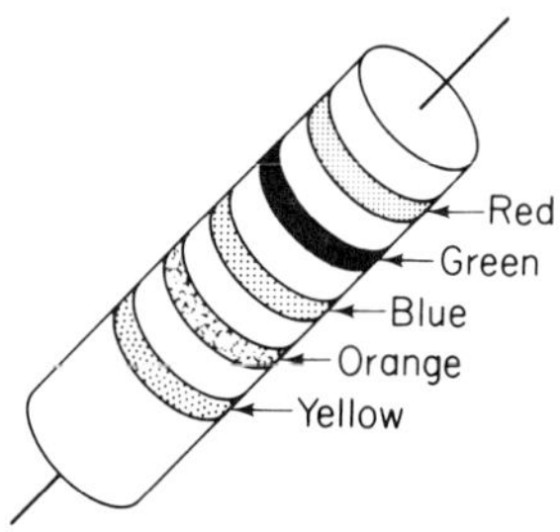

Figure 2-25. Read the Code

This resistor has an ohmic value of 25,000,000 ohms, a tolerance of 3 percent, and a failure rate of 1/1000th of 1 percent per 1000 hours. Not a very likely resistor to be sure, but that is what the bands indicate.

What is the resistance and tolerance of the resistors shown in Figure 2-26?

Resistor A is read from bottom to top.

↑ Gold = 5%	
Black = none	Resistor A = 47 ohms ± 5%
Violet = 7	(Black in the 3rd band indicates that
Yellow = 4	no zeros will be added.)

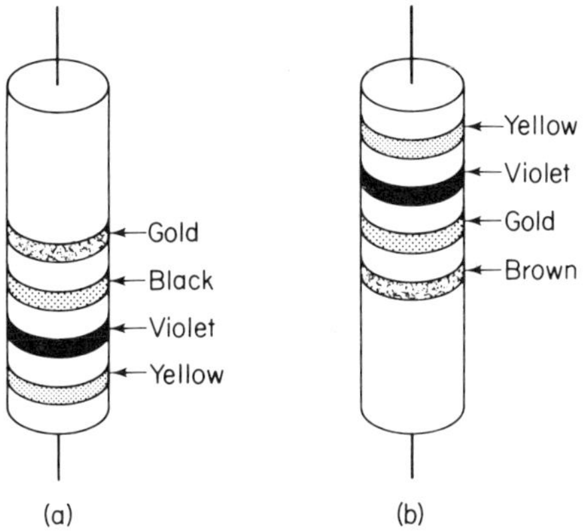

Figure 2-26. Read Resistance and Tolerance

Resistor B is read from top to bottom.

↓ Yellow = 4
Violet = 7
Gold = multiply by 0.1
Brown = 1% Resistor B = 4.7 ohms ± 1%

What is the minimum and maximum acceptable resistance for the resistor shown in Figure 2-27?

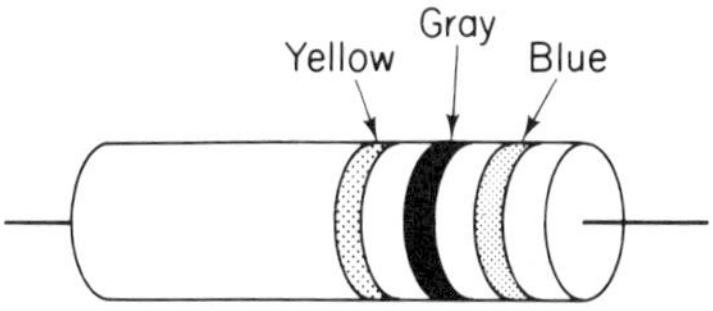

Figure 2-27. Calculate Minimum and Maximum Resistance

The code reads 680,000 ohms, but the absence of a fourth band gives us a 20 percent tolerance. Twenty percent of 680,000 ohms is 136,000 ohms. Minimum acceptable resistance is 680,000 − 136,000 or 544,000 ohms. Maximum acceptable resistance is 680,000 + 136,000 or 816,000 ohms. Unrealistic? Not at all. This is a standard size resistor, and 20 percent tolerance is more common than any other tolerance value.

the body-end-dot system

This is an old system that evolved before a standard for color coding systems was adopted. It is no longer used on new resistors, but you may encounter some old equipment with these resistors still intact. It deserves an honorable mention for its years of faithful service.

Instead of bands, specified parts of the resistor were colored. These parts were: the *body,* one *end,* and a *dot* on the center of the body. The colors were read in that order; body-end-dot. This system is illustrated in Figure 2-28.

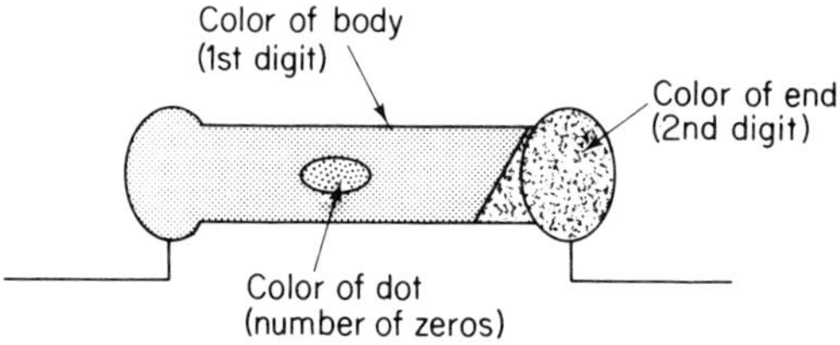

Figure 2-28. Significance of Colored Areas

The colors used in this system are the same that you have learned previously. Sometimes, a fourth color appears on the opposite end. If so, it is either gold or silver to represent a five or ten percent tolerance, respectively. More often than not the tolerance was 20 percent, which left the tolerance end with a color the same as that of the body. The resistor depicted in Figure 2-29 is 1.5 million

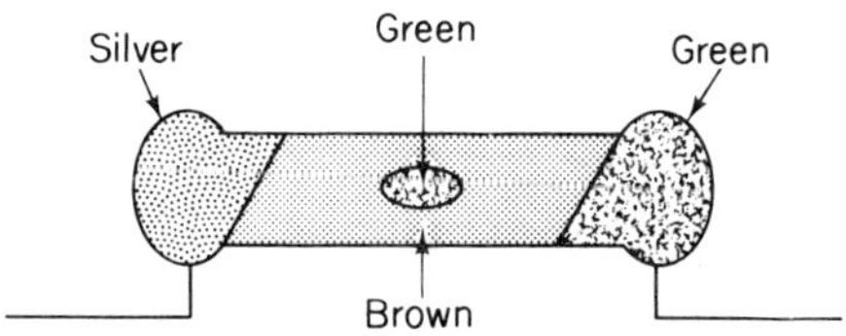

Figure 2-29. Body-End-Dot Coded Resistor

ohms with 10 percent tolerance. You will notice that after you have mastered one coding system, the other is no problem.

There are a few additional direct current (dc) circuit components that you will encounter in the near future. This seems to be an appropriate time to introduce them.

SCHEMATIC SYMBOLS FOR CIRCUIT COMPONENTS

It is not our intent to burden you with a complete list of dc circuit components. In fact, a complete list today would be less than a complete list tomorrow. Although such a list might be interesting, it would serve little purpose. Our objective is to cover several components in order to develop a feel for the symbols, schematic use, and functions of the important ones while learning the principles of dc circuits.

At this point we shall show a list of schematic symbols with the common name of the component that each represents. And to avoid later confusion, let's clarify our meaning of *circuit component.* In this book, it will mean any device represented in a circuit by a *single schematic symbol* with the exceptions of grounds and wire connections. Many of the circuit components contain other circuits that are much more complex than the schematic on which they appear. Still, if the internal function of the device is incidental to the circuit, it is quite proper to represent it as a single component.

This list appears in Figure 2-30. You will notice some components there that we have already covered. They are included for review and ease of cross-reference.

As you have already observed in studying electrostatic charges, and again in resistor sizes, electrical technicians must be proficient in handling numbers—from very small fractions to very large whole numbers. Many times, both of these extremes must be manipulated in the same problem. Not only is this awkward at times, but it also tends to promote errors. For instance, one could lose a zero from either of the following numbers with very little effort:

0.0000000000000000000000142
142000000000000000000000

There is a shorthand method of handling these numbers that will both increase the ease of calculation and decrease the chance of inadvertent errors.

POWERS OF TEN

This is known as *engineer's shorthand* or *scientific notation* and is well worth the effort to develop proficiency in its use. It not only applies to electricity but also to any other use of numbers. In fact, from this point on, you need never perform a calculation with any number smaller than 1 nor larger than 10.

power of a decimal position

Since our arabic numbering system is based on a radix of ten (0,1,2,3,4,5,6,7,8,9), each decimal position represents a power of ten.

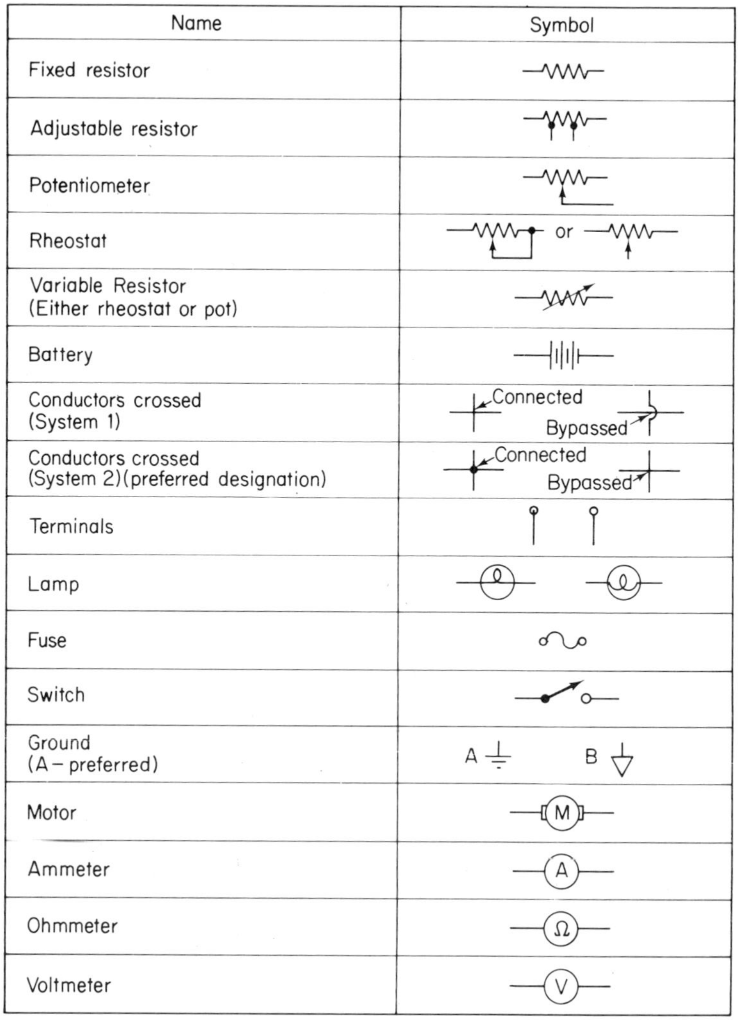

Figure 2-30. Component Names and Symbols

The quotient of any number divided by ten can be obtained by moving the decimal point of that number one place to the left. The converse is true for obtaining a product of any number and ten. This is effected by moving the decimal point in the number one place to the right.

Example: 4976.42/10 = 497.642 and 4976.42 × 10 = 49764.2.

You probably do this already—it is using the powers of ten.

Have you thought of moving the decimal point from place to place in order to make the number less cumbersome? People do this when performing calculations on the slide rule. "But," you say, "the number wouldn't be the same. It wouldn't represent the same quantity." It would be if you kept track of the decimal point. To help retain a firm grasp on that elusive point, let's imagine that moving it to the left is going in a positive direction and to the right is a negative direction. We can now start at the decimal point, assign it position zero, and assign a power of ten to each decimal position. Positions to the left will be positive powers while positions to the right are negative powers. This is illustrated in Figure 2-31.

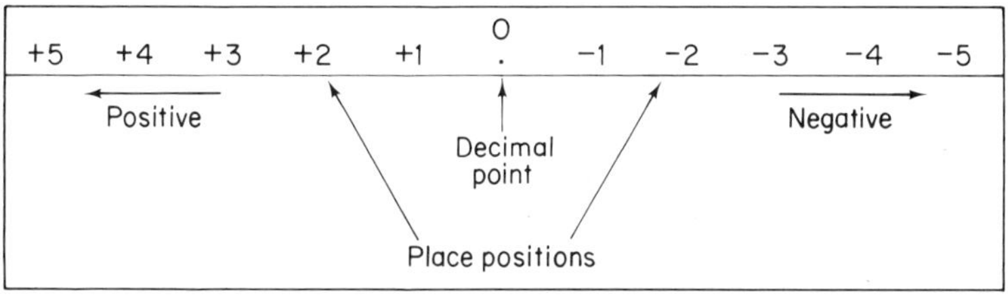

Figure 2-31. Powers of Decimal Positions

Powers are expressed like this:

1. 10^5 (read "10 to the positive fifth power").
2. 10^{-5} (read "10 to the minus fifth power").

Referring back to Figure 2-31, we can see that 10^5 is located five positions to the left of the decimal point. On the other hand, 10^{-5} is five places to the right of the decimal point. Thus, $10^5 = 10 \times 10 \times 10 \times 10 \times 10 = 100{,}000$.

We see from this that the positive power simply indicates the number of times ten is to be multiplied by itself. Since negative powers represent fractions, the negative power indicates the number of times that ten is divided into one in order to create that fraction. Thus, $10^{-5} = 1 \div 10 \div 10 \div 10 \div 10 \div 10 = 0.00001$.

Perhaps a simpler way to express the same idea is this: the power of ten indicates the place position in respect to position zero. The chart in Figure 2-32 was constructed on this premise.

You agree that this is fine for numbers which evenly multiply or divide by ten, but you ask, "What about the other possible numbers?"

We hasten to answer that question.

Power of ten	Numerical value
10^{12}	1000000000000
10^{11}	100000000000
10^{10}	10000000000
10^{9}	1000000000
10^{8}	100000000
10^{7}	10000000
10^{6}	1000000
10^{5}	100000
10^{4}	10000
10^{3}	1000
10^{2}	100
10^{1}	10
10^{0}	1
10^{-1}	0.1
10^{-2}	0.01
10^{-3}	0.001
10^{-4}	0.0001
10^{-5}	0.00001
10^{-6}	0.000001
10^{-7}	0.0000001
10^{-8}	0.00000001
10^{-9}	0.000000001
10^{-10}	0.0000000001
10^{-11}	0.00000000001
10^{-12}	0.000000000001

Figure 2-32. Value of Powers of Ten

expressing numbers as powers of ten

Any number can be expressed as a number between one and ten times a power of ten. Thus, 6,000,000 $= 6 \times 10^{6}$; 42,000 $= 4.2 \times 10^{4}$; and 0.000039 $= 3.9 \times 10^{-5}$.

How was this determined? Take the 6,000,000 for instance. Where is the decimal point? It is understood to be to the right of the sixth zero; like this: 6,000,000.←(Decimal point.) Now, move that decimal point until the number is between one and ten; like this: 6.000000. This shows our number to be six times some power of ten. Write it thus: 6×10.

To determine whether the power is positive or negative, ask, "Which way did the decimal point move?" In this case, it moved *left;* so, our power is *positive.* Count the number of positions that the decimal point moved, and this gives the power of ten. In this case, it moved six positions. Therefore, 6,000,000 $= 6 \times 10^6$.

Let's examine the fraction, 0.000039. Move the decimal point until the number is between one and ten; thus, 00003.9. The number is: 3.9×10. The *power is negative* because the decimal point moved to the *right.* How far did it move? Five positions. Therefore, $0.000039 = 3.9 \times 10^{-5}$.

It is a simple process, and you need never again wrestle with unwieldy numbers. To get back from the power to the number, just reverse the above process. For example:

$$7.91 \times 10^6 = 7{,}910{,}000$$

$$2.4 \times 10^{-7} = 0.00000024$$

In the first case, the positive power locates our number to the left of the zero position. The power (6) says that the decimal is six places from the zero position. Move the decimal point to the right six places, and fill the empty positions with zeros; thus: 7910000.

The negative power locates the number to the left of the zero position. The 7 says that the decimal is seven places from its rightful position. Move the decimal point seven places to the left and fill the empty positions with zeros; thus: 0.00000024.

To emphasize the utility of the usage of powers of ten, we shall solve a few problems with otherwise cumbersome numbers.

solving problems with powers of ten

To divide powers, the power of the divisor is subtracted from the power of the dividend. For example: $10^6/10^3 = 10^{6-3} = 10^3$ and $10^6/10^{-3} = 10^{6-(-3)} = 10^{6+3} = 10^9$.

To multiply powers, the powers of the multiplier and multiplicand are added together. For example: $10^4 \times 10^3 = 10^{4+3} = 10^7$; $10^2 \times 10^5 \times 10^6 = 10^{2+5+6} = 10^{13}$; and $10^{-2} \times 10^4 = 10^{4+(-2)} = 10^2$.

Perform the indicated operations using powers of ten.

$$\frac{690{,}000 \times 0.0035}{4200}$$

Express each number as a number between one and ten times a power of ten. The problem then becomes:

$$\frac{6.9 \times 10^5 \times 3.5 \times 10^{-3}}{4.2 \times 10^3}$$

Combine the powers:

$$\frac{6.9 \times 3.5 \times 10^{2}}{4.2 \times 10^{3}} = \frac{6.9 \times 3.5 \times 10^{-1}}{4.2}$$

Multiply top numbers:

$$\frac{6.9 \times 3.5 \times 10^{-1}}{4.2} = \frac{24.15 \times 10^{-1}}{4.2}$$

Move the decimal point and change the power in order to keep the number within bounds of one to ten:

$$\frac{24.15 \times 10^{-1}}{4.2} = \frac{2.415 \times 10^{0}}{4.2}$$

Since $10^0 = 1$, that term can now be dropped. This leaves $2.415/4.2 = 0.575$. This is less than one; so move the decimal point, and replace the power of ten: 5.75×10^{-1}.

From the above process, it should be clear that normal arithmetic operations are performed with the numbers. At the same time, the powers are manipulated in order to keep the numbers simple. The operations on the powers were previously described: *subtract powers when dividing; add powers when multiplying.*

Here is another example: $\dfrac{8{,}400{,}000 \times 6{,}000{,}000{,}000}{4{,}200 \times 30{,}000}$. Express in powers of ten: $\dfrac{8.4 \times 10^{6} \times 6 \times 10^{9}}{4.2 \times 10^{3} \times 3 \times 10^{4}}$. Combine the powers: $\dfrac{8.4 \times 6 \times 10^{6+9-3-4}}{4.2 \times 3} = \dfrac{8.4 \times 6 \times 10^{8}}{4.2 \times 3}$. Divide by inspection:

$\dfrac{\overset{2}{\cancel{8.4}} \times \overset{2}{\cancel{6}} \times 10^{8}}{\cancel{4.2} \times \cancel{3}}$. Multiply: $2 \times 2 \times 10^{8} = 4 \times 10^{8}$.

This should be enough practice with the basic procedures. We shall get more practice as we move through the book. Before we leave the subject of mathematics, note that in addition to the normal $\times$, multiplication will be indicated by one of the following methods:

1. Two quantities written together with no space between them. Such as: IR (read I times R); AB (read A times B); or XY (read X times Y).
2. Two quantities enclosed in parentheses with no space between center parentheses: $(A + B)(B + C)$ (read $A + B$ multiplied by $B + C$).

You already know from the standard resistor chart, Figure 2-18, that resistance values vary from a fraction of an ohm to millions of ohms. Some other common units of measure have even a wider range. Even the powers of ten shorthand is too bulky for easy use on schematics. As you might have expected, a system of *unit prefixes* and their symbols have been developed for expressing *very large and very small quantities.*

UNITS OF MEASURE

You have already encountered the units of measurement shown in Figure 2-33.

Unit	Measurement of	Abbreviations
Coulomb	Charge	none
Volt	Voltage	V
Second	Time	s
Ohm	Resistance	Ω
Mho	Conductance	℧
Ampere	Current	A
Watt	Power	W

Figure 2-33. Common Units of Measurement

Some of the other units of measurement that will be encountered as we move along are shown in Figure 2-34.

Unit	Measurement of	Abbreviations
Hertz	Frequency	Hz
Farad	Capacitance	F
Henry	Inductance	H

Figure 2-34. Units of Measurement to be Used Later

In electrical work, some of these units are simply not flexible enough. For example, the second as a measure of time becomes awkward when a horse running around a race track is being timed. The stop watch divides the second into smaller increments of time for this purpose. In electricity millions of specific actions take place in the space of one second. The prefixes along with their symbols go a long way toward mitigating this problem.

unit prefixes

The unit prefixes have a direct relation to powers of ten. By attaching one of these prefixes to a unit, we may enlarge or diminish the basic unit to the level that best suits our purpose. For instance, when dealing with very short time intervals, we might want to express time as a millionth part of a second. If so, we would attach the prefix *micro* to the unit, the second. We are then speaking in terms of *microseconds.* Figure 2-35 lists the prefixes along with their symbols and power-of-ten relationships.

Prefix	Symbol	Power of ten
tera	T	10^{12}
giga	G	10^{9}
mega	M	10^{6}
myria	my	10^{4}
kilo	k	10^{3}
hecto	h	10^{2}
deka	da	10^{1}
deci	d	10^{-1}
centi	c	10^{-2}
milli	m	10^{-3}
micro	μ	10^{-6}
nano	n	10^{-9}
pico	p	10^{-12}

Figure 2-35. Unit Prefixes

There are a few other prefixes, but you are not likely to encounter them in electricity. Of the thirteen listed in Figure 2-35, you are certain to use the prefixes *mega, kilo, milli, micro, nano,* and *pico.* The prefix *pico* is a relative newcomer to the field of electricity. It is replacing the *micromicro* whose symbol is $\mu\mu$. You are likely to encounter the *obsolete* micromicro from time to time.

association of units and prefixes

Mega is normally associated with resistance, voltage, frequency, and power. It is common terminology to use *megohms, megavolts, megahertz,* and *megawatts.* In each case, the quantity expressed is 1,000,000 times the size of the basic unit. The symbol for megaunits is the normal basic unit symbol preceded by the mega symbol. The symbol for megaohms is symbolized by MΩ; megavolts is symbolized by MV; megahertz become MHz; and megawatts is symbolized by MW.

Kilo is also a term normally associated with resistance, voltage, frequency, and power. The terms then become *kilohms, kilovolts, kilohertz,* and *kilowatts.* In these cases, the quantities that are expressed are 1000 times as large as the basic unit. The symbols are the normal unit symbol preceded by the kilo symbol. The above four units are symbolized by kΩ, kV, kHz, and kW, respectively.

Milli is the term most frequently used with current, voltage, power, inductance, and time. The expressions are *milliamperes, millivolts, milliwatts, millihenries,* and *milliseconds.* These units are equal to a thousandth part of the basic unit. The symbols are mA, mV, mW, mH, and mS, respectively.

Micro is the term most frequently attached to the units for current, voltage, power, capacitance, and time. These units are a millionth part of the basic unit. The expressions and symbols are: microamperes $= \mu$A; microvolts $= \mu$V; microwatts $= \mu$W; microfarad $= \mu$F; and microsecond $= \mu$S, respectively.

Nano is a term rarely encountered, except in connection with time. The term *nanosecond* is very common in electric circuits. It is an interval of time equivalent to a billionth part of a second. The symbol in this case is nS. Of course, the possibility of seeing nano used with other units should not be completely ruled out.

The term *pico* is most frequently associated with capacitance. The term used is *picofarad.* This is a trillionth part of a *farad,* the basic unit. The symbol is pF. When dealing with the pico, always be on the alert for the micromicro, which has the same meaning. A *micromicrofarad* (equal to a picofarad) is symbolized as $\mu\mu$F. Sometimes the farad is symbolized as fd which is really meant as an abbreviation for the word farad.

Being able to convert back and forth from very large to very small units is a great asset to the electrical technician. For that matter, it is even more of an asset to the electrical student.

procedure for converting units

The table in Figure 2-36 indicates the operation required in converting from one unit to another.

The table in Figure 2-36 is summarized in the chart of Figure 2-37. The numbers and arrows indicate in which direction and by how many positions one should move the decimal point.

Figures 2-36 and 2-37 were included as aids to get you started.

From	To	Multiply by
pico	micro	10^{-6}
pico	milli	10^{-9}
pico	unit	10^{-12}
micro	pico	10^{6}
micro	milli	10^{-3}
micro	unit	10^{-6}
milli	pico	10^{9}
milli	micro	10^{3}
milli	unit	10^{-3}
unit	pico	10^{12}
unit	micro	10^{6}
unit	milli	10^{3}
unit	kilo	10^{-3}
unit	mega	10^{-6}
kilo	mega	10^{-3}
kilo	unit	10^{3}
mega	kilo	10^{3}
mega	unit	10^{6}

Figure 2-36. Conversion by Multiplying

Converting from ↓	Converting to →						
	mega	kilo	unit	milli	micro	nano	pico
mega	–	3 →	6 →	9 →	12 →	15 →	18 →
kilo	← 3	–	3 →	6 →	9 →	12 →	15 →
unit	← 6	← 3	–	3 →	6 →	9 →	12 →
milli	← 9	← 6	← 3	–	3 →	6 →	9 →
micro	← 12	← 9	← 6	← 3	–	3 →	6 →
nano	← 15	← 12	← 9	← 6	← 3	–	3 →
pico	← 18	← 15	← 12	← 9	← 6	← 3	–

Figure 2-37. Conversion by Moving the Decimal Point

At first, you will use them frequently; later, you will refer to them to clear up an occasional doubt; and eventually, you will have no need for them at all. Use them now to follow through some conversions.

practice on unit conversion

Convert 25 amperes to milliamperes and to microamperes. Proceed as follows: Unit to milli: 25000. The decimal point moved three positions to the right. When the vacant positions are filled with zeros, we have 25,000 milliamperes (or 25,000 mA). Unit to micro: 25000000. The decimal point moved six places to the right. The new unit is 25,000,000 microamperes (or 25,000,000 μA).

Convert 50 ohms to kilohms and to megohms. Follow this procedure: Unit to kilo: 0.050. The decimal point moved to the left three positions. Fill the vacant position with a zero. We have 0.050. Since, in fractions, zeros to the right of the significant figures have no meaning, this becomes 0.05 kilohm (or 0.05 kΩ). Unit to mega: 0.000050. The decimal point moved six places to the left, and the number becomes 0.00005 megohm (0.00005 MΩ).

We could go through an infinite number of problems, but the procedure would be the same. Study these carefully, then practice on some problems of your own design. Don't hasten to leave this area. Use of the powers of ten and the conversion of units should be mastered at this point. Future problems will be constructed on the assumption that this has been done.

This concludes Chapter 2. The review exercise provides a good review and some vital practice.

REVIEW EXERCISES

1. What is the relationship between the quantity of free electrons in a material and that material's quality as a:
   (a) Conductor?
   (b) Insulator?
2. Draw the symbols for an ammeter, a voltmeter and an ohmmeter.
3. Conductance and resistance are ______________ of one another.
4. A battery has an E_{NL} of 12V and an E_L of 11.5V. When furnishing 5A of current, what is its internal resistance?
5. In good conductors, resistance is ______________ proportional to length and ______________ proportional to cross-sectional area.
6. What is the relationship between resistance and temperature in materials with a negative temperature coefficient?
7. Name four types of energy that can readily be converted to electrical energy.

8. The ______________ is a common device for converting mechanical energy to electrical energy.
9. What are the most common devices for converting chemical energy to electrical energy?
10. List the three principal parts of a dry cell.
11. Six dry cells connected in the manner shown in Figure P2-11 will give a terminal voltage of ______________ volts and a ______________ current capacity.

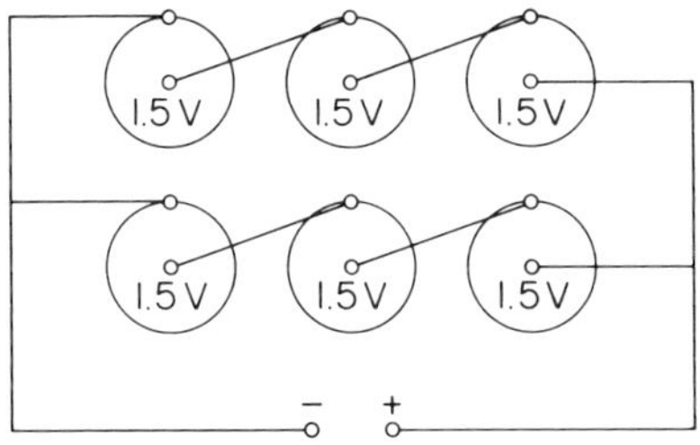

Figure P2-11

12. What are the three principal parts of the lead-acid cell?
13. The hydrometer measures the specific gravity of ______________.
14. A reading of 1.100 on the hydrometer indicates that the battery is fully ______________.
15. Why should only distilled water be added to a battery?
16. The ampere-hour is a unit of ______________ for rating ______________.
17. Most common resistors are constructed either of powdered ______________ or high resistance ______________.
18. What is the difference between physical size and ohmic value of a resistor?
19. Resistors dissipate energy in the form of ______________.
20. What are the three general types of resistors?
21. If resistance needs to be changed from time to time, which type of resistor is best suited?
22. The rheostat is a ______________ control while the potentiometer is a ______________ control.
23. What is meant by a standard size resistor?

24. What important data can be obtained from a resistor color code?
25. Decode all possible data from this resistor's color code:

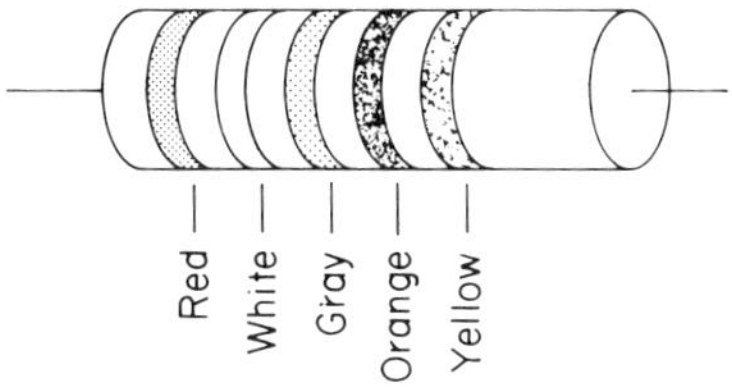

Figure P2-25

26. Read this code:

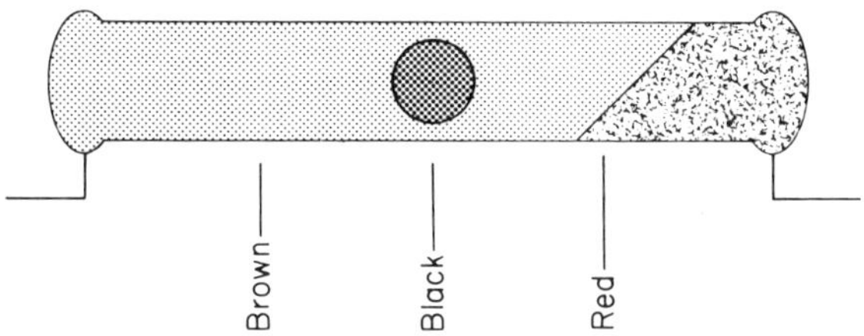

Figure P2-26

27. Identify these symbols:

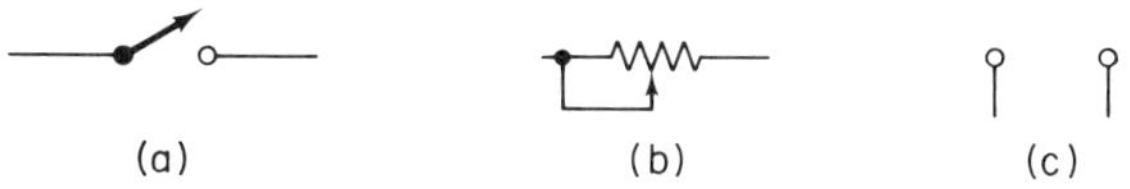

Figure P2-27

28. Express this number as a power of ten; 2,000,000.
29. Solve this problem and express the result as a number between 1 and 10 times a power of ten.

$$\frac{(6 \times 10^4 \times 3 \times 10^2)(2 \times 10^3 \times 4 \times 10^5)}{0.0002}$$

30. Convert these units as indicated.
(a) 3mA to ____________ A; and to ____________ μA.
(b) 4MΩ to ____________ kΩ; and to ____________ Ω.
(c) 6kV to ____________ V; and to ____________ mV.

3

series dc circuits

A *series circuit* is so named because all components are connected end-to-end. The only path for current is through all circuit components, one after the other. The *dc* part of the title is an abbreviation for direct current. The voltage source for a dc circuit allows current in only one direction. Our primary concern is the circuit, but we will examine some electrical instruments along the way.

USE OF DC MEASURING INSTRUMENTS

The *only measurements* that we will be concerned with in this chapter are *voltage, current,* and *resistance.* There are several varieties of meters for each. We will concern ourselves with a basic type for each purpose.

the voltmeter

This instrument measures *voltage,* and its scale is thus calibrated in volts. In common practice, many different voltages may need to be measured in a normal day's work. As a rule, the voltmeter will be capable of measuring three or more ranges of voltages, all on the same meter movement.

For example: A meter may have a range selector switch which selects ranges such as: 0 to 5 volts, 0 to 10 volts, or 0 to 25 volts.

Another meter may have these ranges: 0 to 5 volts, 0 to 50 volts, and 0 to 500 volts.

The presence of a range selector switch makes each meter more flexible and decreases the number of separate instruments required to do the job. A sample voltmeter face is shown in Figure 3-1.

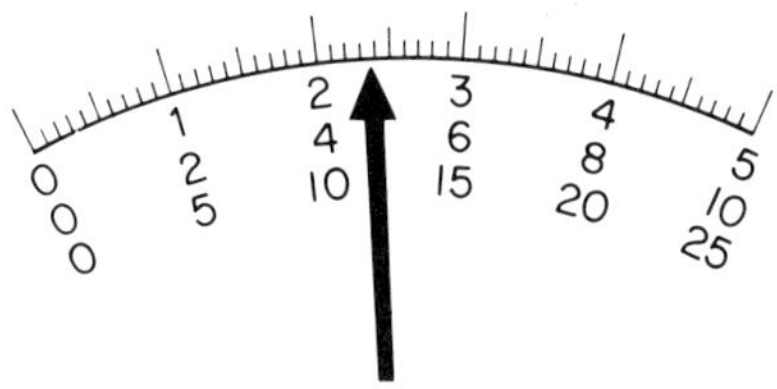

Figure 3-1. Sample Voltmeter Face

The voltmeter draws a small amount of current from the circuit to activate the pointer. The movement of the indicating needle is proportional to the current drawn. Since current is also directly proportional to voltage, it is a simple matter to calibrate the scale and label it in volts.

The schematic *symbol* for the voltmeter is 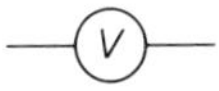 , and it is placed in parallel to the circuit component across which voltage is being measured. Remember that voltage is a difference in potential between two points: two battery terminals, two ends of a resistor, etc.

Figure 3-2 illustrates how the voltmeter is connected to a circuit. Meter M_1 is measuring the voltage across R_1; M_2 the voltage across R_2; and M_3 the terminal voltage of the battery. The voltages thus measured are frequently referred to as voltage drops, and since $E = IR$, they are also called *IR* drops.

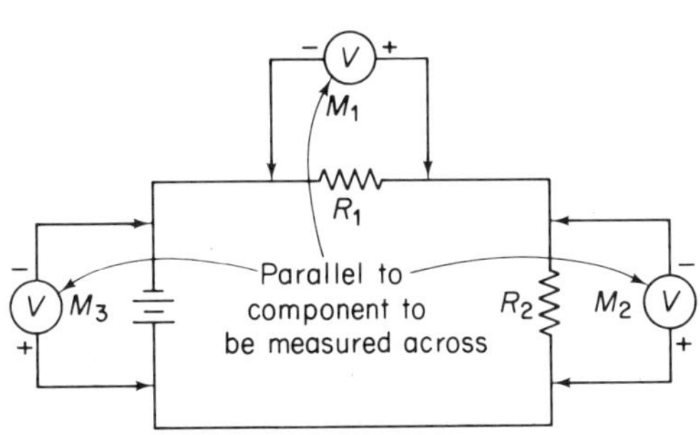

Figure 3-2. Proper Use of Voltmeter

Proper *polarities* must be observed when using the voltmeter. The test leads, which are used to connect the meter to the circuit, are colored to distinguish between the positive lead and the negative lead. To avoid meter damage, the negative lead always must be placed at the point nearest the negative terminal of the voltage source.

The voltmeter is a common, relatively inexpensive instrument but a highly important one. It is imperative that one use it when checking and maintaining electronic equipment.

the ammeter

The ammeter is used to measure *current.* It may be calibrated in amperes, milliamperes, or microamperes. It must be inserted as part of the circuit so that all of the current being measured flows through the meter. A sample ammeter is shown in Figure 3-3.

Figure 3-3. Sample Ammeter (courtesy of Simpson Electrical Company)

This is the ammeter *symbol* —(A)—, and Figure 3-4 shows its use in a circuit.

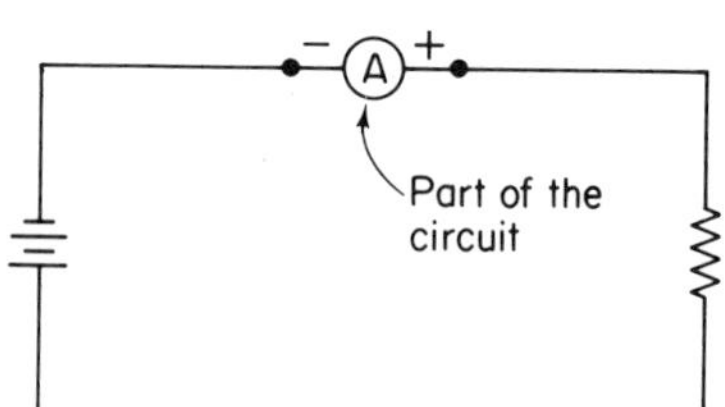

Figure 3-4. Ammeter in a Circuit

Notice that by the scales in Figure 3-3 the ammeter can also measure more than one range of currents in many cases.

Again, *polarity* must be observed because the current must pass through the meter in a given direction to avoid damage to the sensitive movement. This is accomplished by connecting the positive lead to the conductor from the positive terminal and the negative lead to the conductor from the common, or negative, terminal.

To measure current, the circuit is first disconnected as shown in Figure 3-5. Then the circuit is completed by connecting the ammeter into the gap as illustrated in Figure 3-6. After meter connections are made, close the switch.

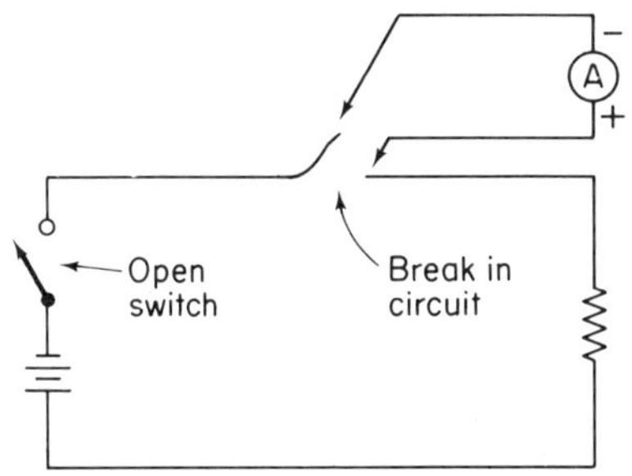

Figure 3-5. Preparing to Measure Current

Now, we need a device for measuring resistance.

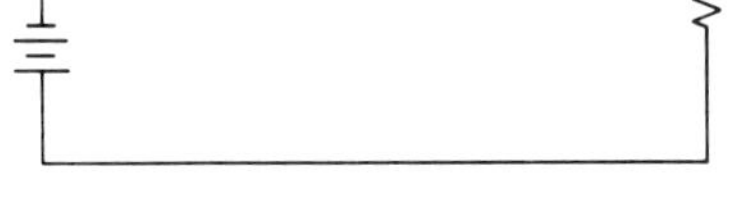

Figure 3-6. Measuring Current

the ohmmeter

One of the first and most important things to learn about the ohmmeter is that it *must not* be used in a *live circuit.* It has its own internal voltage source, and if current is forced through it from an external circuit, damage could result. Not only is this improper use of the meter, but it also produces wildly inaccurate measurements. When the resistor to be measured is installed in a circuit, at least one end of the resistor is disconnected before using the ohmmeter.

This device normally has only one scale, and it is calibrated from *zero on the right* to *infinity on the left.* The pointer rests on the left side near the infinity mark, but the scale is read from right to left ($\infty \leftarrow 0$). The scale is calibrated in ohms, but a switch allows the selection of multiplication factors. On $R \times 1$, the scale is read directly; on $R \times 10$, the scale is multiplied by ten; and on $R \times 100$, the scale is multiplied by one hundred, A sample ohmmeter face is shown in Figure 3-7.

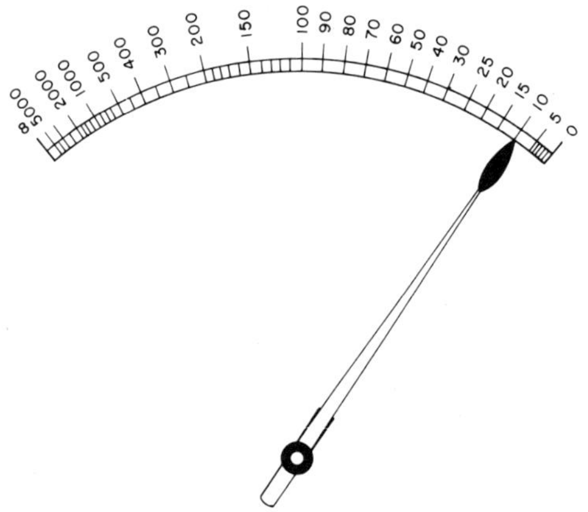

Figure 3-7. Sample Ohmmeter Face

Notice that the scale is nonlinear, and the numbers get crowded together near the left end. The most accurate readings available are between 0 and 100. The selector switch allows more readings to fall in this area. On the $R \times 1$ range, a one thousand-ohm reading would be far to the left. This same one thousand ohms would be read from the 100 point at midscale when the selector is set to $R \times 10$. On $R \times 100$, the present position of the indicator, is one thousand ohms.

The internal voltage source is a small dry cell. Cells tend to weaken for one reason or another, and a weak cell in an ohmmeter will cause inaccurate resistance readings. A *zero control knob* is provided to help compensate for the gradual weakening of the cell. This knob varies a resistance, and hence, the amount of current. Before each resistance measurement, the zero control must be adjusted. The adjustment is made while holding the conducting tips of the ohmmeter leads together. The zero knob is then turned until the pointer lies exactly on the zero mark. This setting guarantees the proper amount of current to take an accurate reading. If the control does not move the needle to zero, the internal cell must be replaced.

The schematic *symbol* for the ohmmeter is $-(\Omega)-$, and its use is illustrated in Figures 3-8 and 3-9.

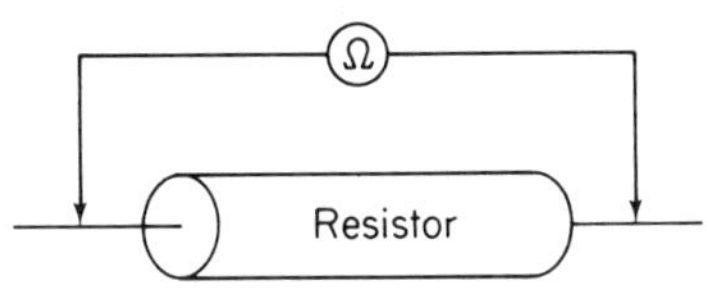

Figure 3-8. Measuring the Ohmic Value of a Resistor

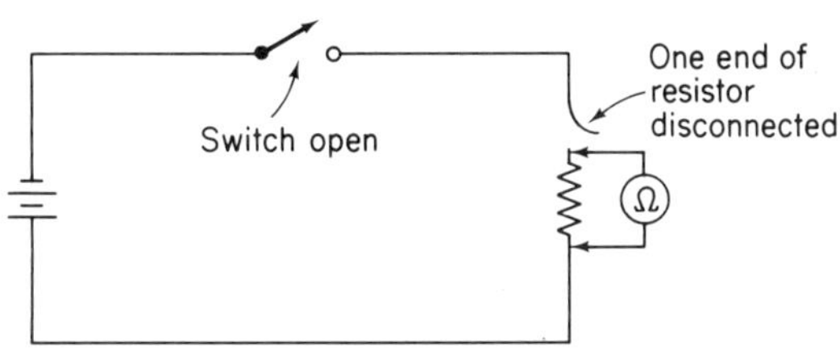

Figure 3-9. Measuring a Resistor in a Circuit

As you have seen, *all three meters operate on current.* This fact has made it possible to construct a single meter that will perform all three functions.

the multimeter

The multimeter is constructed by *combining* a voltmeter, an ammeter, and an ohmmeter into a single unit. This unit has several scales on its face but only one indicator. It has only one meter movement. The multimeter has a variety of uses. One such meter is illustrated in Figure 3-10. In addition to measuring current, voltage and resistance, this particular model measures power ratios.

Notice the four-way selector knob. In one operation, it selects the range and the type of measurement.

Now, it is time to get back to electrical circuits and learn more about the behavior of electricity.

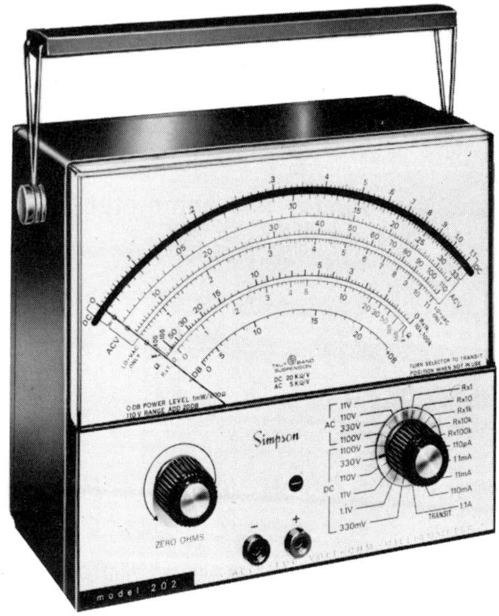

Figure 3-10. Photograph of a Multimeter (Photograph by courtesy of Simpson Electric Company)

SERIES CIRCUITS

Most of the circuits that we have dealt with up to this point have been series circuits. A series circuit has a *single path* for current. This means that the same current can be detected at any point in the circuit. All components in a series circuit are connected in-line (end-to-end) so that the total circuit current exists in each component.

voltage, current, and resistance laws

Ohm's law, which you learned in Chapter 1, applies *not only* to the circuit as a whole, but to each component of the circuit. This is illustrated in Figure 3-11. Notice that I_t (total circuit current) is used in each case because the same current passes through each component.

G. R. Kirchhoff, a German physicist (1824–1887) expanded and clarified Ohm's law with two statements which may be paraphrased as follows:

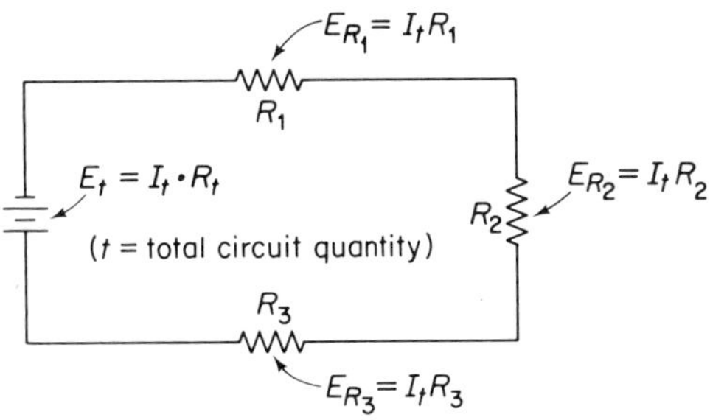

Figure 3-11. Application of Ohm's Law

1. The current entering a point is equivalent to the current leaving that point.
2. The sum of the voltage drops around a closed loop is equal to the applied voltage.

From the first statement we derive a formula stating that current is the same throughout a series circuit. The formula is:

$$I_t = I_1 = I_2 = I_3 = \ldots = I_n$$

Where I_t is total circuit current, and I_1, etc., are currents through each component.

From the second statement comes another important formula:

$$E_a = E_1 + E_2 + E_3 + \ldots + E_n$$

where E_a is the total voltage applied to the circuit, and E_1, etc., are voltages across the individual components.

Kirchhoff intended his statements to apply to all circuits. At this point, we are interested only in their application to series circuits. That is the reason for the above paraphrasing. The formulas are true expressions of *Kirchhoff's laws* as far as series circuits are concerned.

Figure 3-12 illustrates the *current law.* The circuit represented here has a total resistance of 60 ohms with an applied voltage of 120 volts. The total circuit current is therefore two amperes. These same two amperes of current can be read on an ammeter that is placed at any point in the circuit. This means that the current leaving the battery equals the current through R_1 equals the current through R_2 equals the current through R_3 equals current reentering the battery. This law is well founded, easily proven, and completely infallible.

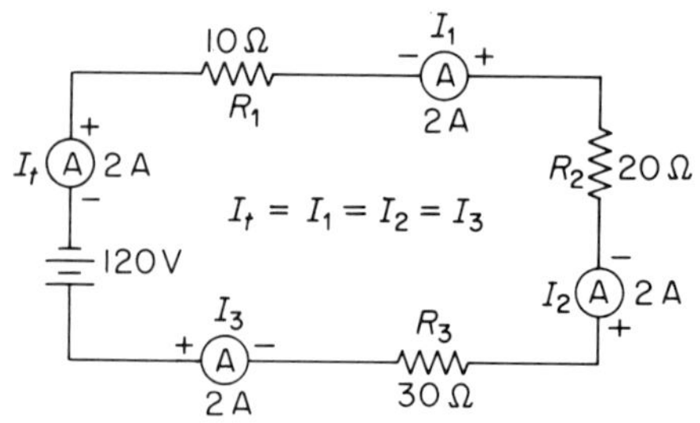

Figure 3-12. Kirchhoff's Current Law

The *voltage law* is illustrated in Figure 3-13.

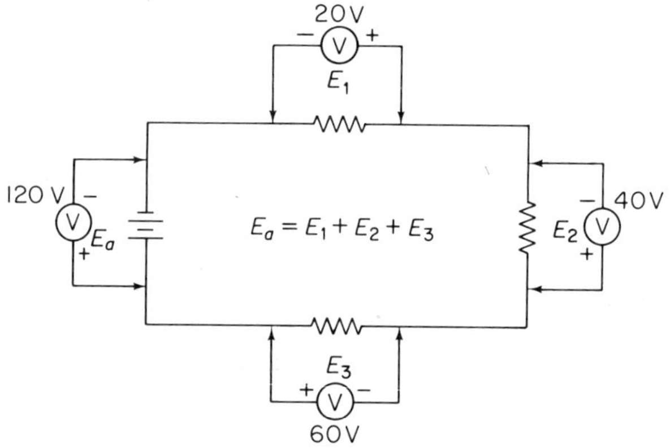

Figure 3-13. Kirchhoff's Voltage Law

In any closed loop (a complete path for current), this law can be verified by making such a series of voltage measurements with a voltmeter. In this case, 120V (E_a) = 20V (E_1) + 40V (E_2) + 60V (E_3).

Let's add another law that has been developed from Kirchhoff's and Ohm's laws: the law for total circuit resistance. The *total resistance* in a series circuit is equal to the sum of the individual resistances; that is

$$R_t = R_1 + R_2 + R_3 + \ldots + R_n$$

where R_t is the total circuit resistance, and R_1, etc., are the resistances of the individual components.

This law is also easily proven by taking a series of resistance measurements as shown in Figure 3-14.

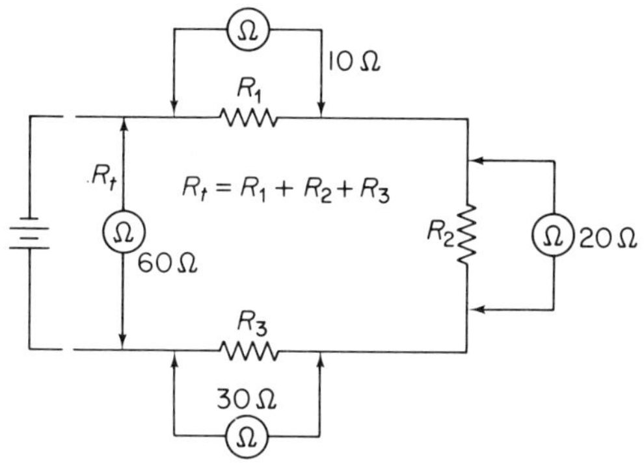

Figure 3-14. Resistance Law

First, the battery is disconnected; then, the resistors are measured one at a time. Finally, the total circuit resistance is measured by placing the ohmmeter across the open ends of the circuit. The results in this circuit are:

$$60\Omega(R_t) = 10\Omega(R_1) + 20\Omega(R_2) + 30\Omega(R_3).$$

Now, do you see how Ohm's law can be applied to any part of a circuit, as well as to the whole circuit?

applying laws to E, I, *and* R

There are three quantities associated with each component. These are: (1) resistance in ohms; (2) voltage in volts; and (3) current in amperes. If any two of these quantities are known, the third can be quickly and easily calculated by applying Ohm's law.

Example 1 (Figure 3-15): Find the voltage drop across R_2.

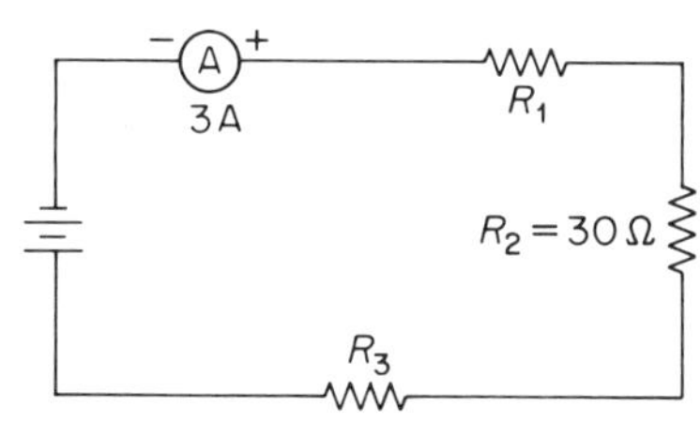

Figure 3-15. Calculate E_{R2}

We don't know the resistance of either R_1 or R_3; nor do we know the amount of voltage applied to the circuit. We do know that R_2 is 30 ohms, and it has three amperes of current through it. From this information, we can calculate E_{R2} as follows: $E_{R2} = I_t \times R_2 = 3A \times 30\Omega = 90V$.

When we know the voltage drop across a resistor and the size of that resistor, we can calculate the current through the resistor.

Example 2 (Figure 3-16): How much current is passing through R_1?

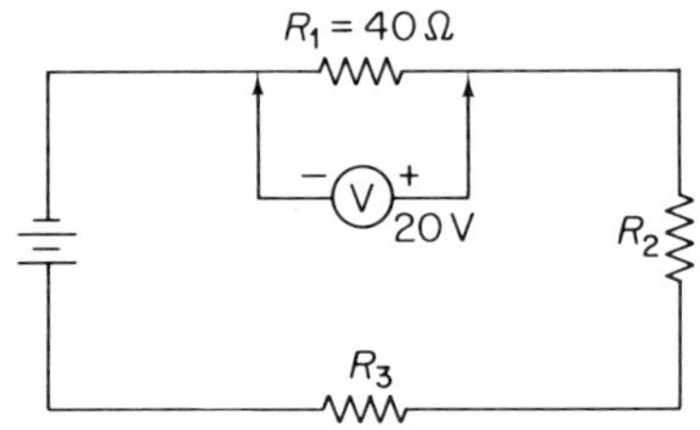

Figure 3-16. Calculate Current through R_1

When we substitute the known values, we have a simple problem.

$$I_{R1} = \frac{E_{R1}}{R_1} = \frac{20V}{40\Omega} = 0.5A$$

Since current is the same throughout the circuit, I_{R1} is also I_{R2}, I_{R3}, and I_t.

Example 3 (Figure 3-17): What is the resistance of R_3?

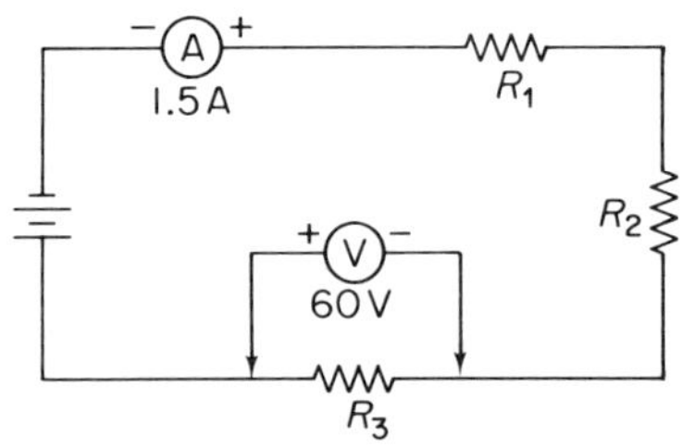

Figure 3-17. Calculate Resistance of R_3

$$R_3 = \frac{E_{R3}}{I_{R3}} = \frac{60V}{1.5A} = 40\Omega$$

Example 4 (Figure 3-18): How much voltage is applied to the circuit?

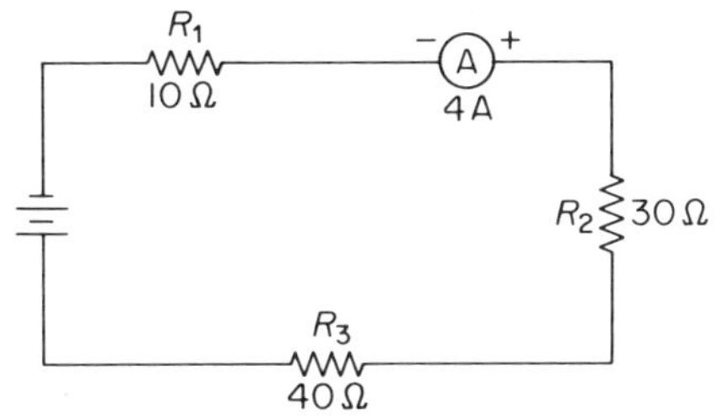

Figure 3-18. Calculate Applied Voltage

Here you have a *choice* of procedures. You may calculate the voltage drop across each resistor and add them together; or you could first find total resistance, then use I_tR_t to solve for E_a. Let's do it both ways and compare the results.

First, the sum of the individual voltage drops:

$$E_1 = 4A \times 10\Omega = 40V$$

$$E_2 = 4A \times 30\Omega = 120V$$

$$E_3 = 4A \times 40\Omega = 160V$$

$$E_a = E_1 + E_2 + E_3 = 40V + 120V + 160V = 320V$$

Now, the total resistance first:

$$R_t = R_1 + R_2 + R_3 = 10\Omega + 30\Omega + 40\Omega = 80\Omega$$
$$E_a = I_tR_t = 4A \times 80\Omega = 320V$$

Comparing the results, we find that by either way, one can determine the applied voltage to be 320 volts.

Example 5 (Figure 3-19): Determine the total circuit resistance, the resistance of R_3, and the voltage drop across each resistor.

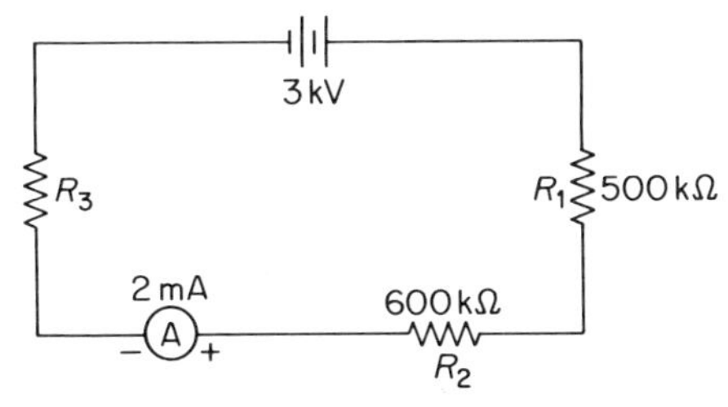

Figure 3-19. Solve for All Unknown Values

We have total voltage and total current. Let's use these values and solve for total resistance. You will note that voltage is given in kilovolts and current in milliamperes. This will give us a chance to use powers of ten in our calculations. Remember that kilo- = 10^3 and milli- = 10^{-3}.

$$R_t = \frac{E_a}{I_t} = \frac{3\text{kV}}{2\text{mA}} = \frac{3 \times 10^3}{2 \times 10^{-3}} = \frac{3 \times 10^{3-(-3)}}{2} = \frac{3 \times 10^6}{2} =$$

$$1.5 \times 10^6 \ (10^6 = \text{mega} = 1{,}000{,}000) = 1.5\text{M}\Omega \ (\text{or } 1{,}500{,}000\Omega)$$

Solving for R_3:

$$R_3 = R_t - (R_1 + R_2) = 1.5\text{M}\Omega - (500\text{k}\Omega + 600\text{k}\Omega)$$

Converting to powers of ten:

$$R_3 = 1.5 \times 10^6 - (0.5 \times 10^6 + 0.6 \times 10^6)\ 1.5 \times 10^6 - 1.1 \times 10^6 = 0.4 \times 10^6 \text{ (which is } 0.4\text{M}\Omega \text{ or } 400\text{k}\Omega)$$

When converting to powers, we manipulate the decimal point to produce the power of ten that will best serve our purpose. In the above case, we needed to add the resistances of R_1 and R_2, and then subtract this sum from 1.5×10^6. Unlike quantities *cannot* be added or subtracted, and different powers of ten are unlike quantities. It was convenient to convert everything to 10^6. This enabled us to add and subtract the basic numbers and retain the common 10^6 without change. Here are a few examples to clarify this point:

Add $2 \times 10^3 + 3 \times 10^4 + 40 \times 10^2$. These are unlike quantities and cannot be added. But they can be added if we first find a common power of ten for all. Any power would do. Let's use 10^4.

$$\begin{array}{rcl} 2 \times 10^3 &=& 0.2 \times 10^4 \\ 3 \times 10^4 &=& 3.0 \times 10^4 \\ 40 \times 10^2 &=& 0.4 \times 10^4 \\ \hline & & 3.6 \times 10^4 \end{array}$$

Add basic numbers (→ 3.6) — Bring down common power (→ 10^4)

Subtract $1.5 \times 10^6 - 20 \times 10^4$. These are unlike quantities again. We must change the power of one of the numbers. Let's use 10^4 again.

$$\begin{array}{r} 1.5 \times 10^6 = 150 \times 10^4 \\ -\ 20 \times 10^4 = \ \ 20 \times 10^4 \\ \hline 130 \times 10^4 \end{array}$$

Subtract basic numbers → (arrow to 130); Bring down common power → (arrow to 10^4)

But we used numbers larger than ten, didn't we? Yes, it was convenient to do so, but it was from choice rather than necessity. Let's do the same problem, only this time we shall use 10^6.

$$\begin{array}{r} 1.5 \times 10^6 = 1.5 \times 10^6 \\ -\ 20 \times 10^4 = 0.2 \times 10^6 \\ \hline 1.3 \times 10^6 \end{array}$$

Compare the results:

$$130 \times 10^4 = 1.3 \times 10^6$$

This little detour was intended to clarify a point on unlike terms and, more important, to emphasize the *flexibility* of using powers of ten to solve circuit problems.

Let's get back to the circuit of Figure 3-19. We found the total resistance to be 1.5MΩ and R_3 to be 400kΩ. We have also set out to find the individual voltage drops.

$$E_1 = I_tR_1 = 2\text{mA} \times 500\text{k}\Omega$$

Converting to powers of ten:

$$E_1 = 2 \times 10^{-3} \times 5 \times 10^5 = 10 \times 10^2$$

$$E_1 = 10 \times 10^2\text{V or } 10^3\text{V } (10^3\text{V} = 1000\text{V or 1kV})$$

Voltage drop across R_2:

$$E_2 = I_tR_2 = 2\text{mA} \times 600\text{k}\Omega$$

Converting to powers of ten:

$$E_2 = 2 \times 10^{-3} \times 6 \times 10^5 = 12 \times 10^2 = 1.2 \times 10^3$$

$$E_2 = 1.2 \times 10^3\text{V (or 1.2kV or 1200V)}$$

Calculating the voltage drop across R_3:

$$E_3 = I_tR_3 = 2\text{mA} \times 400\text{k}\Omega$$

Converting to powers of ten:

$E_3 = 2 \times 10^{-3} \times 4 \times 10^5 = 8 \times 10^2 = 8 \times 10^2$V (or 800V or 0.8kV)

Check the results of the voltage calculations against the 3kV which was given as the applied voltage.

$$E_a = E_1 + E_2 + E_3$$
$$3\text{kV} = 1\text{kV} + 1.2\text{kV} + 0.8\text{kV}$$
$$3\text{kV} = 3\text{kV}$$

Given E_a ↗ ↖ Sum of calculated voltage drops

series circuits as voltage dividers

You probably observed during the above calculations that the *voltage applied* to a series circuit *is divided* among the individual resistors. Did you also notice that the voltage drop across a resistor is always directly proportional to the size of that resistor relative to the total circuit resistance? It should then be possible to set up a ratio and proportion that will enable us to make some calculations without using current.

E_a is to E_1 as R_t is to R_1. Expressing that mathematically, we have: $E_a = E_1R_t/R_1$.

Example 1 (Figure 3-20): The above procedure is illustrated in the solution of Figure 3-20 for E_a.

First, add the resistor values to obtain R_t:

$$R_t = 10\Omega + 20\Omega + 30\Omega = 60\Omega$$

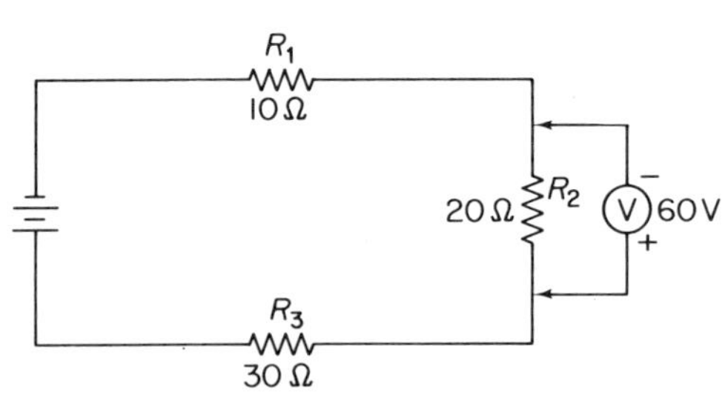

Figure 3-20. Solve for E_a

Substitute known values: $E_a = 3600/20 = 180$V.

Of course this circuit could have been solved by a straight application of Ohm's law, but the above is another useful tool in circuit analysis. With a little practice you will drop the details of the mathematics and solve problems of this type by inspection.

Example 2: It is obvious at first glance that R_2 in Figure 3-20 is one third of R_t. Therefore, the voltage drop across R_2 must be one third of E_a. The voltage is measured as 60V; so, E_a must be 3×60 or 180 volts.

We have discussed voltage, current, and resistance, but little has been said about power.

introduction to power

Frequently we hear statements concerning power that must be liberally interpreted. One such statement is, "The power to this building is 220 volts." Obviously, the meaning is power supply or voltage source since the quantity of power is dependent upon the amount of current as well as voltage.

Voltage is the pressure which causes current through resistance. Power is a measure of the energy used per unit time.

We need a few precise definitions to further illustrate the meaning of power.

Force: That which produces or tends to produce motion.

Energy: The ability to do work.

Work: Production of motion against an opposing force.

Power: The time rate of doing work.

In mechanical power, the *horsepower* is a unit of measurement; it is 550 foot-pounds per second. In other words, when a weight of 550 pounds is lifted against gravity through a distance of one foot in a time of one second, work has been accomplished at the rate of one horsepower. Power then is the time-rate of doing work.

In electricity the unit of measure for power is the watt. When one ampere of current passes through a one-ohm resistor, energy is dissipated at the rate of one watt. What about the time factor? Remember that an ampere is one coulomb passing a point in one second.

power in a dc circuit

The number of watts of power in a dc circuit is equal to the product of the voltage (in volts) times the current (in amperes). Power may also be expressed in terms of current squared times resistance, or voltage squared divided by resistance. Mathematically these three rules are expressed, respectively, as:

$$P = EI$$

$$P = I^2R$$

$$P = \frac{E^2}{R}$$

We have stated that power is measured in watts, and that one watt is used when one ampere passes through a one-ohm resistor. Power is also expressed in terms of kilowatts and milliwatts. Horsepower is another term which expresses a measurement of electrical power; an *electrical horsepower* is 746 watts.

Notice that the title of this section is "Power in a dc Circuit." We have not specified a series circuit because the power laws are the same for all dc circuits. So long as we are dealing with dc, the three formulas listed above may be applied to any circuit or to any component in a circuit.

For example, the rate at which a 100-ohm resistor that has 10 amperes through it consumes energy is calculated as follows:

$$P = I^2R = 10 \times 10 \times 100 = 10{,}000\text{W or } 10\text{kW}$$

The same would be true of this resistor in any type of configuration that caused this amount of current through it.

Total power which is consumed by a dc circuit is the product of E_a and I_t, and this is the same as the summation of the power used by each individual component in the circuit.

Example 1 (Figure 3-21): How much power is consumed by R_2?

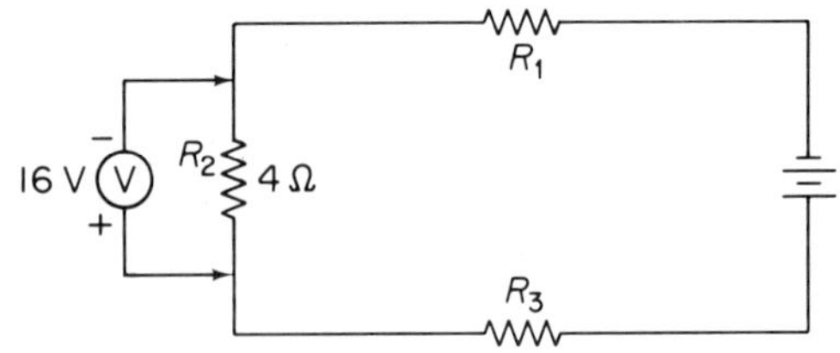

Figure 3-21. Solve for Power of R_2

Solution:

$$R_2 = \frac{E_2^2}{R_2} = \frac{16 \times 16}{4} = 64\text{W}$$

Example 2 (Figure 3-22): What is the total power consumed by the circuit?

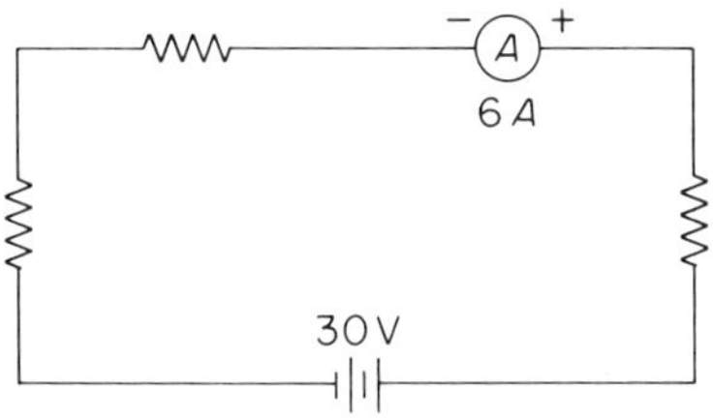

Figure 3-22. Solve for *Pt*

Solution:

$$P_t = E_a \times I_t = 30\text{V} \times 6\text{A} = 180\text{W}$$

Example 3 (Figure 3-23): What is the power used by each resistor, and what is the total circuit power?

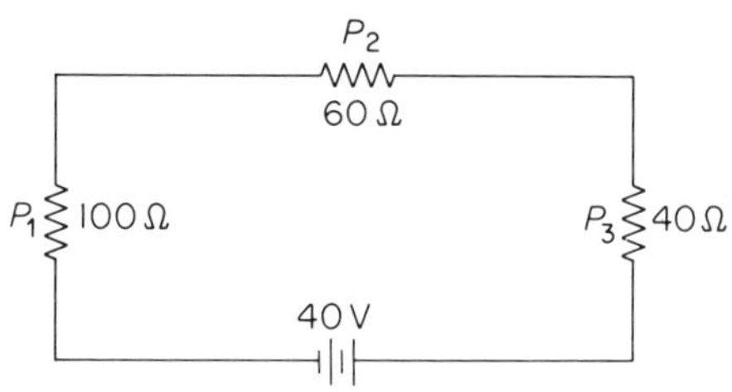

Figure 3-23. Solve for *Pt,* *P*1, *P*2, and *P*3

Solve for P_t:

$$P_t = \frac{E_a^2}{R_t}$$

Find R_t:

$$R_t = 100\Omega + 60\Omega + 40\Omega = 200\Omega$$

Use E_a and R_t:

$$P_t = \frac{40 \times 40}{200} = \frac{1600}{200} = 8\text{W}$$

Solve for P_1, P_2, and P_3 by inspection:

R_1 is one half of R_t; therefore, R_1 should use one half of P_t. P_1 must be 4 watts. That leaves 4 watts for R_2 in conjunction with R_3. R_2 is 60 percent of the combined resistance; so, it must use 60 percent of the 4 watts. P_2 is, therefore, 2.4 watts which leaves 1.6 watts for P_3.

We can *check* these figures by solving for I_t, then using I_t to solve for P_1, P_2, and P_3.

Solve for current:

$$I_t = \frac{E_a}{R_t} = \frac{40}{200} = 0.2\text{A}$$

Solve for P_1:

$$P_1 = I^2R_1 = 0.2 \times 0.2 \times 100 = 4\text{W}$$

Solve for P_2:

$$P_2 = I^2R_2 = 0.04 \times 60 = 2.4\text{W}$$

Solve for P_3:

$$P_3 = I^2R_3 = 0.04 \times 40 = 1.6\text{W}$$

Before getting into more complex circuits, we need to devote a little time to a much used, and sometimes abused, schematic symbol.

the electrical ground

The *ground* is a circuit symbol that *does not* represent a component. It simply signifies either a *common* connection or a point of *zero* reference—sometimes both. When only one grounded point appears on a schematic, it can only appear for reference purposes. When two or more grounded points are shown, the grounds signify not only points of zero potential, but also a common conductor returning to the voltage source. This common conductor return path may be composed of any conducting material. Frequently, it is the metal chassis in which the circuit is housed. Sometimes, it is a buss bar or a strap braided from low resistance wire.

Since ground is the reference point, positive and negative voltages are measured *relative* to this point. Not only are these potentials relative to ground, they are actually defined in terms of ground. For instance, any point in any circuit can be either positive or negative depending upon the two reference points. This is illustrated in

Figure 3-24. The current flows through the resistors from point *A* toward point *D*. Points *B*, *C*, and *D* are all positive with respect to point *A*. Point *B* is positive in respect to point *A*, but negative with respect to points *C* and *D*. Point *C* is positive with respect to points *A* and *B*, but negative with respect to point *D*. Thus, we see that positive and negative voltages are relative terms.

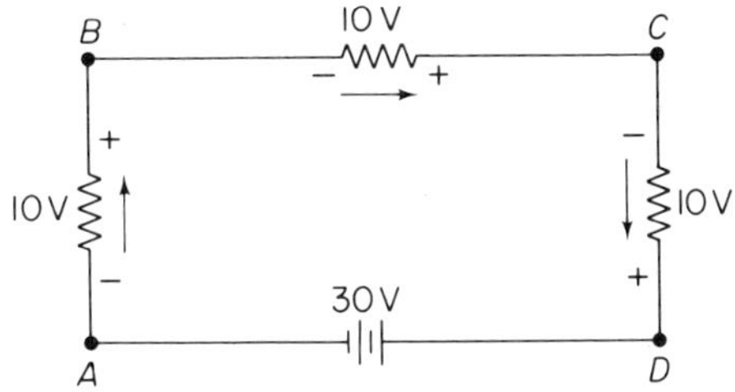

Figure 3-24. Relative Potentials

Think of it like this:

1. There must be two points of reference for there to be a potential difference.
2. The point that has current away from it is negative with respect to the other point.
3. The point that has current toward it is positive with respect to the other point.

Go back to Figure 3-24, select any two points and determine their potential relative to each other.

When the ground is used, the potentials are taken relative to ground. When only one point is specified, the second point is understood to be ground. Test point *X* has 40 volts. This means that point *X* is 40 volts positive with respect to ground.

Ground is the point of zero potential; all voltages are either positive or negative with respect to ground. It must be emphasized that zero (ground) in one circuit has no relation to zero in another circuit, nor to any point outside of that circuit. When a hand or the probe of a meter touches ground in a circuit, one will frequently discover that this is a very lively ground. Figure 3-25 shows voltages with respect to ground.

The only real difference between a positive power supply and a negative power supply is the *direction of current* with respect to ground. This is shown in Figure 3-26.

In circuit *A* the current path is through the resistor to ground,

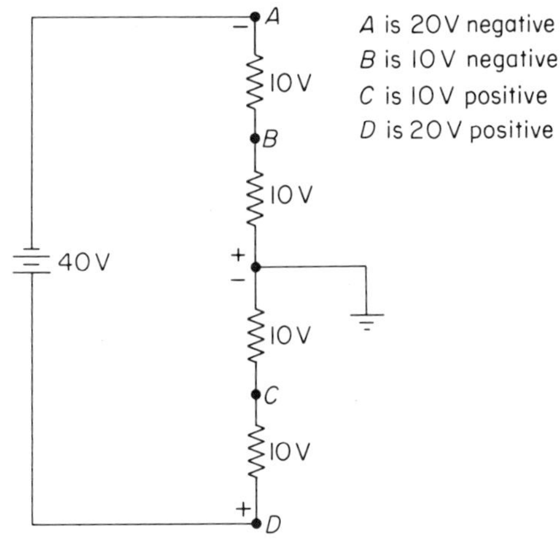

Figure 3-25. Voltages with Respect to Ground

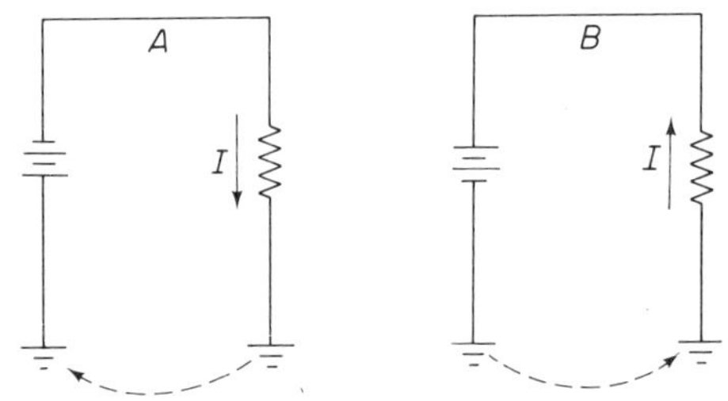

Figure 3-26. Negative and Positive Power Supplies

and back to the battery. This battery is connected as a negative power supply. In circuit *B*, the current path is from the battery through ground, and then through the resistor back to the battery. The battery in circuit *B* is connected as a positive power supply. Some automobiles have negative voltage systems; others have positive voltage systems. The only real difference between the systems is the terminal of the battery that is connected to ground. All grounded points in a given circuit are on the ground conductor and therefore are zero potential. But in the above diagrams, ground in one circuit is not zero with respect to ground in the other.

The tapped resistor may be placed in a series circuit and thus furnish both *positive and negative* voltages from the *same battery*. This is illustrated in Figure 3-27.

The voltmeter between point *A* and ground reads a *negative* 25 volts. The meter between point *B* and ground reads a *positive* 25

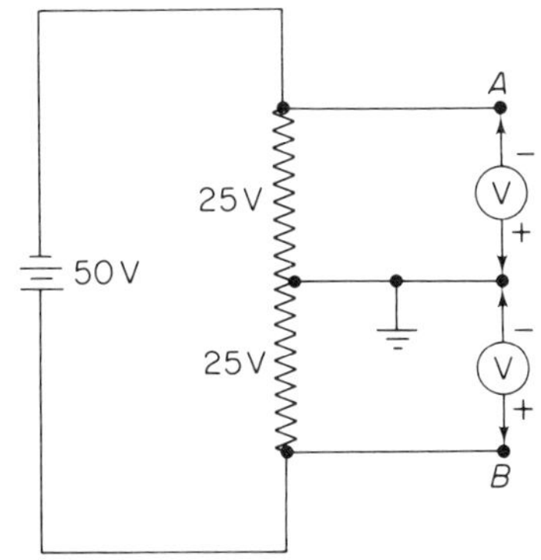

Figure 3-27. Positive and Negative Voltages from Same Battery

volts. Practically all circuits make use of the electrical ground. There is also a circuit component that appears in nearly all circuits.

the circuit protector

The *fuse* is placed in a circuit to protect the components from excessive current. The conductor inside the fuse is intended to melt when a specified level of current is reached. The fuse may be designed to handle any amount of current—from a fraction of a microampere to several thousand amperes. The circuit designer chooses a fuse that will provide safety to the components in his circuits. The idea is to allow the required amount of current to pass, but not much more. Some circuits require surges of high current for short intervals with a much lower operating current. This type of circuit requires a *time lag* fuse. The time lag fuse will allow current surges for brief periods without burning out.

A *series circuit* may be protected by placing a fuse *at any point* in the circuit. (The same current exists through the fuse no matter where it is placed.)

The fuse provides protection by breaking the circuit when it burns out. This has the same effect as the opening of a switch. Figure 3-28 is a fused circuit. When the switch is closed, the circuit is complete, current is present, and the lights burn. When excessive current is drawn, the conducting element inside the fuse burns out. This breaks the circuit just as effectively as opening the switch.

The fuse is a one-time switch. It breaks the circuit when excessive current is present, but in doing so it destroys itself and must be replaced. Compared to other circuit components, the fuse is not

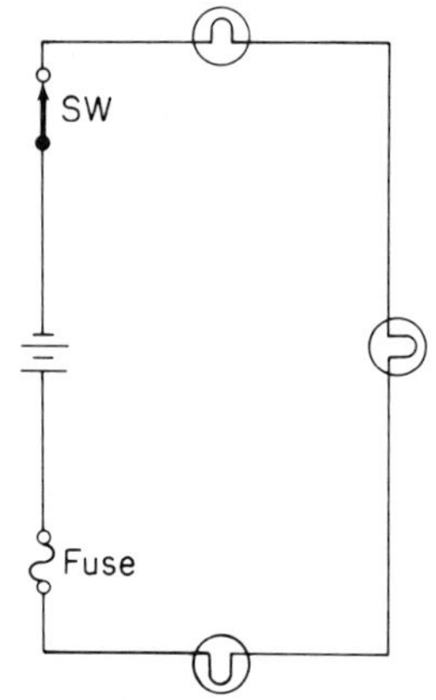

Figure 3-28. A Fuse Protected Circuit

expensive and provides adequate protection for the much more expensive items.

In circuits that are not designed to withstand current surges, replacing fuses can be annoying. This led to the development of the *circuit breaker*. It does the same job as the fuse, but after having broken the circuit, it can be reset (i.e., restored). Circuit breakers are not practical for all circuits, however. They are bulky and rather expensive. Many circuits are still protected by fuses.

circuit failures

Two of the *most common* circuit failures are *opens* and *shorts.* An open is a break in the circuit which causes an undesired interruption of current. The short is an undesirable path which allows current to bypass one or more components. Opens stop the current; shorts cause an increase in current. Figure 3-29 illustrates an open in a circuit.

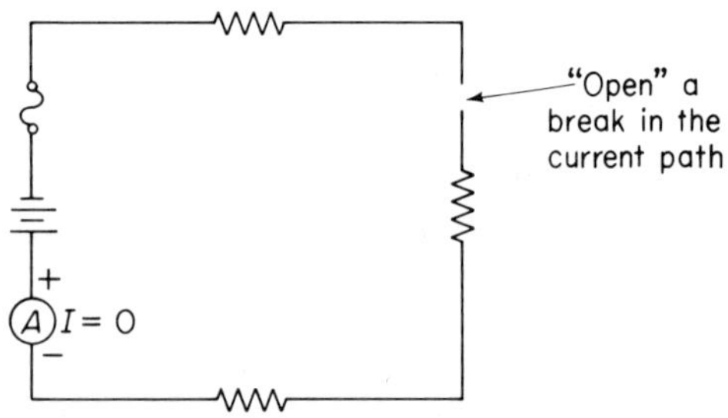

Figure 3-29. An Open Circuit

The open can be caused by many things: a bad solder connection, a burned-out resistor, even a burned-out light bulb. Repair is relatively simple once the open has been located. Usually, repair consists of resoldering a connection or replacing a bad component.

The short is illustrated in Figure 3-30. The most frequent cause

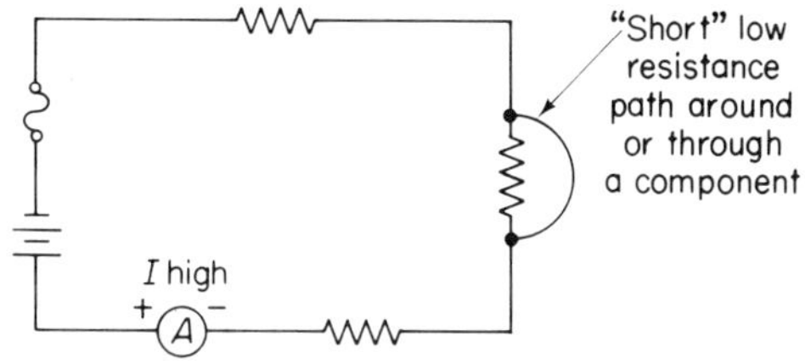

Figure 3-30. A Short Circuit

of a short is a faulty component which provides a low resistance path for current. Damaged or worn insulation on circuit wiring may allow crossed wires to touch, and shorts result.

This schematic shows a *shorted resistor* which *is not* a common occurrence. Shorts are far more likely to occur with other types of components.

Shorts and opens are being covered at the same time because it is not unusual for both to appear in a circuit at the same time. In Figure 3-30, the shorted resistor could cause a current large enough to burn out the fuse. This would produce an open in the fuse as well as the short around the resistor.

An open fuse is an indication of a circuit trouble. The fuse should not be replaced until the trouble has been located and repaired.

The voltmeter is a good tool to use in locating opens and shorts in series circuits. Since no current is in a circuit with an open, a zero reading will be obtained in the circuit in all cases except two:

1. When measuring across the power supply.
2. When measuring across the open.

Either 1 or 2 will give a voltage reading equal to the *applied voltage.* This is shown in Figure 3-31. Notice that the voltmeter actually bridges the gap when it is placed across the open. This allows a small current, but nearly all the voltage is dropped across the meter. The reading is so close to the value of E_a that you immediately know the open is between the meter leads.

With the shorted component the voltmeter reads *zero* because the two meter leads are electrically at the same point. You know

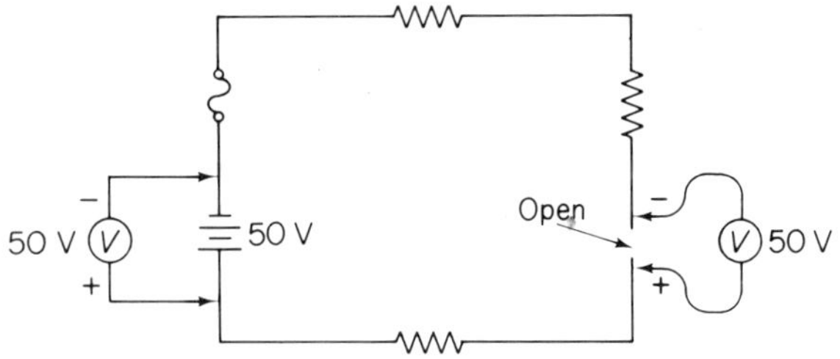

Figure 3-31. Voltmeter in an Open Circuit

that the normal component would show a voltage drop. When you know that a voltage should be read, but the meter reading is zero, that is the shorted component. This is illustrated in Figure 3-32.

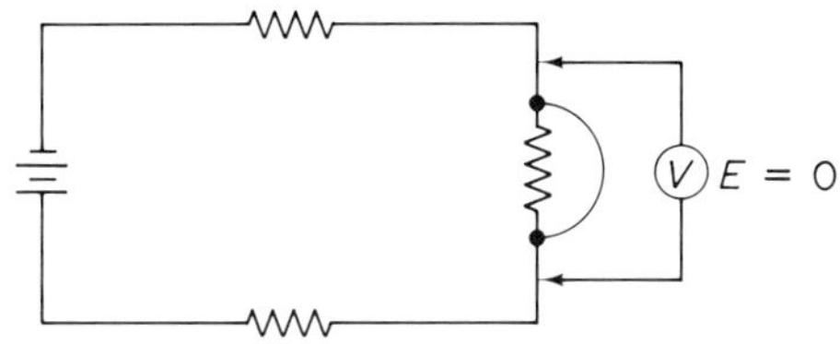

Figure 3-32. Voltmeter Detecting a Short

Another useful instrument in locating opens and shorts is the ohmmeter. The procedure is done in two steps:

1. Check continuity.
2. Measure resistance.

Figure 3-33 shows the *continuity check.* This is a check to determine if the current has a continuous path between two points. The two points in Figure 3-33 comprise the entire circuit, and this is a good starting place. Notice that we removed the voltage by opening the switch before connecting the ohmmeter.

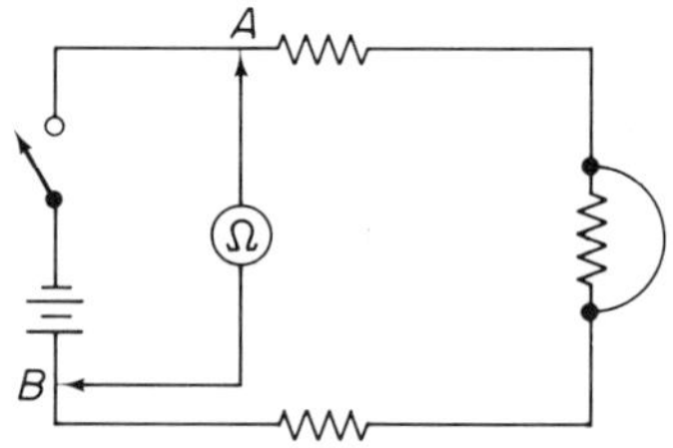

Figure 3-33. Checking Continuity

After continuity has been checked, the resistors are measured and checked against the color code values or the values indicated on the schematic. Any reading on the meter in Figure 3-33 proves that current has a complete path (continuity) from point *A*, all the way around the circuit, to point *B*. Having taken a reading, the possibility of an open circuit is eliminated, but the circuit could still have a short.

As the resistors are measured one by one, all will show a normal reading (within tolerance of schematic values) except the shorted resistor. When this reading is attempted, the meter will indicate either zero or a very low value of resistance. This is shown in Figure 3-34.

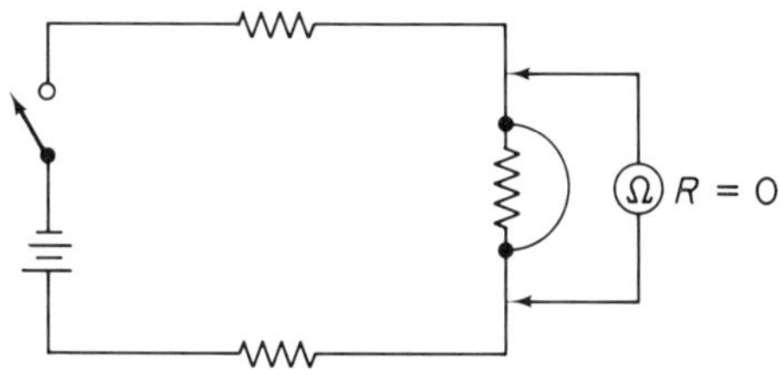

Figure 3-34. Detecting a Short With an Ohmmeter

When the circuit is open, the continuity check will give a resistance reading of infinity (∞). This is an excellent indication that the circuit is open—but where? The resistance measurement from point to point can now be replaced by a modified continuity check. With one lead at the start of the circuit, move the other along from point to point until the infinity reading appears. Secure the lead at that point (see Figure 3-35).

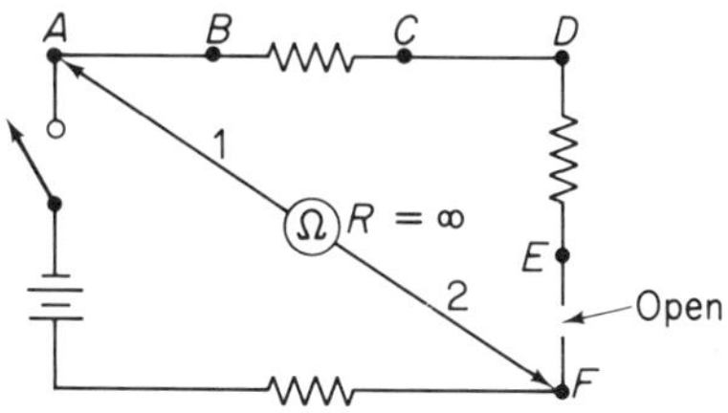

Figure 3-35. Modified Continuity Check

With lead 1 secured to point *A*, lead 2 is moved from point *B*, to *C*, to *D*, to *E*, and then to *F*. At points *B*, *C*, *D*, and *E* the needle *moved,* showing continuity. At point *F* the needle remained on *infinity,* showing an open circuit between points *E* and *F.*

This concludes Chapter 3. The following review exercises are designed to reinforce learning on the key points.

REVIEW EXERCISES

1. What is a dc circuit?
2. The voltmeter in Figure 3-1 is on the 25V scale. What voltage is being measured?
3. Why must polarities be observed when using a dc voltmeter?
4. The ammeter in Figure 3-3 is on the 1.5 ampere range, and the indicator points to the number 6. What current is being measured?
5. The ______________ is connected in series as part of the circuit being measured.
6. The most common device for measuring resistance is the ______________.
7. Why are ohmmeter readings more accurate on the right half of the scale?
8. Why must the ohmmeter be zeroed prior to each use?
9. A series circuit has a single ______________ for current.
10. Write Kirchhoff's laws for voltage and current in a series circuit.
11. Draw a circuit with a 5-Ω, a 10-Ω, and a 15-Ω resistor in series with a 60-V battery.
12. What is the I_t and R_t for the circuit of item 11?
13. Connect the ohmmeter into the circuit of Figure P3-13 to measure R_t:

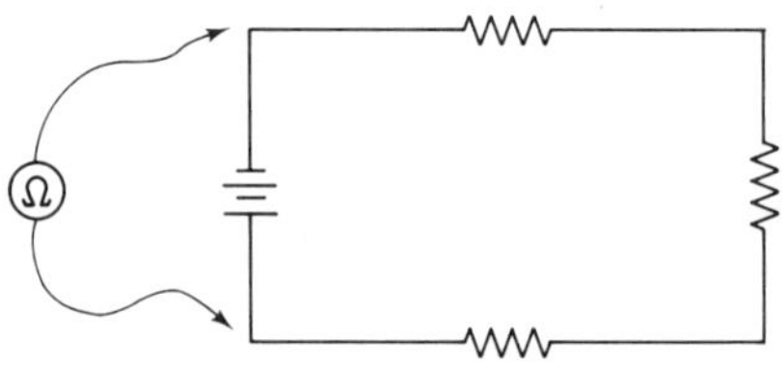

Figure P3-13

14. A series circuit has four resistors (of 5Ω, 10Ω, 20Ω, and 40Ω), and a 150-V power supply. What is the *IR* drop across each resistor?
15. A series circuit has 12-kV resistance with a current of 2mA. What is the total resistance?
16. What must be done to these numbers before they can be added or subtracted?: 3×10^{-3}; 2×10^{-4}; 4×10^{2}. Why?
17. What characteristic of a series circuit causes it to be classified as a voltage divider?
18. In a given circuit, $E_a = 60$V, $R_t = 120$kΩ, and R_x is dropping 10V. What is the resistance of R_x?
19. Power is the ______________________ of doing work.
20. The electrical horsepower is ____________ watts.
21. Write the three power formulas.
22. In any dc circuit, P_t is the ____________ of the power losses.
23. One electrical ground in a circuit signifies a ____________ point. Two or more grounds also indicate a ____________ conductor.
24. A tapped resistor may furnish both positive and negative voltages from the same ____________.
25. A battery with the positive terminal grounded is a ____________ power supply.
26. The most common circuit protector is the ____________.
27. The circuit breaker can be ____________ by throwing a switch or pressing a button.
28. Two common circuit failures are ____________ and ____________.
29. A voltmeter across an open reads ____________; across a short, ____________.
30. An ohmmeter across an open reads ____________; across a short, ____________.
31. A continuity check is performed using an ____________. Its purpose is to verify a ____________ path between two points.
32. The circuit in item 13 has 50V applied and dissipates energy at the rate of 200W. Solve for total circuit: (a) Current; (b) Resistance.

33. See item 32: All resistors are the same size. What is the: (a) resistance of each?; (b) power of each?
34. A 500-W dc motor is drawing 5A of current. What is the amplitude of the applied voltage?

4

parallel and series-parallel dc networks

In this chapter we shall deal with arrangements that are somewhat more complex and a little closer to *practical* circuits than those of Chapter 3. The use of instruments will be covered again as they apply to these circuits.

PARALLEL CIRCUITS

Parallel circuits are so named because they have *two or more branches for current.* These different current paths are electrically parallel to one another and to the power supply. Some texts refer to them as *shunt circuits.* This phrase is also descriptive because a portion of the current is *shunted* through each branch. Figure 4-1 is a schematic of a parallel circuit.

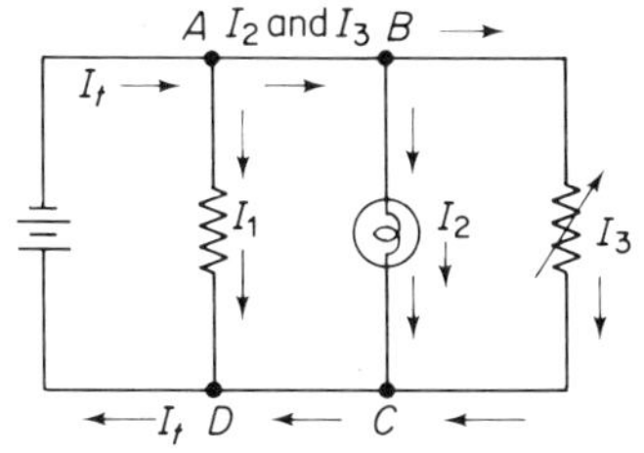

Figure 4-1. A Sample Parallel Circuit

laws for parallel circuits

At first, parallel circuits will feel a bit strange. There are several differences between these circuits and the series circuits. But a parallel circuit is still an *electric* circuit, and its actions can be predicted. Most of the rules have been developed already. Ohm's law and Kirchhoff's laws are still applicable, and, as previously stated, the power laws are applicable as well.

The *primary difference* between a series circuit and a parallel circuit is:

> The series circuit divides its total voltage among the components while a parallel circuit divides its total current among the branches.

Refer back to Figure 4-1 and notice the arrows that represent the direction of current. I_t leaves the negative terminal of the battery. It splits into two parts at point *A* and splits again at point *B.* At point *C,* I_3 recombines with I_2 and goes on to point *D.* At point *D,* I_1 recombines with the other currents, and the total current is united again. I_t goes from point *D* back to the positive terminal of the battery.

We have just illustrated Kirchhoff's current law as it applies to parallel circuits:

$$I_t = I_1 + I_2 + I_3 + \ldots + I_n$$

where I_1, I_2, etc., are the individual branch currents.

Remember that we earlier paraphrased Kirchhoff's statement about current. The actual statement was:

> The algebraic sum of all currents at any junction is equal to zero.

Analyze this statement and see if it isn't actually saying: "The total circuit current is equal to the *sum* of the branch currents."

How about the voltage? Examine Figure 4-2. Notice that all

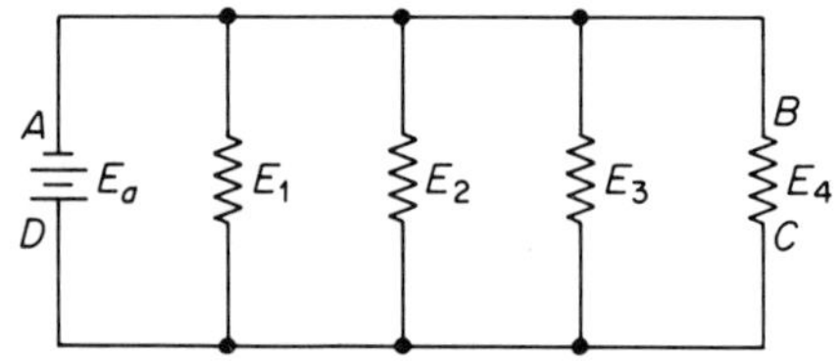

Figure 4-2. Voltage in a Parallel Circuit

parts of the conductor from point *A* to point *B* are connected directly to the negative terminal of the battery. On the other side, all parts of the conductor from point *C* to point *D* are connected directly to the positive terminal. This means that the total circuit voltage (E_a) is applied directly across each branch; that is:

$$E_a = E_1 = E_2 = E_3 = \ldots = E_n$$

Is this also an expression of Kirchhoff's law? Well, Kirchhoff actually said:

> The algebraic sum of the voltage drops around a closed loop is equal to zero.

Doesn't that statement actually imply that the *total* circuit voltage is dropped around *each loop*?

Ohm's law is no problem. The formulas for *E, I,* and *R* can be applied to both the total circuit and the individual components.

Resistance in a parallel circuit appears a bit awkward until you have become accustomed to it. Since E_a is impressed across each branch, Ohm's law ($I = E/R$) shows that the branch with the largest resistance will have the smallest current. When more branches are added, the current will have more paths to divide among, and at the same time, the total current will increase.

When one refers to parallel circuits, conductance is actually a term more appropriate than resistance. In fact, the resistance formula for parallel circuits was derived from the formula for conductance.

Conductance (*G*) for a circuit is:

$$* \; G_t = G_1 + G_2 + G_3 + \ldots + G_n$$

where G_1, G_2, etc., are the branch conductances.

Since resistance and conductance are reciprocals of each other, resistance becomes:

$$* \; \frac{1}{R_t} = \frac{1}{R_1} + \frac{1}{R_2} + \frac{1}{R_3} + \ldots + \frac{1}{R_n}$$

The above formula is known as the *reciprocal* formula, and in this form it is a bit difficult to handle. If we were to remove the fraction from the left member of the equation, it would yield a more usable form. This can be done by taking the reciprocal of both sides. The result is:

* These two formulas are identical: $G = 1/R$ and $R = 1/G$.

$$R_t = \frac{1}{\frac{1}{R_1} + \frac{1}{R_2} + \frac{1}{R_3} + \cdots + \frac{1}{R_n}}$$

This formula is probably the most commonly used resistance formula for parallel circuits. It is commonly known as the *Reciprocal of the Sum of the Reciprocals;* it enables calculation of the total circuit resistance in one operation, regardless of the number of branches.

When the circuit has only two branches, this formula may be used, but a simpler one is normally employed. It is called the *Product Over the Sum,* and looks like this:

$$R_t = \frac{R_1 R_2}{R_1 + R_2}$$

Before we get bogged down in a list of rules and regulations, let's analyze some parallel circuits.

ANALYSIS OF PARALLEL CIRCUITS

Example 1 (Figure 4-3): What is the total current in the circuit?

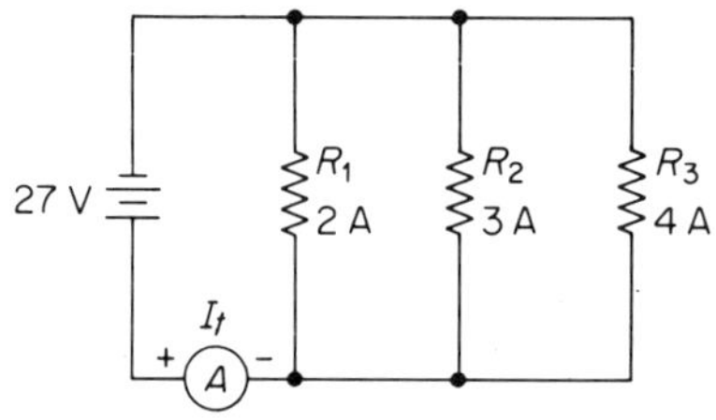

Figure 4-3. Solve for I_t

Solution:

$$I_t = I_1 + I_2 + I_3 = 2A + 3A + 4A = 9A$$

Example 2 (Figure 4-3): What is the total circuit resistance?

$$R_t = \frac{E_a}{I_t} = \frac{27V}{9A} = 3\Omega$$

Example 3 (Figure 4-3): Which resistor is the smallest in the circuit?

Solution by inspection:

R_3 has the largest current; therefore, it is the smallest resistor.

Example 4 (Figure 4-3): What is the actual size of R_3?

$$R_3 = \frac{E_a}{I_3} = \frac{27V}{4A} = 6.75\Omega$$

How can this be? The total circuit resistance is 3 ohms, yet the smallest resistor is more than twice that amount. Remember how our formula was derived. The total resistance of a parallel circuit is always less than the smallest branch resistance.

Example 5 (Figure 4-3): What is the power in this circuit?

Solution:

$$P_t = E_a I_t = 27V \times 9A = 243W$$

Example 6 (Figure 4-3): What is the power in each branch? What is the sum of these powers?

Solution:

$$P_1 = E_a I_1 = 27V \times 2A = 54W$$
$$P_2 = E_a I_2 = 27V \times 3A = 81W$$
$$P_3 = E_a I_3 = 27V \times 4A = 108W$$
$$\text{Sum} = 54W + 81W + 108W = 243W$$

The solutions to examples 5 and 6 prove that the total circuit power is equal to the sum of the individual powers. This is the same rule which we applied in series circuits.

Example 7 (Figure 4-4): Use the product over the sum formula to solve for total circuit resistance.

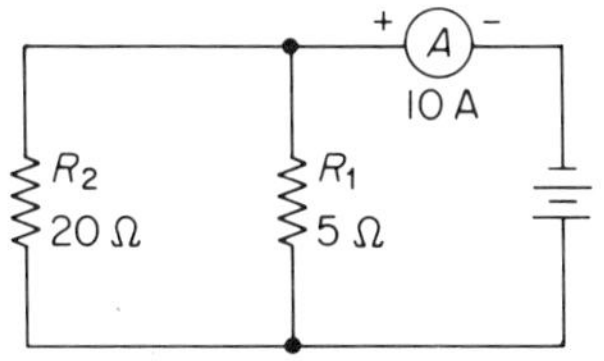

Figure 4-4. Solve for R_t (two branches)

Solution:

$$R_t = \frac{R_1 R_2}{R_1 + R_2} = \frac{20 \times 5}{20 + 5} = \frac{100}{25} = 4\Omega$$

Example 8 (Figure 4-4): What is the current through each branch?

Solution by inspection:

R_2 has four times the resistance of R_1. Therefore, R_1 has four times the current of R_2. The total circuit current is 10 amperes. This divides in a ratio according to relative resistor sizes. Eight amperes flow through R_1, and two amperes flow through R_2.

Example 9 (Figure 4-4): How much voltage is applied?

Solution:

$$E_a = I_t R_t = 10\text{A} \times 4\Omega = 40\text{V}$$

Example 10 (Figure 4-4): Use branch resistances and applied voltage to solve for branch currents.

Solution:

$$I_1 = \frac{E_a}{R_1} = \frac{40\text{V}}{5\Omega} = 8\text{A}$$

$$I_2 = \frac{E_a}{R_2} = \frac{40\text{V}}{20\Omega} = 2\text{A}$$

Compare these results with the solution in Example 8.

Example 11 (Figure 4-4): What is the power in this circuit?

Solution (one of several):

$$P_t = E_a I_t = 40\text{V} \times 10\text{A} = 400\text{W}$$

Example 12 (Figure 4-5): Use the reciprocal of the sum of the reciprocals formula to solve for total circuit resistance.

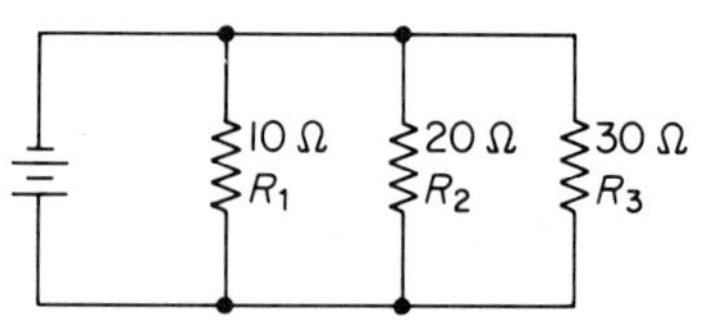

Figure 4-5. Solve for *Rt* (more than two branches)

Solution:

$$R_t = \frac{1}{\frac{1}{R_1} + \frac{1}{R_2} + \frac{1}{R_3}} = \frac{1}{\frac{1}{10} + \frac{1}{20} + \frac{1}{30}}$$

Take the lowest common denominator and combine the reciprocals:

$$R_t = \frac{1}{\frac{6+3+2}{60}} = \frac{1}{\frac{11}{60}}$$

Take the reciprocal:

$$R_t = \frac{60}{11} = 5.45\Omega$$

Some circuits seem to invite *solution by inspection.* Here are some examples.

1. Two equal resistors in parallel produce an equivalent resistance that is one-half the value of one of the resistors, e.g., two 20-ohm resistors in parallel = 10 ohms. Two 10-ohm resistors in parallel = 5 ohms.
2. Three equal resistors in parallel produce an equivalent resistance that is one-third the value of one of the resistors. Three 15-ohm resistors = 5 ohms. Three 45-ohm resistors = 15 ohms. For any number of equal resistors in parallel, $R_t = R/N$, where R = resistance of one resistor, and N = the number of resistors.

Nice, simple, complete circuits lend themselves very well to a rigid application of all our laws—but how about portions of circuits?

solving parallel networks

It is sometimes necessary to solve for *portions* of a circuit in order to predict the results of adding such a portion into a circuit. We shall refer to these incomplete circuits as networks. For instance, the resistive network in Figure 4-6 is not a complete circuit because the power supply is missing.

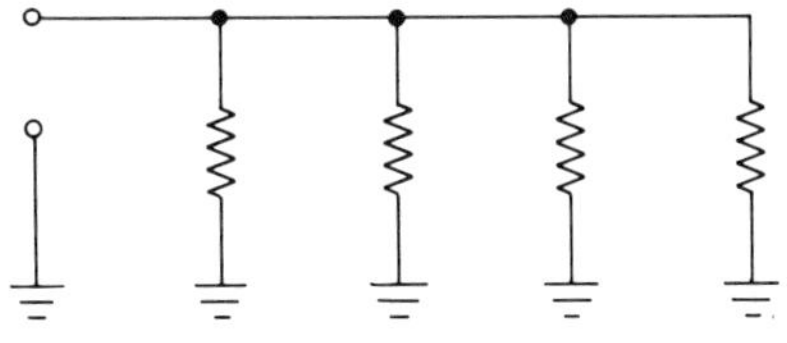

Figure 4-6. Parallel Resistive Network

An easy way to solve a resistive network for total resistance is *to assume a convenient value of voltage.* Any voltage will do, but it is best to choose one that can be easily divided by the value of each resistor. Use the actual resistor values with the assumed voltage to calculate the current in each branch. Add these hypothetical currents to get total circuit current. Divide I_t into the assumed voltage, and solve for actual network resistance. This technique is illustrated in Figure 4-7. (Of course, other solutions for resistance may be used for these networks.)

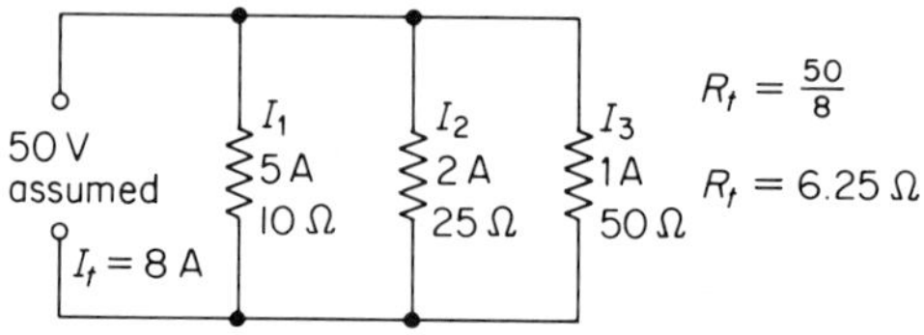

Figure 4-7. Using Assumed Voltage

Example 1 (Figure 4-8): Use assumed voltage to find the network resistance.

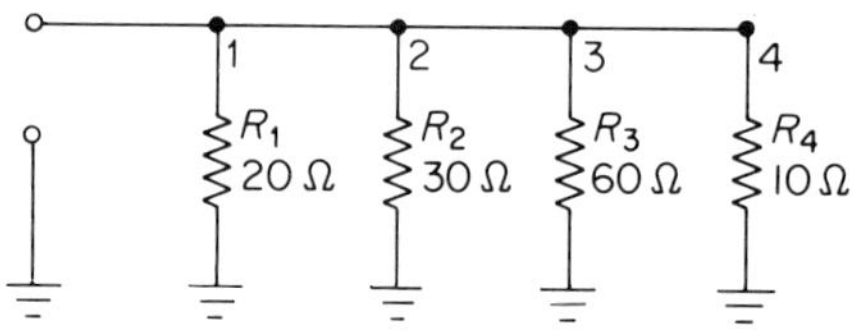

Figure 4-8. Solve for *Rt*

Assume a voltage:

60 volts will serve well.

Calculate current in each branch:

$$I_1 = \frac{60}{20} = 3\text{A}$$

$$I_2 = \frac{60}{30} = 2\text{A}$$

$$I_3 = \frac{60}{60} = 1\text{A}$$

$$I_4 = \frac{60}{10} = 6\text{A}$$

Add for total circuit current:

$$I_t = 3\text{A} + 2\text{A} + 1\text{A} + 6\text{A} = 12\text{A}$$

Solve for R_t:

$$R_t = \frac{E_a}{I_t} = \frac{60V}{12A} = 5\Omega$$

Thus far we have considered only circuits of a fixed character. Would the characteristics change if the circuit configuration were altered?

effect of adding or removing branches

All pure parallel circuits, such as those we create on paper, have all resistance grouped into the resistive components. The main line to and from the power supply is considered as zero resistance. Adding or removing branches with these circuits will change total resistance and total current, but it will have no effect on either the resistance or current of the other branches. In other words, each branch becomes an independent circuit with the entire power supply at its disposal.

Example 1 (Figure 4-9): How much change will occur in this circuit when the switch is closed?

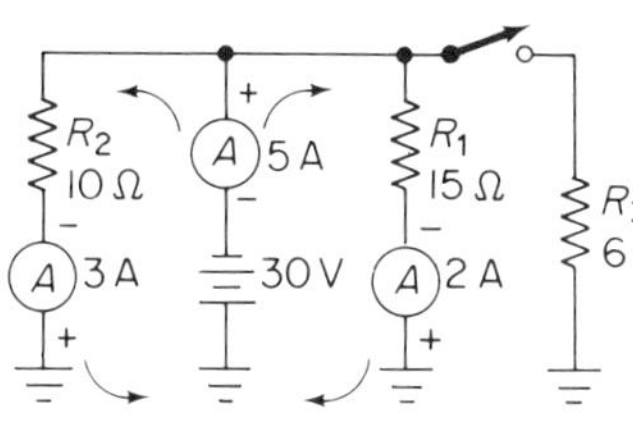

Figure 4-9. Changing Number of Branches

Presently, the circuit consists of *two* branches; one of 15 ohms and the other of 10 ohms. Total circuit resistance is 6 ohms, and total current is 5 amperes.

Close the switch:

Now the circuit has a *third* branch of six ohms. The 30V (E_a) across this resistor produces an additional five amperes of current:

$$I_t = 2A + 3A + 5A = 10A$$

$$R_t = \frac{6\Omega \times 6\Omega}{6\Omega + 6\Omega} = 3\Omega$$ (The original 6 ohms in parallel with the six-ohm branch that was added.)

Since the resistance of the first two branches does not change and the voltage across them remains constant, the current through them is not *affected.*

Open the switch:

The six-ohm resistor branch is now removed, and the circuit is the same as it was shown to be in the schematic of Figure 4-9.

We specified that these conditions exist for pure parallel circuits. In practice, however, *there are no pure parallel circuits.* The line conductor resistance becomes a significant factor as more and more branches are added. Each branch draws additional current, and the line current gets rather heavy. This heavy current causes a noticeable voltage drop across the minute resistance of the line conductor. The voltage across each branch is diminished, and branch currents are affected.

For this reason, it is unwise to place too many extensions on one wall plug. Houses are wired with parallel circuits. Each added lamp, toaster, heater, etc., adds another branch for current and increases the line current (I_t). When the line current gets high enough to burn the insulation from the wires, fires will sometimes result despite the protection afforded by fuses or circuit breakers.

parallel circuit failures

Opens and shorts are a bit more difficult to locate in a parallel circuit than they are in a series circuit.

The voltmeter cannot be used for a short because the short blows the fuse and opens the circuit. Without voltage to the circuit the voltmeter is useless. Neither can it be used to locate an open unless the open happens to be in the main line rather than in one of the branches. Since E_a is applied across each branch, an open branch produces the same voltage reading that is found across a good branch.

The ohmmeter then is the principal tool in locating parallel circuit defects. The continuity check serves no purpose because continuity will exist as long as a single branch is intact. The *procedure is* as follows:

1. Remove power from the circuit.
2. Open a lead in the suspected branch.
3. Measure the resistance of that branch.

4. Compare the measured resistance with color-coded values or values shown on the schematic.
5. If the results from step 4 are favorable, the branch is good. Reconnect the lead.

Steps 2 through 5 are repeated until the bad branch is located. If the branch is *shorted,* it will produce a *full-scale* deflection of the indicator; this is a zero reading. If the branch is *open,* there will be *no deflection* of the needle; this reading is infinity. Figure 4-10 illus-

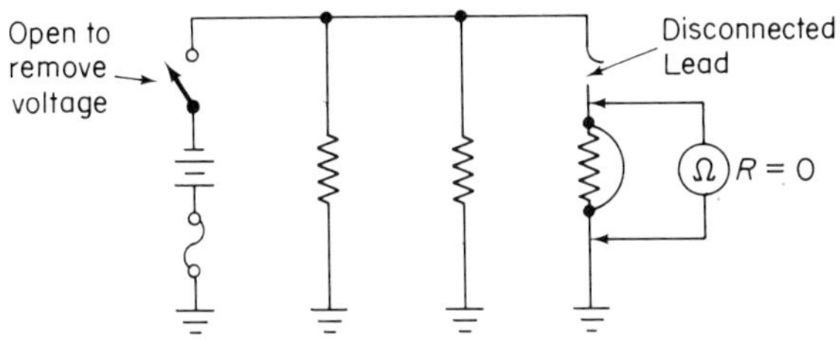

Figure 4-10. Ohmmeter Across a Short

trates the detection of a short, and Figure 4-11, the detection of an open.

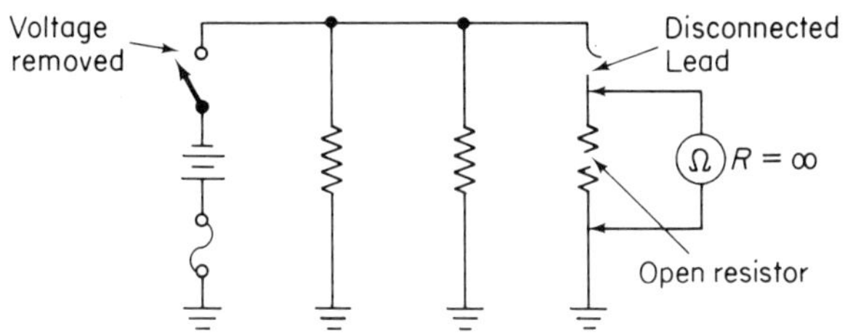

Figure 4-11. Ohmmeter Across an Open

If one attempts a measurement without disconnecting the branch, the ohmmeter will be unable to detect either a short or an open. Figure 4-12 shows what happens with the short.

The broken line shows the path of current. Since this is a path of no resistance, the meter shows a full-scale deflection. The same indication would result from one's trying to measure any branch. A short is obviously present—but where is it? Which branch? There is no way to tell without disconnecting the branches one at a time.

Figure 4-13 illustrates the result of one's trying to locate an open without disconnecting the branch.

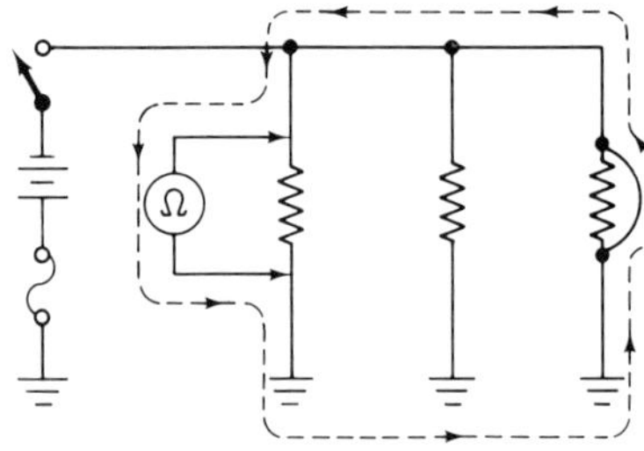

Figure 4-12. Result of Short and All Branches Connected

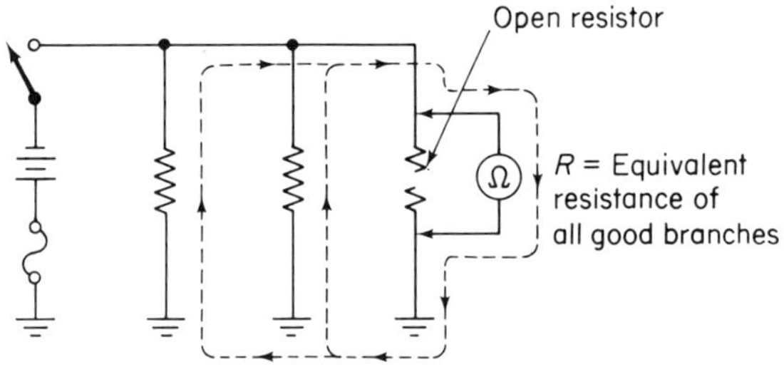

Figure 4-13. Results of Open with All Branches Connected

The current from the meter takes a path through all good branches as shown by the broken lines. The meter actually measures the equivalent (parallel) resistance of all of the good branches. This same reading would occur across each branch in the circuit.

large and small values in the same circuit

Example 1 (Figure 4-14): What is the applied voltage?

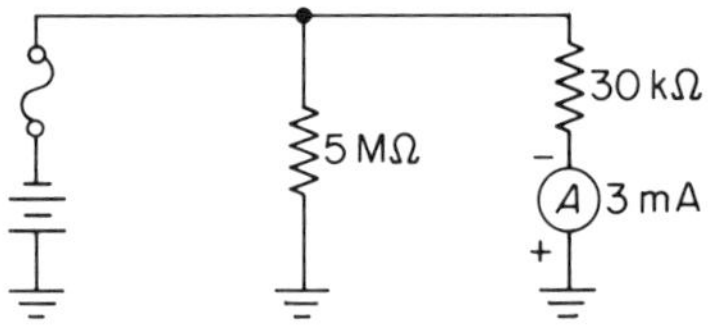

Figure 4-14. Large Resistor—Small Currents

Solution:

$$E_a = IR = 3\text{mA} \times 30\text{k}\Omega$$

Change to powers of ten:

$$E_a = 3 \times 10^{-3} \times 30 \times 10^{3} = 3 \times 30 = 90\text{V}$$

Example 2 (Figure 4-14): How much current is present in the 5MΩ resistor?

Solution:

$$I = \frac{E}{R} = \frac{90V}{5M\Omega}$$

Change to powers of ten:

$$I = \frac{90V}{5 \times 10^6} = \frac{90 \times 10^{-6}}{5} = 18 \times 10^{-6} = 18\mu A$$

Example 3 (Figure 4-14): What is the total current in the circuit?

Solution:

$$I_t = I_1 + I_2 = 18\mu A + 3mA$$

Move decimal point to express currents in like units:

$$18\mu A = 0.018mA$$

Add:

$$I_t = 0.018mA + 3mA = 3.018mA$$

This could have been calculated in microamperes as follows (The end result is the same.):

$$3mA = 3000\mu A$$

$$I_t = 18\mu A + 3000\mu A = 3018\mu A$$

$$3018\mu A = 3.018mA$$

Most circuits encountered in electronics are neither pure series nor pure parallel, but some combination of the two.

SERIES-PARALLEL (SP) CIRCUITS

Because *all complex circuits* are composed of some components in series and others in parallel, they are commonly known as *series-parallel* (SP) circuits. A series-parallel circuit has one or more components in series (using total current), plus two or more parallel branches. The simplest form of a series-parallel circuit is shown in Figure 4-15.

Total circuit current leaves the negative terminal of the battery, passes through the series component (R_1), and then I_t divides between the parallel branches (R_2 and R_3).

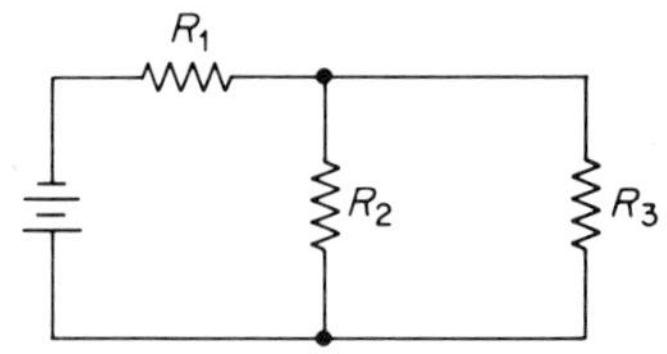

Figure 4-15. Series-Parallel Circuit

procedures for analyzing SP circuits

Analysis for these circuits is best accomplished in *two parts:*

1. parallel section;
2. series section.

While in the parallel section of the circuit, use the rules that we learned in connection with parallel circuits. In the series section, apply the rules for series circuits. Figure 4-16 illustrates the two sections.

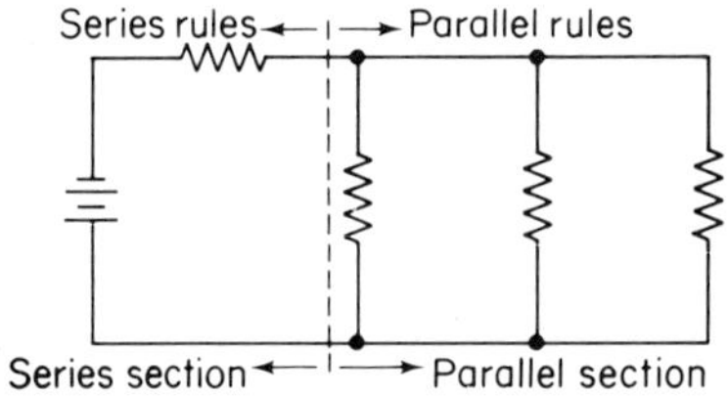

Figure 4-16. Sections of an SP Circuit

Total circuit current in a SP circuit exists in the series section and is divided among the branches. Part of the applied voltage drops across the series components, and the remainder appears across each parallel branch. The circuit resistance is obtained by using parallel circuit resistance formulas to reduce the parallel section to an equivalent resistance. R_t is the sum of the equivalent parallel resistance plus the series resistance.

Ohm's and Kirchhoff's laws are still adaptable and may be applied to the whole circuit as well as to any part thereof.

The starting point for a problem on series-parallel circuits is flexible and varies with the circuit. There is no "best" starting point common to all problems. First, inspect the circuit. When you locate a spot where two values are known, you have found a good starting point. After one calculation is performed, the results generally reveal enough information to allow additional calculations to be made. When the circuit is handled in small steps, there may be several steps, but they are all relatively simple.

solving for SP circuit problems

Example 1 (Figure 4-17): Determine the equivalent resistance (R_e) of the parallel section.

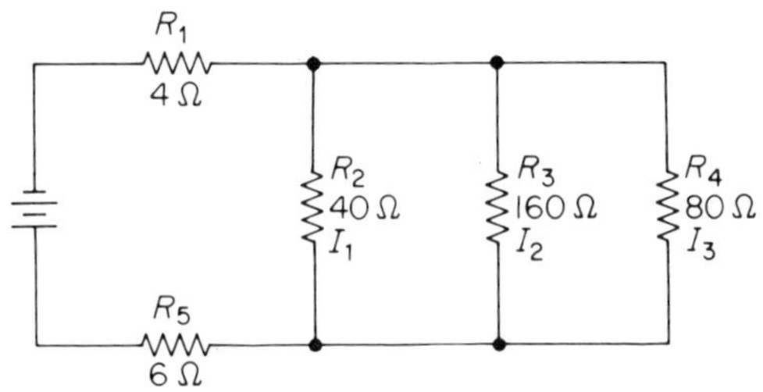

Figure 4-17. Series-Parallel Resistance

Solution:

$$R_e = \frac{1}{\frac{1}{R_2} + \frac{1}{R_3} + \frac{1}{R_4}} = \frac{1}{\frac{1}{40} + \frac{1}{160} + \frac{1}{80}}$$

Take the lowest common denominator and add the reciprocals:

$$R_e = \frac{1}{\frac{4+1+2}{160}} = \frac{1}{\frac{7}{160}}$$

Take the reciprocal and divide for R_e:

$$R_e = \frac{160}{7} = 22.8\Omega$$

Example 2 (Figure 4-17): What is the series resistance (R_s)?

Solution:

$$R_s = R_1 + R_5 = 4\Omega + 6\Omega = 10\Omega$$

Example 3 (Figure 4-17): Combine R_e and R_s to obtain R_t.

Solution:

$$R_t = R_s + R_e = 10\Omega + 22.8\Omega = 32.8\Omega$$

Example 4 (Figure 4-18): How is the voltage distributed in this circuit?

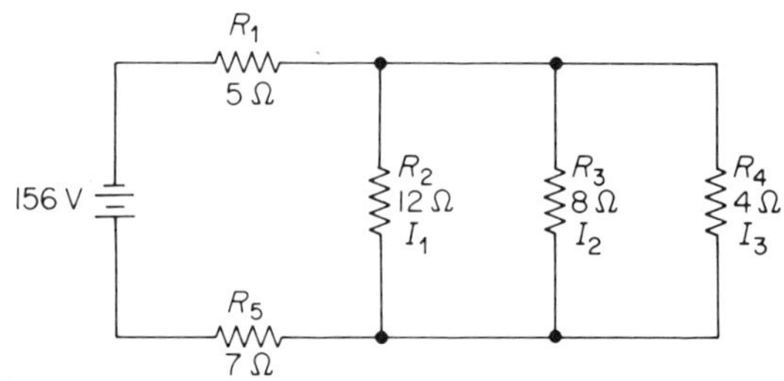

Figure 4-18. Determine Voltage Distribution

Solution:

First determine R_t:

$$R_t = R_s + R_e = 5\Omega + 7\Omega + \frac{1}{\frac{1}{12} + \frac{1}{8} + \frac{1}{4}} = 5\Omega + 7\Omega + \frac{1}{\frac{11}{24}}$$

$$R_t = 5\Omega + 7\Omega + 2.18\Omega = 14.18\Omega$$

Solve for I_t:

$$I_t = \frac{E_a}{R_t} = \frac{156\text{V}}{14.18\Omega} = 11\text{A}$$

Solve for voltage drop across each series component:

$$E_1 = I_t R_1 = 11\text{A} \times 5\Omega = 55\text{V}$$
$$E_5 = I_t R_5 = 11\text{A} \times 7\Omega = 77\text{V}$$

Solve for voltage across the parallel branches (E_p):

$$E_p = E_a - E_s = 156\text{V} - 55\text{V} - 77\text{V} = 24\text{V}$$

This same voltage (24V) appears across each branch.

Example 5 (Figure 4-18): What is the power in the circuit?

Solution:

$$P_t = I^2 R_t = 11 \times 11 \times 14.18 = 1715.78\text{W} \text{ (or 1.7kW)}$$

Example 6 (Figure 4-19): Determine the applied voltage of this circuit.

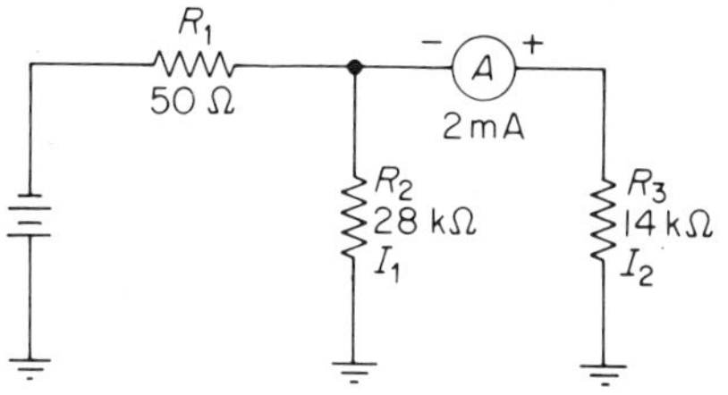

Figure 4-19. Handling Large and Small Values

Solve for E_p:

$$E_p = I_2R_3 = 2\text{mA} \times 14\text{k}\Omega$$

Change to powers of ten:

$$E_p = 2 \times 10^{-3} \times 14 \times 10^3 = 28\text{V}$$

Solve for I_1:

$$I_1 = \frac{E_p}{R_2} = \frac{28\text{V}}{28\text{k}\Omega}$$

Change to powers of ten:

$$I_1 = \frac{28}{28 \times 10^3} = \frac{28 \times 10^{-3}}{28} = 10^{-3} = 1\text{mA}$$

Solve for I_t:

$$I_t = I_1 + I_2 = 1\text{mA} + 2\text{mA} = 3\text{mA}$$

Solve for E_s:

$$E_s = I_tR_1 = 3\text{mA} \times 50\Omega = 3 \times 10^{-3} \times 50 = 150 \times 10^{-3} = 150\text{mV}$$
(or 0.15V)

Solve for E_a:

$$E_a = E_s + E_p = 0.15\text{V} + 28\text{V} = 28.15\text{V}$$

Example 7 (Figure 4-19): What is the power in branch 2?

Solution:

$$P_3 = (I_2)^2 \times R_3 = (2\text{mA})^2 \times 14\text{k}\Omega$$

Change to powers of ten:

$$P_3 = (2 \times 10^{-3})(2 \times 10^{-3})(14 \times 10^3) = 56 \times 10^{-3} = 56\text{mW}$$
(or 0.056W)

predicting results of circuit changes

In the SP circuit any change affects all parts of the circuit. This is a result of the voltage distribution. Changes in branch resistance, number of branches, series resistance, or applied voltage all affect the voltage distribution and, consequently, the amount of current through each component. This is illustrated in Figure 4-20.

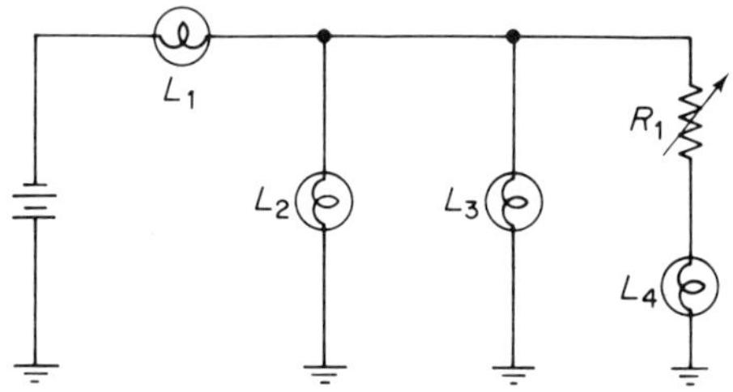

Figure 4-20. Effect of Varying Resistance

Example 1 (Figure 4-20): Assume that all four lights are burning with *equal brilliance.* How will each light change when R_1 is increased?

Solution by inspection:

1. R_1 increases.
2. Current in that branch decreases.
3. Total circuit current decreases.
4. Less voltage drops across L_1.
5. More voltage drops across each branch.
6. L_1 gets dimmer.
7. L_2 and L_3 grow brighter.
8. L_4 grows dimmer. (Even though the branch voltage increased, the larger resistor allows less branch current. This means less voltage across the light.)

Example 2 (Figure 4-21): What happens to the brilliance of these lights when a branch is added by closing the switch?

Solution by inspection:

1. The new branch draws some current.
2. Total circuit current increases.
3. More voltage drops across L_1.

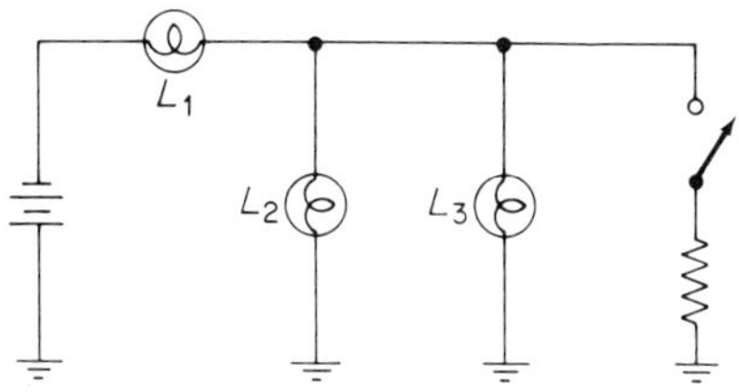

Figure 4-21. Effect of Adding a Branch

4. Less voltage drops across each branch.
5. L_1 gets brighter.
6. L_2 and L_3 grow dimmer.

When branches are added to an existing circuit, the new portion of the circuit will be powered by the first circuit. You have already noticed some effects of this operation. If the original circuit is called a voltage divider, and the added circuit is called a load, we have a *loaded voltage divider.*

LOADED VOLTAGE DIVIDERS

The subject of *voltage dividers* has come up several times, but a full explanation is not possible before developing the laws of SP circuits. The voltage dividers that we have seen in connection with series circuits don't serve much purpose until we connect them to other circuits. When this has been done, we have a series-parallel circuit. Figure 4-22 shows a voltage divider circuit before the load is added.

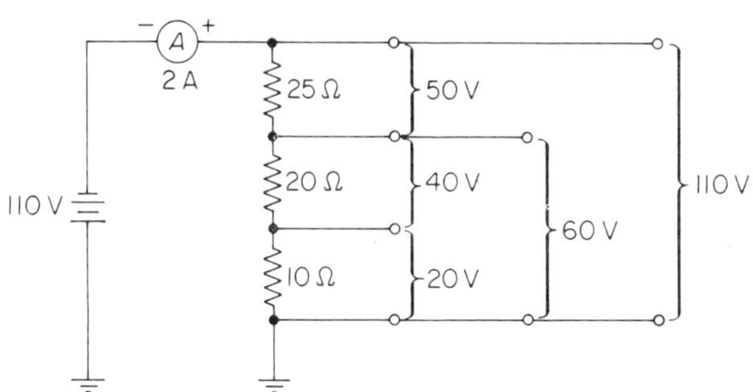

Figure 4-22. A Voltage Divider without a Load

results of loading the circuit

When the tapped voltage is put to use, the total circuit current changes, and the available voltage changes. Figure 4-23 shows this trend.

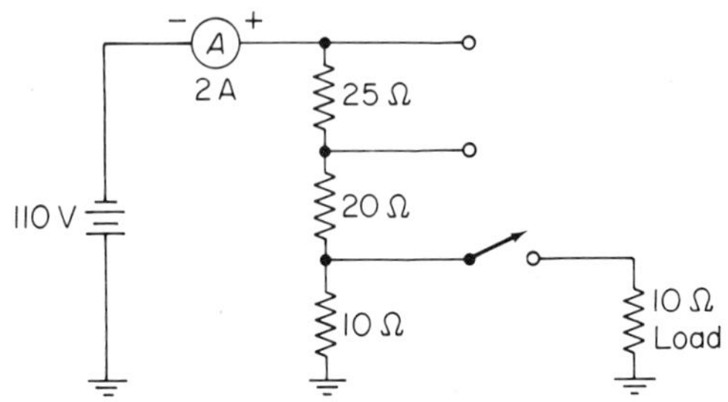

Figure 4-23. Results of Loading a Voltage Divider

When the switch has been closed, a 10-ohm load resistor is placed in parallel with the 10-ohm section of the voltage divider. The equivalent resistance of this section drops to 5 ohms. Total circuit resistance decreases from 55 ohms to 50 ohms. Total circuit current increases from 2 amperes to 2.2 amperes. There are 45 ohms of resistance in the series section, and 99 volts drop across this section. This leaves only 11 volts that will appear across the two parallel resistors.

Obviously, there is a big difference between *apparent* output of the *unloaded* voltage divider and *actual* output of the *loaded* voltage divider. The voltage divider is a flexible and useful device, but it has to be designed with the load requirements clearly in mind. The definition of an *electric load* is another case of conflicting terminology. Some authors positively state that load is synonymous with current; others are equally sure that load means the device that current flows through. Still others will argue that load is the power. It really boils down to a situation in which one chooses a definition to suit himself and stands ready to explain his meaning when using the term. However, with the term, "loaded voltage divider," we have no problem. Everyone would agree that it is a voltage divider that supplies voltage for another circuit.

Voltage dividers are designed in a *reverse procedure.* First you determine the type of circuit which you wish to be powered from the voltage divider. When this is done, you have ascertained the amount of voltage needed and the amount of current that will be drawn when the proper voltage is supplied. Thus, it is a simple matter to design a voltage divider that will furnish the required voltage.

designing voltage dividers

Problem 1: You have a circuit which requires 180 volts, and it will draw seven milliamperes of current. Your power supply is 250 volts. Design a voltage divider that will enable your circuit to function properly.

Analyze the problem:

1. You need a *series resistor* to drop the difference between the desired 180 volts and the available 250 volts.
2. You need a *resistor in parallel* with the load which will carry all current in excess of seven milliamperes. Draw a circuit and label the known specifications such as has been done in Figure 4-24.

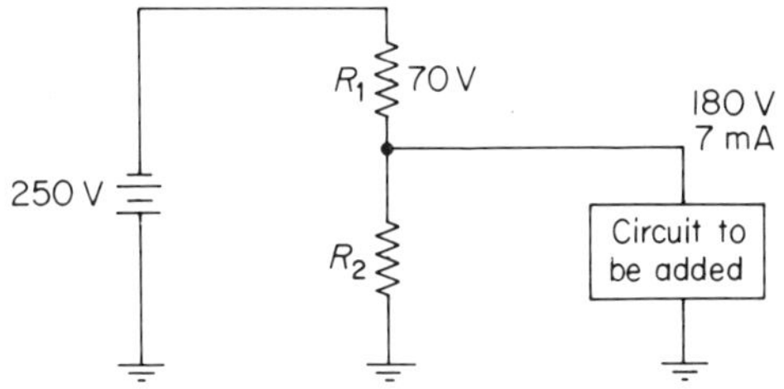

Figure 4-24. Designing a Voltage Divider

Determine what size resistor to use for R_1:

If the size is greater than 10kΩ, the load current alone will drop more than the 70 volts required. Select the next standard resistor lower than 10kΩ from the list in Figure 2-18—this is 8.2kΩ. Designate 8.2kΩ as the resistance of R_1.

Determine I_t: Since R_1 is now 8.2kΩ, and it must drop exactly 70V, how much current is required?

$$I_t = \frac{70\text{V}}{8.2\text{k}\Omega} = 8.54\text{mA}$$

Since seven milliamperes flow through the load, this will leave 1.54mA to go through the parallel resistor (R_2).

Solve for R_2:

$$R_2 = \frac{180\text{V}}{1.54\text{mA}} = 117\text{k}\Omega$$

The closest *standard* resistor to 117kΩ is 120kΩ. The answer seems to be a choice between a 120kΩ standard resistor and a variable resistor that can be adjusted to 117kΩ. If the circuit being added is an average circuit, the 120kΩ will be well within tolerance, and it will cost less than a variable resistor. However, the variable resistor does have a role in voltage divider circuits as we shall see in the next section.

potentiometers and rheostats

The *potentiometer* is a handy device where voltage needs to be varied from time to time. It is a critical item when building voltage dividers for close tolerance circuits. Figure 4-25 shows its use as a voltage divider.

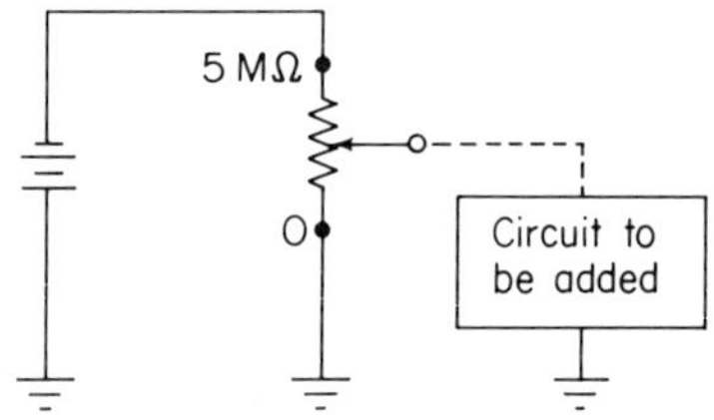

Figure 4-25. Potentiometer as a Voltage Divider

This function is similar to the volume control circuit in a radio or television receiver. The apparatus is very simple. The center arm is connected to the load circuit. The variable control is moved about until the desired level of voltage is obtained. This potentiometer is variable from 0 to 5MΩ and can tap off any level of voltage from 0 to E_a.

The *rheostat* is a current control and is built to withstand relatively high current values. It is usually placed in series with the component that needs the controlled current. This is shown in Figure 4-26.

Moving the arm toward *A* allows more current; moving it toward *B* reduces the current.

When a variety of loads are powered by a single voltage divider, the circuit becomes slightly more complicated.

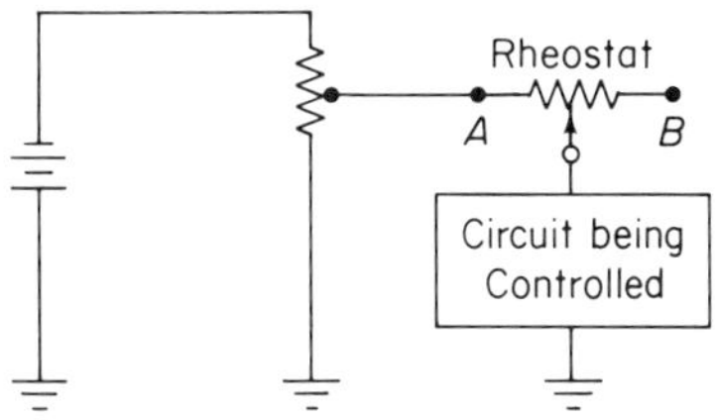

Figure 4-26. Rheostat Controlling Current

multiple loads

The multiple load voltage divider presents no great problem as long as we remember that we are dealing with the SP circuit, and each change in circuit configuration *affects every component.* To reemphasize this fact, let's solve the circuit in Figure 4-27 for voltages E_1, E_2, and E_3 with respect to ground.

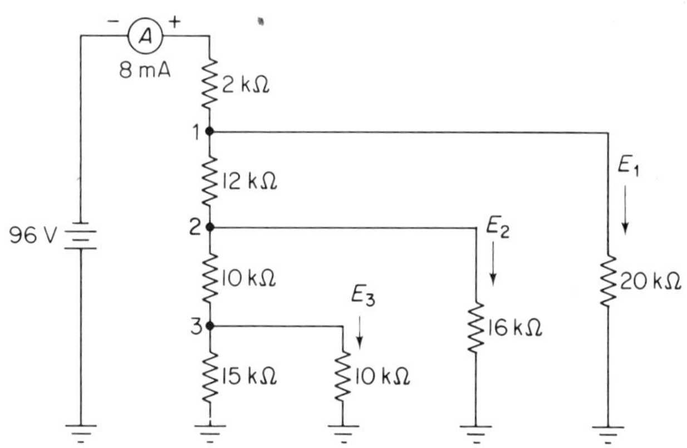

Figure 4-27. Multiple Load Voltages

Easy values have been selected to simplify the solution. The principle is the important thing, not the mathematics.

$$E_s = 8\text{mA} \times 2\text{k}\Omega = 16\text{V}$$

$$E_1 = 96\text{V} - 16\text{V} = 80\text{V}$$

At point 1, the 8mA will divide equally because, from that point to ground, there is a resistance of 20kΩ in each path. The 4mA through the 12kΩ resistor drops 48V; leaving 32V for E_2.

At point 2, the 4mA divides equally through two paths of 16kΩ each. The 2mA through the 10kΩ resistor drops 20V, leaving 12V for E_3.

At point 3, the 2mA divides again with 1.2mA going through the 10kΩ resistor. This leaves 0.8mA passing through the 15kΩ resistor. Since all currents are toward ground, all voltages are negative with respect to ground.

The above conclusions may be readily verified by application of Ohm's and Kirchhoff's laws.

Problem: Design a voltage divider from a 400-volt power supply that will satisfy all three of these loads:

1. 5mA at 30V.
2. 10mA at 200V.
3. 20mA at 250V.

Solution:

The 400-V supply exceeds our *greatest* voltage requirement by 150V. We need a series resistor to drop this difference. Each load needs separate dropping resistors in order to reduce the voltage to a usable value. The circuit will appear somewhat like the one in Figure 4-28.

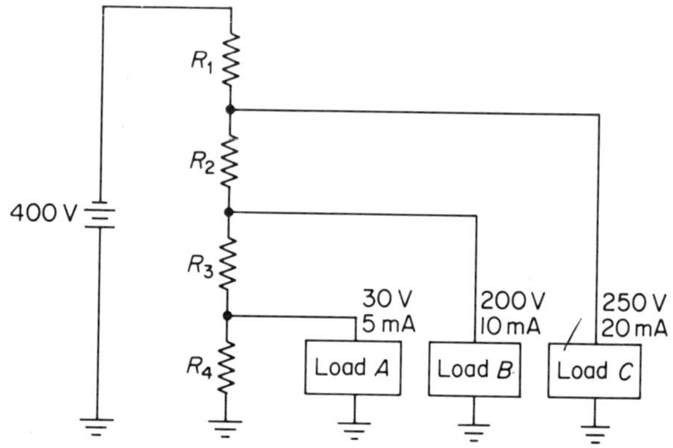

Figure 4-28. Planning for Specified Loads

Resistor requirements:

1. R_4 is a shunt. It must be of a size that shunts excess current around load *A*.
2. R_3 must drop 170V.

3. R_2 must drop 50V.
4. R_1 must drop 150V.

No involved calculations are called for—just a little straight thinking.

1. Since load *A* draws 5mA at 30V, it must have a resistance of 6kΩ.
2. Let's match that by placing a 6kΩ resistor at R_4.
3. R_4 will now carry 5mA. The summation of these currents is 10mA through R_3.
4. R_3 must drop 170V. $R_3 = 170V/10mA = 17k\Omega$.
5. R_2 must carry this 10mA plus the 10mA of load *B*. R_2 must also drop 50V. $R_2 = 50V/20mA = 2.5k\Omega$.
6. The 20mA through R_2, plus the 20mA through load *C*, goes through R_1. With 40mA, R_1 must drop 150V. $R_1 = 150V/40mA = 3.75k\Omega$.

Admittedly this is a bit haphazard, but it will work. Remember that such work with common electric circuits does not require extreme accuracy. There are two points which we failed to mention: (1) resistors in these exact sizes may not be available; and (2) we must keep an eye on the power ratings of the resistors which we use. Both of these problems can be solved by one purchase. Instead of buying four separate resistors, one need only acquire a tapped resistor with three adjustable bands. The only factor remaining for concern is power rating. The series portion of the resistor (R_1) must be capable of dissipating six watts of power (40mA $\times$ 150V). Most adjustable resistors can handle much more power than this.

To remind ourselves that not all circuits have such wide tolerances, let's consider a means of measuring the precise value of a resistor.

BRIDGE CIRCUITS

The subject of bridge circuits is a course in itself. They are *precision devices* and find many uses in taking precise resistance measurements. Figure 4-29 illustrates a bridge circuit.

balanced bridge

The circuit in Figure 4-29 is commonly referred to as a *balanced bridge* or a *Wheatstone bridge.* It consists of a power supply, a gal-

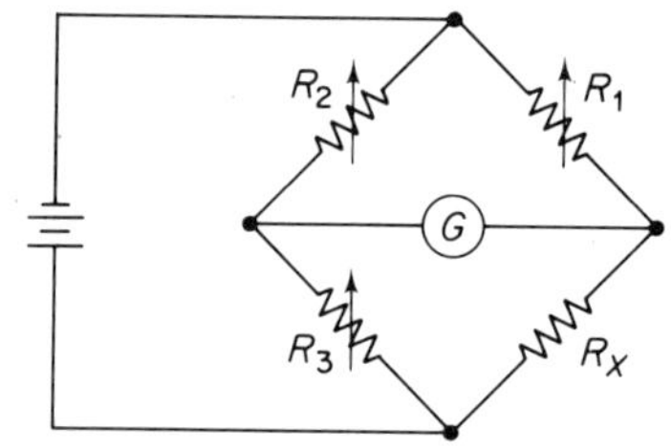

Figure 4-29. Bridge Circuit

vanometer, three variable resistors (R_1, R_2, and R_3), and a fourth unknown resistor to be measured. The *galvanometer* is a sensitive current measuring device that measures current in either direction.

The three variable resistors are calibrated in exact ratios, and R_1, R_2, and R_3 are adjusted until there is no current through the galvanometer. At this time, the exact size of R_x can be read from a calibrated dial. The resistance ratios are:

$$R_1/R_x = R_2/R_3$$
$$R_x = R_1R_3/R_2$$

Some complex circuits fit into patterns similar to the bridge circuit except for the fact that their currents are not balanced.

unbalanced bridge

At times, one encounters circuits that *do not* fit into any of our three patterns (series, parallel, or series-parallel). One such circuit is the *unbalanced bridge* illustrated in Figure 4-30.

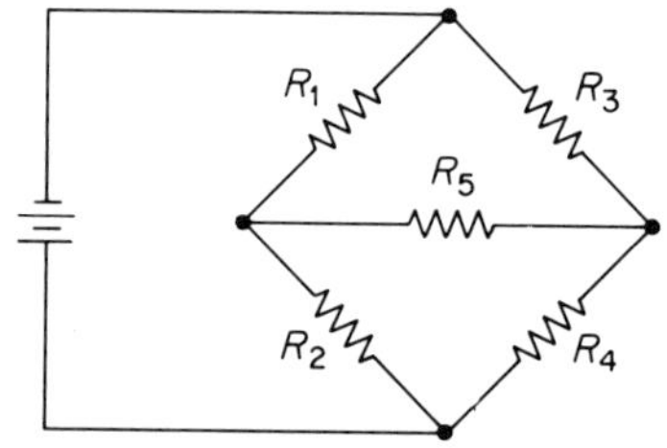

Figure 4-30. An Unbalanced Bridge

The current direction for each resistor should be obvious except for the current through R_5. This current may go in either direction depending upon the relative sizes of the other four resistors. This being the case, it is difficult to determine which resistors are parallel to each other. This circuit is not adaptable to the conventional methods of solution, but there are ways to handle it.

solving bridges by Kirchhoff's law

Solving this type of circuit by Kirchhoff's law requires *formulating the equations* and solving for several unknown values at the same time. Kirchhoff's laws provide a guide for designing the equations. Let's review:

> the algebraic sum of the voltage drops around any complete loop is equal to zero, and the algebraic sum of the currents at a junction is equal to zero.

Circuits must be labeled with all known values, and symbols must be established for *unknown* values. Voltage drops across resistors are assigned negative or positive values depending on the direction of their current. If the direction of current is unknown, assume an arbitrary direction. Voltage drops across batteries are positive or negative depending upon the direction in which the drop crosses the battery. For instance, *when crossing a resistor with the current, the* IR *drop is negative; against the current, the* IR *drop is positive. When crossing a battery with the current, the voltage is positive; against the current, the voltage is negative.* Using these values of voltages and *IR* drops, a separate formula is designed for each complete loop.

Let's practice on a simple circuit.

Example (Figure 4-31): Find the voltage across and the current through each resistor.

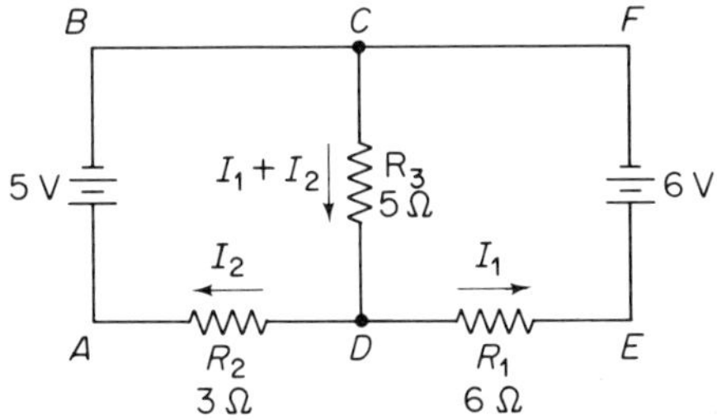

Figure 4-31. Multiple Power Supplies

Establish formulas for each loop. (Points *A, B, C, D, E,* and *F* will help keep track of the current paths.):

Loop *A, B, C, D, A:* (*A*) $5 - 5(I_1+I_2) - 3I_2 = 0$

Loop *E, F, C, D, E:* (*B*) $6 - 5(I_1+I_2) - 6I_1 = 0$

Voltage drops are in terms of *IR* because we *cannot* yet say how many volts are dropped across a resistor. The battery voltages and resistor sizes are the only values that we are sure of.

Place known values on *one side* of the equality sign and unknown values *on the other:*

Formula (A): $5 = 5(I_1+I_2) + 3I_2$

Formula (B): $6 = 5(I_1+I_2) + 6I_1$

Clear the formulas by combining:

$$\text{(A)}\quad 5 = 5I_1 + 5I_2 + 3I_2$$
$$5 = 5I_1 + 8I_2$$
$$\text{(B)}\quad 6 = 5I_1 + 5I_2 + 6I_1$$
$$6 = 11I_1 + 5I_2$$

Place the formulas together and solve for I_1:

$$\text{(A)}\quad 5 = 5I_1 + 8I_2$$
$$\text{(B)}\quad 6 = 11I_1 + 5I_2$$

Multiply *A* by 5 and *B* by 8:

$$\text{(A)}\quad 25 = 25I_1 + 40I_2$$
$$\text{(B)}\quad 48 = 88I_1 + 40I_2$$

Subtract *A* from *B:*

$$\text{(B)}\quad 48 = 88I_1 + 40I_2$$
$$\text{(A)}\quad \underline{25 = 25I_1 + 40I_2}$$
$$23 = 63I_1$$
$$I_1 = 23/63 = 0.365\text{A}$$

Use the simplest formula, substitute for I_1, and solve for I_2:

$$5 = 5I_1 + 8I_2$$
$$5 = 5 \times 0.365 + 8I_2$$
$$5 = 1.825 + 8I_2$$
$$5 - 1.825 = 8I_2$$
$$I_2 = \frac{3.175}{8} = 0.397\text{A}$$

Both solutions resulted in *positive* values for current. This indicates that the *assumed direction* for both currents was correct. Had either result been negative, it would show that we assumed the wrong direction for that current.

Referring back to Figure 4-31, write in the current values, and solve for voltage drops across each resistor.

$$E_1 = 0.365\text{A} \times 6\Omega = 2.19\text{V}$$
$$E_2 = 0.397\text{A} \times 3\Omega = 1.19\text{V}$$
$$E_3 = (0.365\text{A} + 0.397\text{A})5\Omega = 3.81\text{V}$$

When these voltage drops are added up, it appears that we have less than the applied voltage. However, the effective applied voltage is the sum of the three voltage drops. The difference between this and the 11 volts in the two batteries is caused by bucking currents through the opposing power supplies.

Let's try the unbalanced bridge circuit in Figure 4-32 by following the same procedures. Solve for all values of current and voltage.

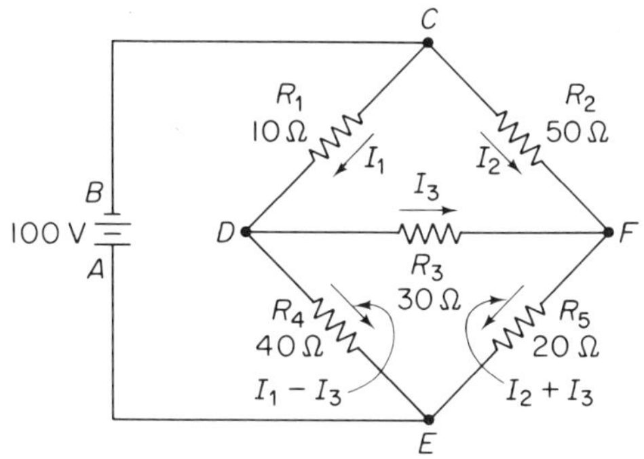

Figure 4-32. Solving the Unbalanced Bridge

The circuit has already been labeled with all known values, and symbols have been established for the currents. We actually have *five* values of unknown currents, but all five can be accurately expressed in combinations of I_1 I_2, and I_3. We have assumed that current through R_3 is from *D* to *F*. If this is right, I_3 will have a positive value; if not, I_3 will be negative.

Since there are *three loops,* we will need *three equations.* Let's designate them (A), (B), and (C). The loops for each equation are (the letter in parentheses refers to a loop):

(A) A, B, C, D, E, A
(B) A, B, C, F, E, A
(C) A, B, C, D, F, E, A

Set up and simplify equation (A):

$$100 - 10I_1 - 40(I_1 - I_3) = 0$$
$$100 - 10I_1 - 40I_1 + 40I_3 = 0$$
$$100 = 50I_1 - 40I_3$$
$$\text{(A)} \qquad 10 = 5I_1 - 4I_3$$

Set up and simplify equation (B):

$$100 - 50I_2 - 20(I_2 + I_3) = 0$$
$$100 - 50I_2 - 20I_2 - 20I_3 = 0$$
$$100 - 70I_2 - 20I_3 = 0$$
$$100 = 70I_2 + 20I_3$$
$$\text{(B)} \qquad 10 = 7I_2 + 2I_3$$

Set up and simplify equation (C):

$$100 - 10I_1 - 30I_3 - 20(I_2 + I_3) = 0$$
$$100 - 10I_1 - 30I_3 - 20I_2 - 20I_3 = 0$$
$$100 - 10I_1 - 20I_2 - 50I_3 = 0$$
$$100 - 10I_1 + 20I_2 + 50I_3$$
$$\text{(C)} \qquad 10 = I_1 + 2I_2 + 5I_3$$

Equation (C) contains *all three* unknown currents. We'll use it along with (B) to eliminate I_2. The results, when used with (A), will then reveal the actual value of either I_1 or I_3.

Write (B) and (C) together:

$$\begin{aligned} &\text{(B)} \quad 10 = \phantom{I_1 + {}} 7I_2 + 2I_3 \\ &\text{(C)} \quad 10 = I_1 + 2I_2 + 5I_3 \end{aligned}$$

Multiply (B) by 2 and (C) by 7:

$$\begin{aligned} &\text{(B)} \quad 20 = \phantom{7I_1 + {}} 14I_2 + 4I_3 \\ &\text{(C)} \quad 70 = 7I_1 + 14I_2 + 35I_3 \end{aligned}$$

Subtract (B) from (C):

$$\begin{aligned} &\text{(C)} \quad 70 = 7I_1 + 14I_2 + 35I_3 \\ &\text{(B)} \quad 20 = \phantom{7I_1 + {}} 14I_2 + 4I_3 \\ \hline &\text{(D)} \quad 50 = 7I_1 \phantom{{} + 14I_2} + 31I_3 \end{aligned}$$

This result is given a designation, equation (D), because we shall need it for our solution. Equation (A) contains the same currents as equation (D). Write them together and solve for I_3. Why not I_1?: No special reason; I_3 just appears to have an easier solution.

$$\begin{aligned}
&(D) \quad 50 = 7I_1 + 31I_3 \\
&(A) \quad 10 = 5I_1 - 4I_3
\end{aligned}$$

Multiply (D) by 5 and (A) by 7:

$$\begin{aligned}
&(D) \quad 250 = 35I_1 + 155I_3 \\
&(A) \quad 70 = 35I_1 - 28I_3
\end{aligned}$$

Subtract (A) from (D):

$$\begin{aligned}
&(D) \quad 250 = 35I_1 + 155I_3 \\
&(A) \quad 70 = 35I_1 - 28I_3 \\
&\overline{\quad\quad 180 = \quad\quad\quad 183I_3}
\end{aligned}$$

Solve for I_3:

$$\begin{aligned}
183I_3 &= 180 \\
I_3 &= \frac{180}{183} = 0.98\text{A}
\end{aligned}$$

I_3 shows a positive value, which means that its direction is from *D* to *F*—as we assumed it to be.

Substitute I_3 into (B) and solve for I_2:

$$\begin{aligned}
(B) \quad 10 &= 7I_2 = 2I_3 \\
10 &= 7I_2 + 2(0.98) \\
10 &= 7I_2 + 1.96 \\
10 - 1.96 &= 7I_2 \\
7I_2 &= 8.04 \\
I_2 &= 8.04/7 = 1.15\text{A}
\end{aligned}$$

Substitute I_3 into (A) and solve for I_1:

$$\begin{aligned}
(A) \quad 10 &= 5I_1 - 4I_3 \\
10 &= 5I_1 - 4(0.98) \\
10 &= 5I_1 - 3.92 \\
5I_1 &= 10 + 3.92 \\
5I_1 &= 13.92 \\
I_1 &= \frac{13.92}{5} = 2.78\text{A}
\end{aligned}$$

In summary:

$$I_1 = 2.78\text{A}$$
$$I_2 = 1.15\text{A}$$
$$I_3 = 0.98\text{A}$$
$$I_1 - I_3 = 2.78 - 0.98 = 1.8\text{A}$$
$$I_2 + I_3 = 1.15 + 0.98 = 2.13\text{A}$$

Use the above values for currents, and solve for the voltage drop across each resistor.

$$E_1 = 2.78\text{A} \times 10\Omega = 27.8\text{V}$$
$$E_2 = 1.15\text{A} \times 50\Omega = 57.5\text{V}$$
$$E_3 = 0.98\text{A} \times 30\Omega = 29.4\text{V}$$
$$E_4 = 1.80\text{A} \times 40\Omega = 72\text{V}$$
$$E_5 = 2.13\text{A} \times 20\Omega = 42.6\text{V}$$

The voltage drops *appear* to be reasonable, but just to be sure, let's *check* them. How? Against Kirchhoff's voltage law. The voltage drops around each loop should be equal to the applied voltage (100V).

Loop (A):	Loop (B):	Loop (C):
27.8V	57.5V	27.8V
72.0V	42.6V	29.4V
99.8V	100.1V	42.6V
		99.8V

The greatest *error* is *two-tenths* of one percent. It would be a very rare circuit that required greater accuracy. In fact, it would take a rare instrument to measure anywhere near this accuracy. Even this small margin of error could be reduced by carrying out the calculations to three or four decimal places.

Kirchhoff's law is indispensable to us, but we should have more than one method of solving complex networks.

equivalent networks

In actual practice, the unbalanced bridge is encountered quite often. Unbalanced bridge problems can be solved by reducing the circuits to simpler *equivalent networks.* For instance, the networks in Figure 4-33(a) and (b) are identical.

Notice that the top portion of the network (R_1, R_2, and R_3) forms the Greek letter delta (Δ). If we separate the delta from R_4 and R_5, we have the configuration of Figure 4-34.

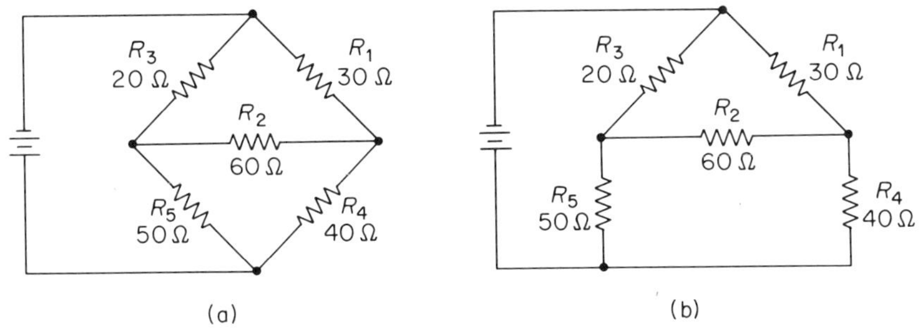

Figure 4-33. Identical Bridge Circuits

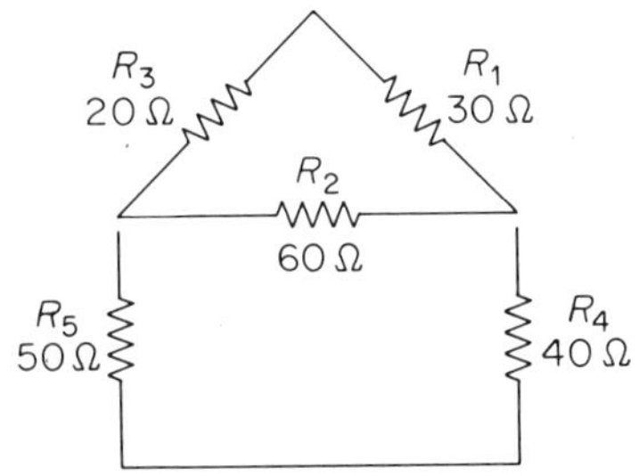

Figure 4-34. Simplifying the Circuit

If the equivalent resistance of the delta can be determined, the circuit solution will be easy. Networks (a) and (b) in Figure 4-35

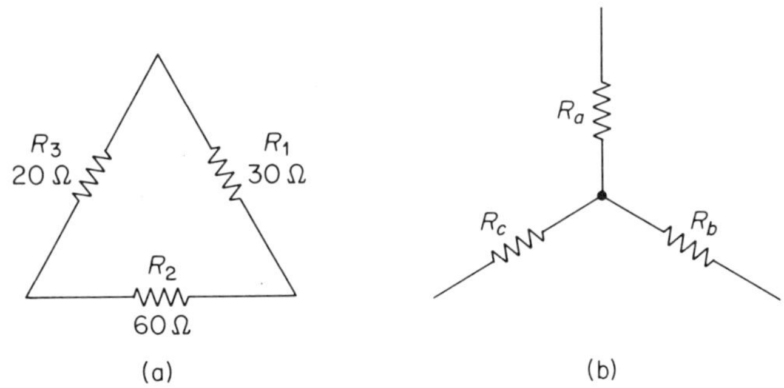

Figure 4-35. Equivalent Circuits

can be made equivalent to one another by calculating proper values of resistance to assign to (b).

Network (b) is called either a "star" or "Y" network because of its shape. We'll refer to (a) as Δ and to (b) as Y. In order to convert from Δ to Y, resistánce values must be calculated for R_a, R_b, and R_c. Start by drawing the Y inside the Δ as shown in Figure 4-36.

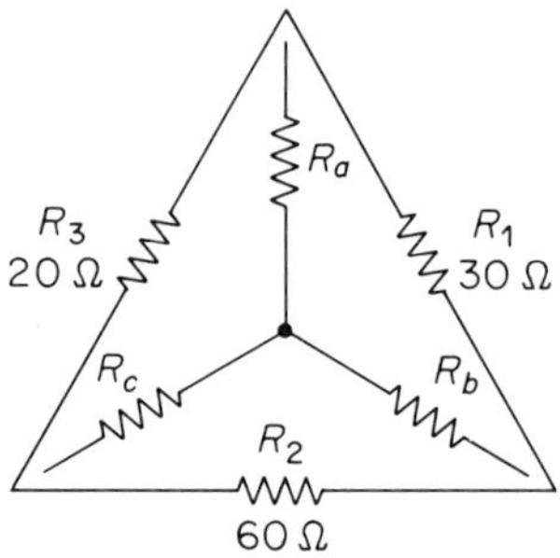

Figure 4-36. Converting △ to *Y*

The resistance for any Y-resistor is equal to the product of its adjacent Δ resistances divided by the summation of the Δ resistances. Reducing this to three formulas, we have:

$$R_a = \frac{R_1 R_3}{\Sigma^R \Delta}$$

$$R_b = \frac{R_1 R_2}{\Sigma^R \Delta}$$

$$R_c = \frac{R_2 R_3}{\Sigma^R \Delta}$$

The Greek letter sigma (Σ) means "the summation of." Therefore, $\Sigma^R \Delta = R_1 + R_2 + R_3$. Calculate the values for R_a, R_b, and R_c in Figure 4-36.

Solving for R_a:

$$R_a = \frac{R_1 R_3}{\Sigma^R \Delta} = \frac{30 \times 20}{110} = \frac{600}{110} = 5.5\Omega$$

Solving for R_b:

$$R_b = \frac{R_1 R_2}{\Sigma^R \Delta} = \frac{30 \times 60}{110} = \frac{1800}{110} = 16.4\Omega$$

Solving for R_c:

$$R_c = \frac{R_2 R_3}{\Sigma^R \Delta} = \frac{20 \times 60}{110} = \frac{1200}{110} = 10.9\Omega$$

Use the above values and draw the equivalent of the bridge circuit in Figure 4-33(a) as shown in Figure 4-37.

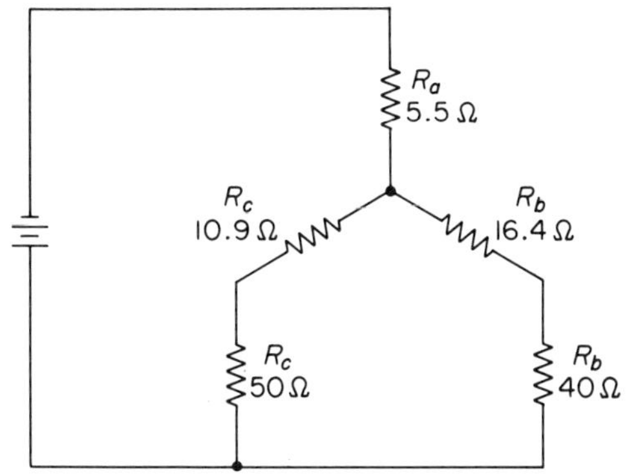

Figure 4-37. SP Equivalent of Figure 4-33

This is much better, but if it still looks unpredictable, rearrange the resistors into a more familiar pattern. This is illustrated in Figure 4-38.

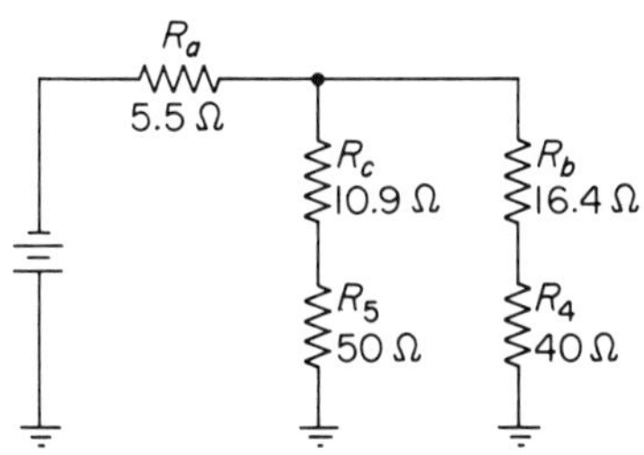

Figure 4-38. Rearrangement of Figure 4-37

This is a relatively simple series-parallel circuit. From this point on to a full solution, there is not much challenge. However, it can be reduced by three more steps. Simplify the SP circuit:

$$R_{\text{branch } 1} = 10.9\Omega + 50\Omega = 60.9\Omega$$
$$R_{\text{branch } 2} = 16.4\Omega + 40\Omega = 56.4\Omega$$

Redraw as in Figure 4-39:

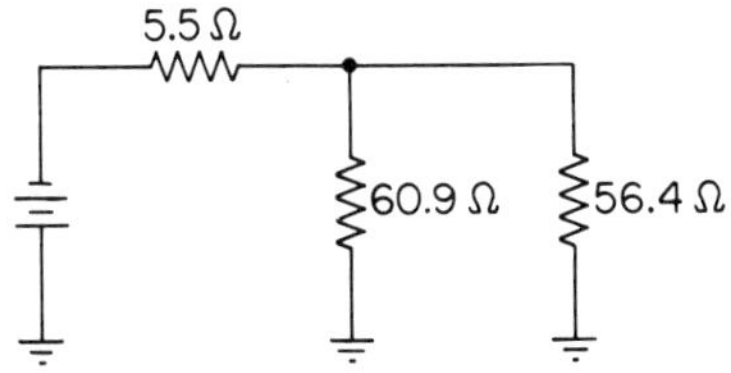

Figure 4-39. Simplest form of SP Equivalent

Reduce the SP circuit to a series circuit:

$$R_e = \frac{60.9 \times 56.4}{60.9 + 56.4} = \frac{3434.76}{117.3} = 29.3\Omega$$

Redraw as in Figure 4-40.

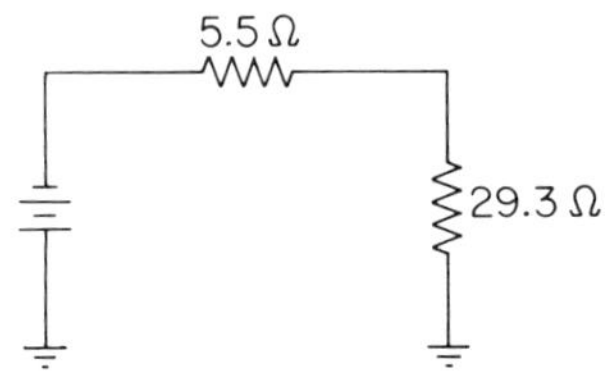

Figure 4-40. Series Equivalent of Figure 4-39

The next simplification produces the value of R_t for Figure 4-33 and gives us the circuit in Figure 4-41.

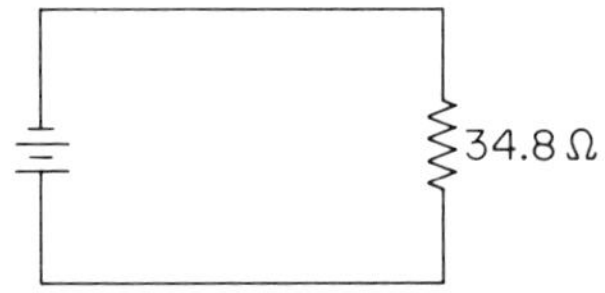

Figure 4-41. Simplest Form of Series Equivalent

In some cases, it might be convenient to convert a Y network to a Δ network. If you understand the conversion of a Δ to a Y, the reverse process will cause no problem. Using Figure 4-42 as a reference, the resistance of any Δ resistor is equal to $R_aR_b + R_bR_c + R_aR_c$ divided by its opposite Y resistance.

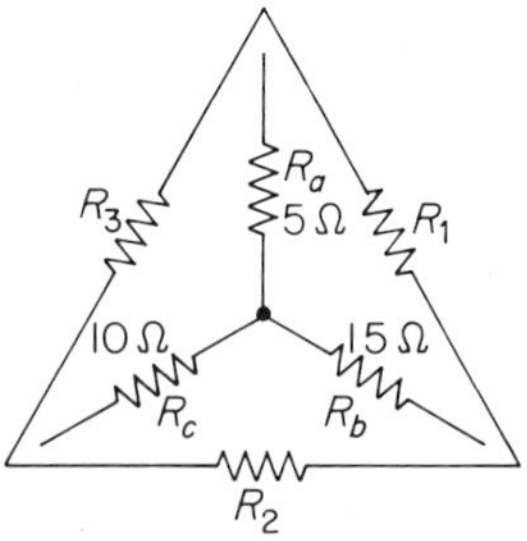

Figure 4-42. Converting Y to △

The three values of Δ produced by applying the above statement to the figure values are:

$$R_1 = \frac{(75 + 150 + 50)}{10} = \frac{275}{10} = 27.5\Omega$$

$$R_2 = \frac{(75 + 150 + 50)}{5} = \frac{275}{5} = 55\Omega$$

$$R_3 = \frac{(75 + 150 + 50)}{15} = \frac{275}{15} = 18.3\Omega$$

Therefore, networks (a) and (b) of Figure 4-43 are electrical equivalents of one another.

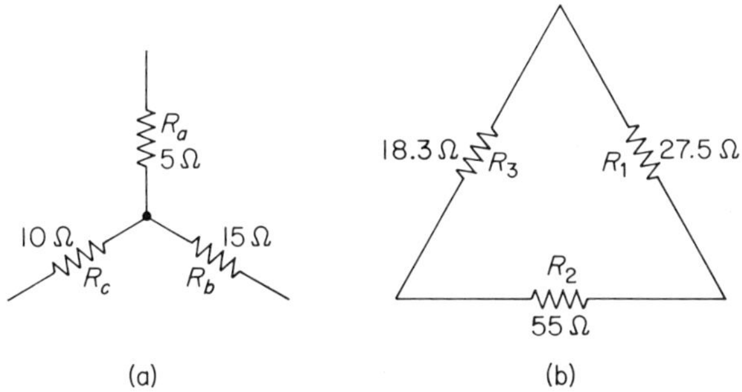

Figure 4-43. *Y*-△ Equivalents

This chapter has developed the principles of circuits to an extent that should enable you to view any dc circuit with a certain sense of familiarity. What you need most is practice in problem solving while the principles are still fresh in your mind. The following review exercises will help you toward that end.

REVIEW EXERCISES

1. A parallel circuit has two or more paths for ______________. The ______________ divides among branches inversely proportional to the branch resistance.
2. Kirchhoff's voltage and current laws for a parallel circuit are ______________ and ______________.
3. Write the two resistance formulas for a parallel circuit.
4. The reciprocal formula was derived from the formula for ______________.

5. A parallel circuit of two branches (10kΩ and 30kΩ) is drawing 10mA of current. How much current is passing through each resistor? What is P_t for the circuit?
6. Solve the network in Figure P4-6 for R_t using the assumed voltage method.

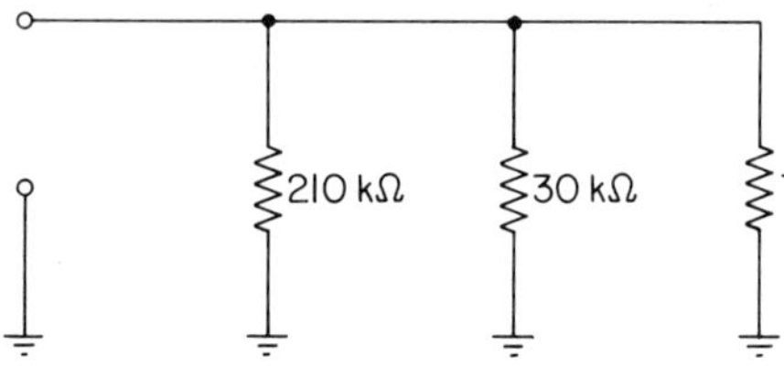

Figure P4-6

7. When a branch is added to an existing parallel circuit, total current ________, and total resistance ________.
8. Existing branches are not affected by ________ or removing branches.
9. Why is the voltmeter ineffective in locating shorts and opens in a parallel circuit?
10. Why must the branch be disconnected in order for the ohmmeter to detect an open or a short?
11. The SP circuit has one or more components in ________ and one or more ________ branches.
12. Write a resistance formula for R_t of a SP circuit of four branches.
13. Analyzing the SP circuit should be accomplished in two sections; the ________ section and the ________ section.
14. Solve the circuit in Figure P4-14 for E_a, R_t, P_t, and I_t.
15. Adding a branch to an existing SP circuit will cause I_t to ________, I_{branches} to ________, R_t to ________, E_s to ________, and E_p to ________.
16. Varying a resistor within an existing branch affects the distribution of ________ to all components.
17. A loaded voltage divider is delivering ________ to another circuit.

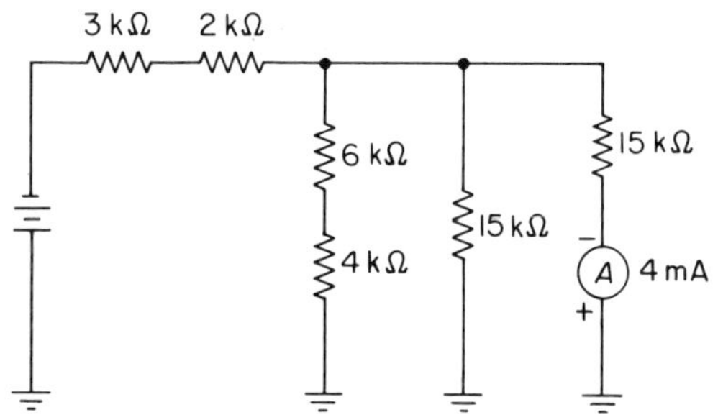

Figure P4-14

18. Adding a load onto a voltage divider is equivalent to adding a ____________ to any SP circuit. It causes a redistribution of ____________.
19. The voltage divider must be designed to fit the ______ ______ requirements.
20. The potentiometer enables a voltage divider to be constructed with ____________ outputs.
21. When current is a critical factor, the variable component is a ____________ connected in ____________ with the load.
22. A chief use for a balanced bridge is precision ____________ measurements.
23. Circuit configurations frequently take the form of ______ ______ bridges.
24. When solving network problems by Kirchhoff's law, a(an) ____________ must be generated for each ____________.
25. Reduce this equation to its simplest form:

$$-6I_1 - 8(I_1 - I_2) - 4(I_1 - I_2) = 0$$

26. Solve these equations for I_1:

$$3I_1 - 2I_2 = 10; \quad 12I_1 + 6I_2 = 12$$

27. Complex circuits are sometimes more easily solved by constructing ____________ networks.
28. Write the three formulas for converting network *A* to network *B* in Figure P4-28.
29. Write the three formulas for converting network *C* to network *D* in Figure P4-29.

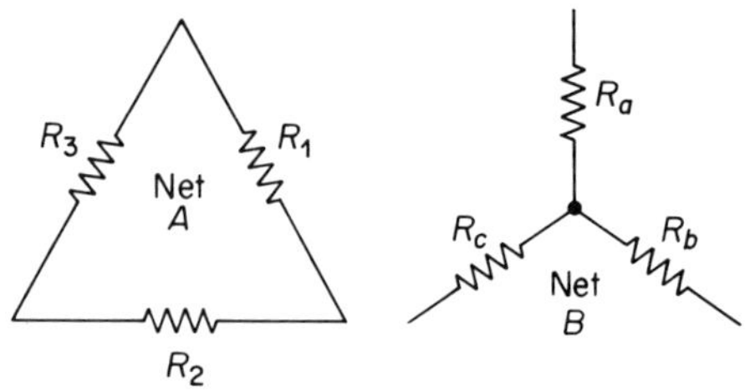

Figure P4-28

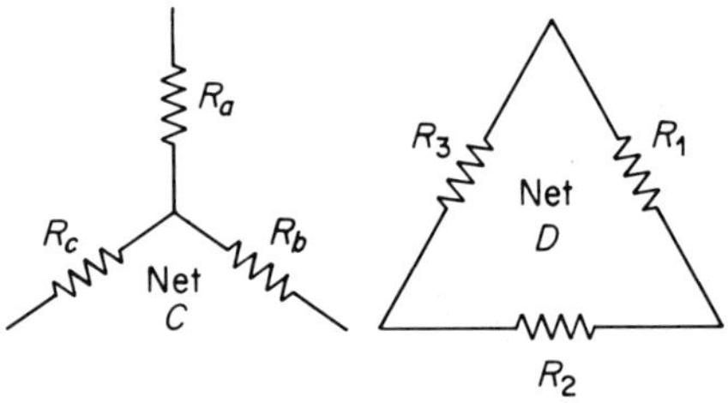

Figure P4-29

30. In Figure P4-28, $R_1 = 10\Omega$, $R_2 = 20\Omega$, and $R_3 = 30\Omega$. What is the resistance of: R_a?; R_b?; R_c?
31. In Figure P4-29, $R_a = 10\Omega$, $R_b = 20\Omega$, and $R_c = 30\Omega$. What is the resistance of: R_1?; R_2?; R_3?

5

interaction of magnetism and electricity

Magnetism *is not* just a subject to fill space; neither is it merely an interesting subject in which one may while away the time. Scientists long ago discovered that magnetism exhibited some properties that resembled the effects of electricity. Later, it was discovered that electric and magnetic properties *complement* one another. Today, *magnetism and electricity* are so entwined that it is *impossible* to separate the two. Magnetism is used to produce electromotive force, and every time electrons move, magnetism is produced. Therefore, any fact that you can learn concerning magnetism enhances your understanding of electricity. Conversely, an extensive knowledge of magnetic properties is an essential part of your knowledge of basic electric principles.

MAGNETIC PROPERTIES

Long before the Christian era, the ancient Greeks had discovered that a certain type of iron ore exhibited a strange attraction for bits of iron. This ore was found in the district of Magnesia in Asia Minor. That particular type of iron ore, *magnetite* (Fe_3O_4), provided the root for the word magnet.

About 2600 B.C. the Chinese found a use for magnetite. They are credited with discovering that when a piece of this ore is suspended and left free to swing, it always points in a particular direction. When they analyzed this action, they discovered that one end pointed to the north and the other to the south. By marking the north seeking end of the suspended ore, the Chinese created the first *magnetic compass.* Thus, magnetite became lodestone (or leading stone). (See Figure 5-1.)

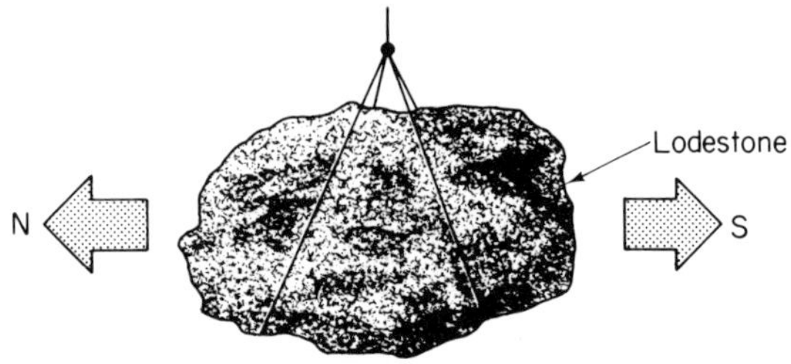

Figure 5-1. The First Compass

terminolgy

Substances that exhibit the properties described above are called *magnets.* The materials that are attracted to magnets are commonly called magnetic materials. The phenomenon associated with magnets and magnetic materials is called *magnetism.* The science of magnetism is a vast and interesting field.

The above definition of magnetic materials places all substances in one of two categories. They are either magnetic or nonmagnetic. This, of course, is a nonscientific as well as a broad classification. It will suffice for a time, but later materials will have to be classified according to their reaction to magnetic influence. Magnets themselves have their own classification.

classification of magnets

Any magnet can be classified as either *natural* or *artificial.* The lodestone (or magnetite) is a natural magnet. It exists in a natural state. Lodestone can presently be found in the United States, Norway, and Sweden. These natural magnets are crude, generally weak, and serve little purpose in the modern world. The earth itself is a huge natural magnet, and it is an exception to the rule of a natural magnet serving no useful purpose. The magnetic properties of the earth are used constantly.

Artificial magnets are magnets into which magnetism has been *induced* by some artificial means. They are generally composed of iron, steel, nickel, or cobalt. Several ways of influencing these materials to become magnets will be covered a bit later. The artificial magnets are far superior to natural magnets. They are more compact, more uniform, and much stronger.

Some types of materials magnetize easily, but lose this magnetism rapidly when the magnetizing influence is removed. These are called *temporary* magnets. They are composed of materials such as silicon steel or soft iron, and these materials are sometimes used in a powdered state.

Other materials are difficult to magnetize, but retain the magnetism for long periods of time. They are classified as *permanent* magnets. Permanent magnets are composed of hard materials such as steel, cobalt, nickel, and permalloy.

The ability of a material to hold its magnetism is known as *retentivity.* Soft iron has a low retentivity while hard steel has a high retentivity. The magnetism that is left in a material after the magnetizing influence has been removed is called *residual* magnetism.

MAGNETIC FIELDS

An invisible force field exists about a magnet, which in many ways resembles the electrostatic field. Its definition is also much the same. A *magnetic field* is the area about a magnet where its *influence can be felt.* This field has a definite direction, and its intensity can be measured.

flux

The magnetic field is composed of imaginary lines of force which are called *flux lines* (or simply *flux*). These lines are heavily concentrated around the ends of a magnet and fan out in all directions. They are illustrated in Figure 5-2, which shows a piece of glass placed over a magnet. When this glass is sprinkled with iron filings, flux patterns can be observed. A gentle tapping of the glass will allow the flux lines to arrange the filings according to the flux concentration as portrayed here. It must be kept in mind that this is a flat plane demonstration only. Actually the field surrounds the magnet as the air surrounds the earth. The *symbol for flux* is the Greek letter phi (Φ).

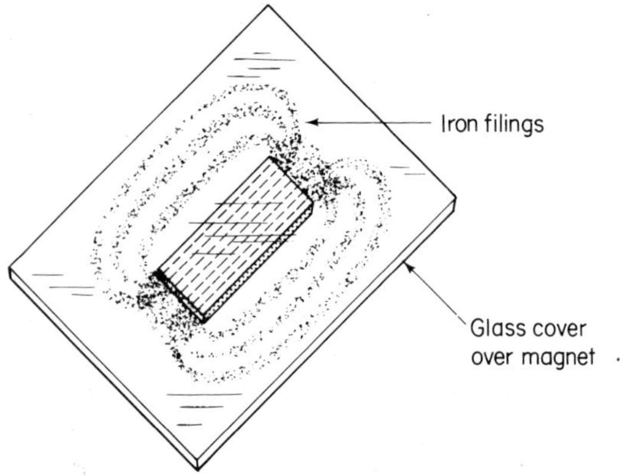

Figure 5-2. Concentration of Flux

magnetic poles

Since a free swinging, suspended magnet always aligns itself with the magnetic lines of the earth, it was only natural that the ends of the magnet would be named *poles.* The end which *seeks* (points to) the *north* pole is called a *north pole.* The *south seeking* end is called a *south pole.* The source of these names is illustrated in Figure 5-3. Examine the drawing and remember that the north pole is the pole that seeks the north pole of the earth, and the south pole is the pole that seeks the south pole of the earth.

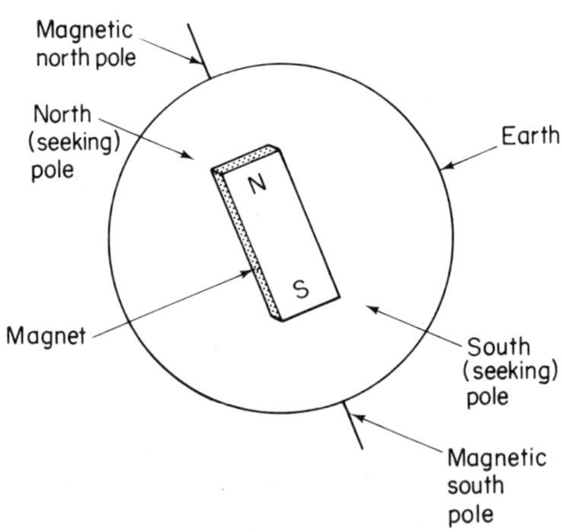

Figure 5-3. Naming of the Magnetic Poles

These magnetic poles exert a force on one another similar to the forces between electrostatic charges. *Like poles repel* one another; *unlike poles attract one another.* This is shown in Figure 5-4.

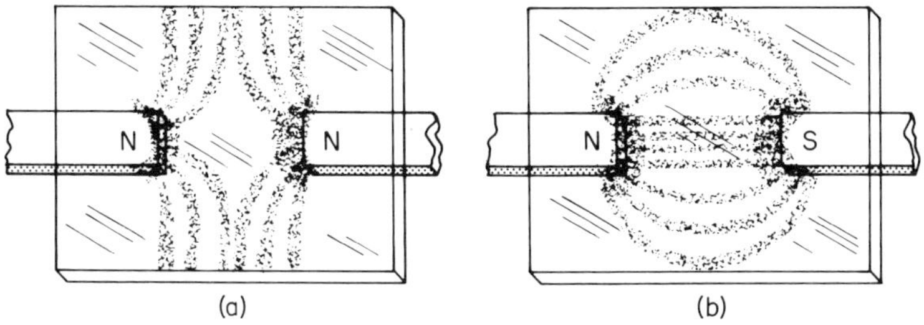

Figure 5-4. Forces Between Poles

In Figure 5-4(a) the pattern of filings clearly shows the repulsion of the flux lines from the two north poles. Exactly the same pattern would be created with two south poles. The force of attraction is just as apparent in Figure 5-4(b). These drawings also illustrate another important point: lines of force in the same direction repel each other. In fact, flux lines *cannot be made to cross.* If lines could be forced to cross, such an action would very likely release a great quantity of energy similar to that released when the atom is split. Considerable research has been conducted with this possibility in mind.

Again, as with the electrostatic field, the direction of the magnetic field was determined by agreement. It was decided to assign it the direction that an isolated north pole would take when placed in the field. Since like poles repel and unlike poles attract, the isolated north pole would move *away from the magnet's north pole* and *toward the magnet's south pole.* Each flux line forms a complete loop: it leaves the north pole; swings around and enters the south pole; and then travels through the magnet back to the north pole. This is illustrated in Figure 5-5. Notice how the lines exist in perfectly straight configurations through the magnet from south to north. After leaving the north pole, they get as far as possible from one another for their trip back to the south pole. Remember that this is a two-dimensional drawing. If you were to view the flux from the end of the magnet, you would see a picture that would appear similar to the water spraying from a lawn sprinkler.

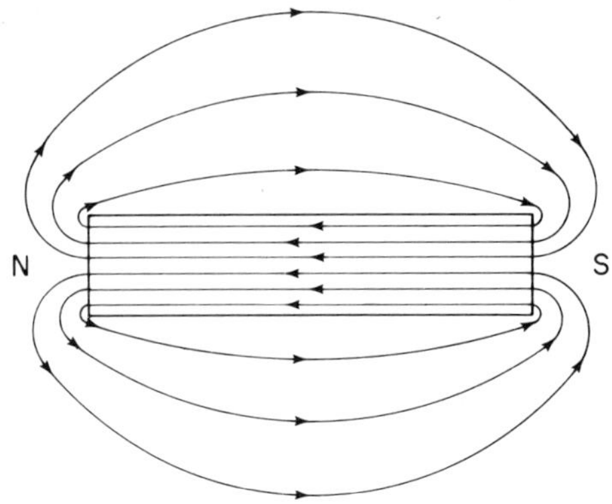

Figure 5-5. Flux Loops in and Around a Magnet

Coulomb's law

Charles A. Coulomb, the physicist, who was introduced earlier in connection with his work with electrostatic charges, experimented extensively with magnetic poles. One of the problems he set about to solve was the isolation of a magnetic pole. Since all magnets have both a north and a south, it is difficult to get the poles separated far enough to observe the influence of either pole. Coulomb used long, slender needles as magnets. In this manner, though the poles were not completely isolated, they were separated far enough for him to formulate some definite ideas. He discovered not only that like poles repel each other and unlike poles attract each other but also a definite pattern to these forces of repulsion and attraction.

He assigned each magnet a definite pole strength relative to a standard magnet. The pole strength was expressed in terms of *unit poles.* He established the fact that the force between poles was *inversely* proportional to the *square* of the distance between them. The results of his investigation were expressed as $F = m_1 m_2 / d^2$.

Where: F is force (attraction or repulsion) in dynes, m_1 and m_2 are pole strengths in unit poles, and d^2 is the square of the distance between the poles in centimeters. (It is assumed in this case that the substance between the poles is air.)

Of course, some substances pass lines of force with greater ease than others. So, when the medium is not air, another factor is involved. The ease with which lines of flux pass through a substance is called *permeability.* The symbol for permeability is the Greek letter mu (μ). The value of μ ranges from less than one for nonmagnetic materials to several thousand for certain alloys. The table in Figure 5-6 shows approximate permeabilities for several different substances. This table shows that flux passes through permalloy 100,000 times as easily as it passes through bismuth. It should also illustrate that the flux lines will pass through any known substance. When we take permeability into account, Coulomb's law becomes $F = m_1m_2\mu/d^2$.

This can be made clearer by solving a few problems.

Substance	Permeability	Substance	Permeability
Bismuth	0.999824	Nickel	1000
Nitrogen (gas)	0.999999995	Iron	7000
Aluminium	1.000022	Permalloy	100000
Air	1.00000029		

Figure 5-6. Approximate Permeabilities

calculating force between poles

The unit pole used by Coulomb is described as a pole of such strength that when two of these poles are separated by one centimeter of air, they will exert a force of one dyne between them. The unit pole is illustrated in Figure 5-7.

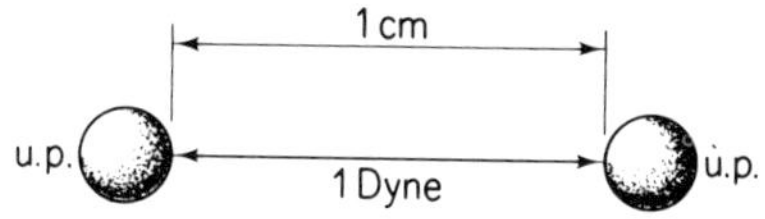

Figure 5-7. Unit Poles

This is the unit we speak of when we refer to pole strength. The abbreviation we will use is u.p.

Example 1 (Figure 5-8): What is the force between the two isolated poles depicted in this drawing?

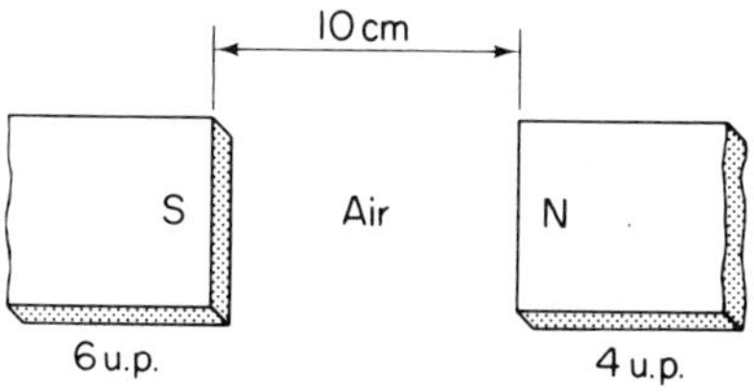

Figure 5-8. Calculate Force

Procedure:

The medium is air which has a permeability of approximately one; thus μ drops from the formula, and we use $F = m_1 m_2 / d^2$.

Substitute values and solve:

$F = 6 \times 4/10 \times 10 = 24/100 = 0.24$ dynes. (Since both poles are south, this is a force of repulsion.)

Example 2 (Figure 5-9): How much force is acting between the isolated poles? What is this force, attraction or repulsion?

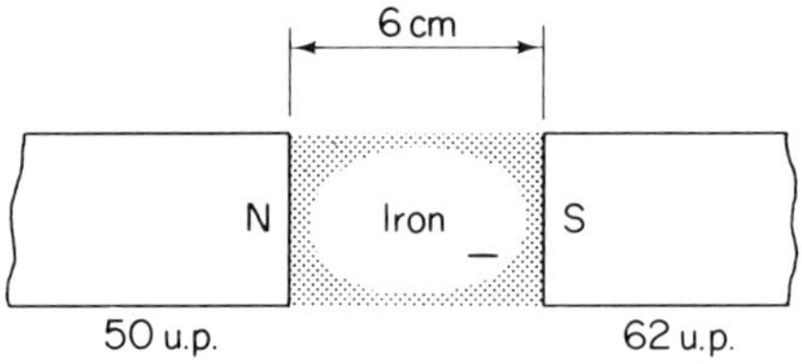

Figure 5-9. How Much Force?

Solution:

$F = m_1 m_2 \mu / d^2 = 50 \times 62 \times 7000/6 \times 6 = 21{,}700{,}000/36 = 2.17 \times 10^6/3.6 = 0.6 \times 10^6$ (or 6×10^5) dynes of attraction.

If this appears to be a large force, keep in mind four important factors:

1. The poles are rather strong.
2. The distance between them is very short.
3. The path for flux is very easy.
4. The dyne is an extremely small force.

When the poles are those of common magnets, the interaction among the poles causes the problem to become more complex. This fact is illustrated in Figure 5-10.

Let's assume that each pole in Figure 5-10 has a strength of 400 u.p. This gives us four sets of figures.

We have a force of repulsion between north poles; $F = 400 \times 400/10 \times 10 = 1600$ dynes.

There is a force of repulsion between south poles; $F = 400 \times 400/50 \times 50 = 64$ dynes.

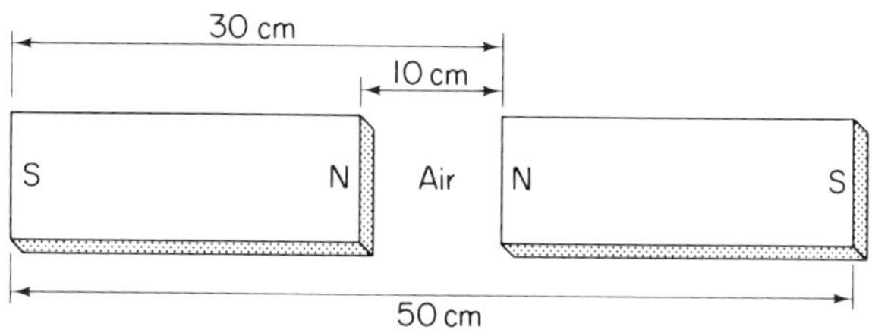

Figure 5-10. Interaction Among Poles

There is a force of attraction between the two pairs of north and south poles; $F = 2(400 \times 400)/30 \times 30 = 355.4$ dynes.

The net force is $1600 + 64 - 355.4 = 1308.6$ dynes of repulsion.

If a magnet is pivoted as shown in Figure 5-11, it can be rotated by bringing another magnetic pole into the vicinity.

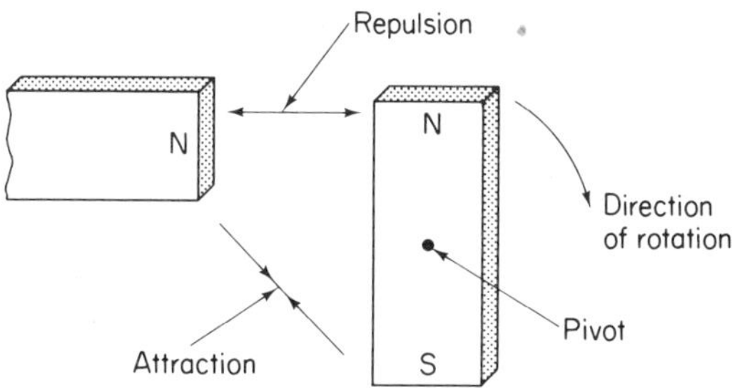

Figure 5-11. Rotation of Pivoted Magnet

Remember this drawing. It illustrates the principle on which the design of our most popular meter movements are based.

The above problems should have illustrated that flux becomes less concentrated as the distance from the pole increases. With proper instruments it would be possible to chart the field intensity.

field strength

We have seen that flux lines leave the north pole, fan out, and re-enter the south pole. If an isolated unit pole were placed at any point in a magnetic field, some force would act on it. *Field strength* (or *intensity*) is expressed in terms of this force. *Field* strength is measured in a unit called the *oersted.* An oersted is the field strength that will exert one dyne of force on a unit pole. The symbol for field strength is H, and the formula is $H = F/m$, where H is field strength in oersteds, F is force in dynes, and m is pole strength in unit poles.

Example 1: An isolated pole of 30 u.p. is placed in a field with a resulting force of 260 dynes. What is the field strength at that point?

Solution:

$$H = F/m = 260/30 = 8.67 \text{ oersteds.}$$

The field strength within the magnet is referred to as *flux density.* It is a measure of the number of flux lines per unit area. One flux line is a *maxwell,* and one maxwell per square centimeter is a *gauss.* The gauss is a unit for measuring flux density, and it can be measured on a *gauss meter.* The symbol for flux density is the Greek letter beta (β).

$$\beta = \Phi/A$$

Where: β is flux density in gauss, Φ is flux lines (maxwells), and A is cross sectional area in square centimeters.

Example 2: A magnet with a cross-sectional area of 40 square centimeters contains 5×10^5 maxwells. What is the flux density?

$$\beta = \Phi/A = 5 \times 10^5/4 \times 10 = 1.25 \times 10^4 \text{ gauss.}$$

Now that we have seen some of the effects of magnetism, it is time to examine some magnetic theories.

MAGNETIC THEORIES

A theory is a *logical idea* that is advanced in an effort to explain some phenomenon; in this case, the phenomenon of magnetism. We will examine the two most popular theories. You may take your choice or reject both as you see fit. Both are well thought out and represent some good, logical analysis.

Weber's theory

This theory has been around a long time and is based on simple experiments that can be performed with a minimium of equipment.

A magnet can be created by *stroking* a magnetic material in a constant direction with a pole of another magnet. This seems to indicate that particles of the material have been induced to align themselves in the direction of the strokes. If this is true, the particles of metal must have already possessed *magnetic polarities.* The overall magnetic properties of the piece of metal are neutral because the poles of the individual particles are arranged in a random pattern.

If we think of these particles as molecules, we have the essentials of Weber's theory, which is also called the *molecular* theory, of magnetism. This is illustrated in Figure 5-12.

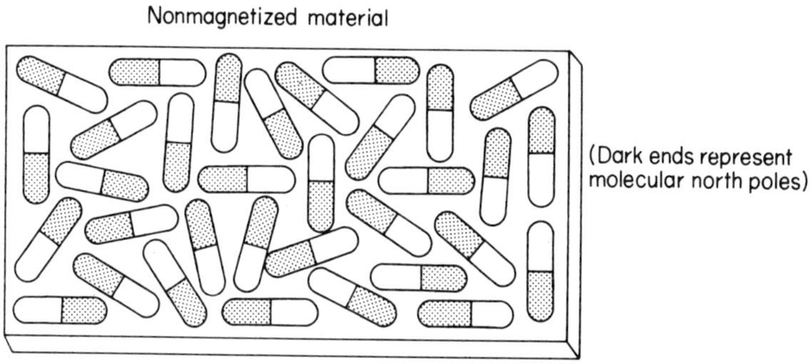

Figure 5-12. Random Arrangement of Molecular Poles

The repeated stroking action *aligns the molecules* as shown in Figure 5-13.

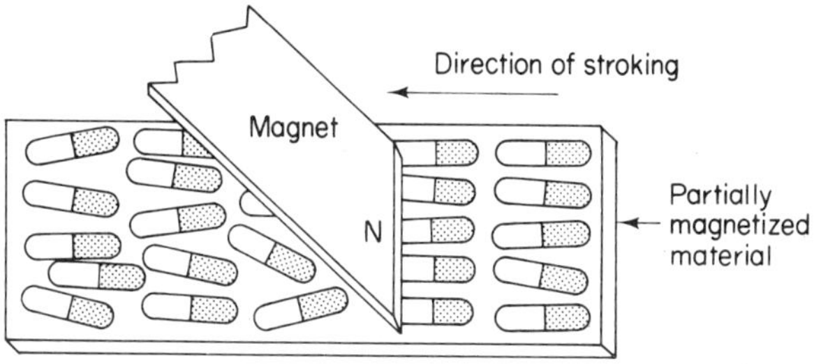

Figure 5-13. Creating a Magnet

When nearly all molecules are aligned, the material is fully magnetized as shown in Figure 5-14.

The piece of metal is now a magnet, and it exhibits the combined effect of all the molecular magnets acting in unison.

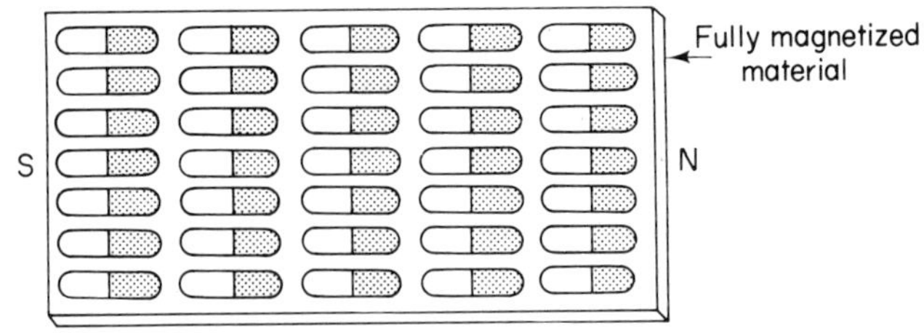

Figure 5-14. Molecules of a Magnet

Another experiment, which seems to support this theory, can be conducted by breaking a bar magnet into small pieces. As each piece is broken from the magnet, it becomes a separate magnet with a north and a south pole. This is illustrated in Figure 5-15. If

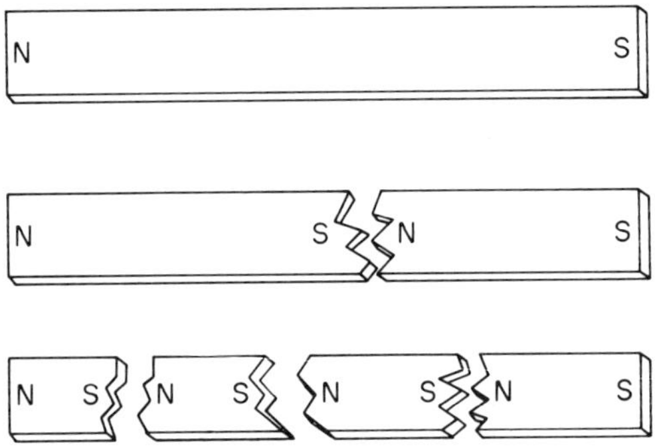

Figure 5-15. Polarities of Broken Magnets

this process could be carried on until a single molecule of metal could be broken off, that molecule should be a magnet with a north and a south pole.

When magnets are exposed to excessive heat or vibration, they tend to lose their magnetism. This also appears to support the molecular theory. Either heat or vibration will increase molecular activity within the material and thus tend to destroy the aligned pattern of the molecules.

domain theory

This more modern theory of magnetism is based on the *spin of the electrons* that orbit the nuclei of the atoms within the material. Comparing an atom to a solar system, we have the nucleus representing the sun and each electron corresponding to an orbiting planet. It has been proved that each electron has a magnetic field about it. It is not too unlikely that each electron spins on its own axis much as the planets do. This is shown in Figure 5-16.

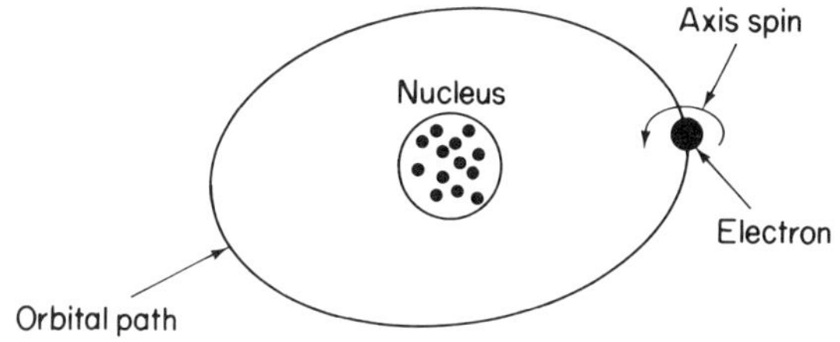

Figure 5-16. Axis Spin of Rotating Electron

The domain theory holds that a *nonmagnetized* atom has an *equal number* of electrons spinning in *both directions.* For lack of more definite references, we shall use clockwise and counterclockwise for these directions. An atom, then, is magnetically neutral when half of its atoms are spinning clockwise and half are spinning counterclockwise. When more than half of the electrons are spinning in one direction, the atom is magnetized.

When a number of magnetized atoms are grouped in a piece of iron, the magnetic forces interact. The magnetic fields of adjacent atoms align themselves in parallel patterns; that is, their magnetic poles are oriented in the same direction.

When this group of magnetized atoms reaches a certain concentration, the group becomes a *domain.* A domain consists of about 10^{14} atoms. It is estimated that about 10 million domains can be contained in a cubic millimeter. Each domain exhibits a strong magnetic field in a specific direction, but normally, the random arrangement of the domains cancels the overall magnetic effect of a material.

An external magnetizing field can cause these domains to *align their fields* in one direction. This harmonious relationship causes the piece of metal to exhibit the combined effect of the magnetic fields of all the domains.

Since both theories have been advanced in an attempt to explain the same set of facts, most arguments in favor of one also support the other. When better theories are formulated, we will abandon the old in favor of the new. If one of these becomes thoroughly proven, it will become a law.

In many electromagnetic devices, magnets and magnetic materials are arranged into magnetic circuits.

MAGNETIC CIRCUITS

In these circuits, *the magnet becomes the power supply;* the *magnetic material* is comparable to the *conductors* and resistors; and the *flux* can be compared to *current.* Before we can fully understand the action of magnetic circuits, we need a better analysis of magnetic materials.

magnetic materials

We promised a better *classification* of magnetic materials. Instead of simply magnetic and nonmagnetic, all materials actually fall into one of *three* categories. These are:

1. diamagnetic.
2. paramagnetic.
3. ferromagnetic.

Materials are categorized according to their capability to respond to a magnetizing influence. This is called *magnetic susceptibility,* and it is a ratio of magnetism produced to the force that produces it.

Diamagnetic materials have strong negative susceptibilities. Bismuth is among the strongest with a susceptibility of negative 13×10^{-6} (−0.000013). Materials in this category show no reaction to ordinary magnetic fields. This accounts for the fact that they are generally called *nonmagnetic.* When exposed to very strong fields, they do react, but in a peculiar manner. They are repelled by the field, and if free to move, they will move out of the field. Copper, silver, gold, and mercury are some other examples of diamagnetic materials. The permeability of these diamagnetic materials is always less than one.

Paramagnetic materials have a positive, although very weak, susceptibility. The susceptibilities range from a positive 10×10^{-6} to 100×10^{-6} (0.00001 to 0.0001). This means that they can be weakly attracted by very strong magnetic fields. Most of the elements, and many compounds, are in the paramagnetic category. Some examples are aluminum, chromium, and platinum. The permeabilities of these materials are slightly greater than unity. Since paramagnetic materials show no reaction to fields of common strength, they are also commonly called nonmagnetic.

Ferromagnetic materials have very high positive susceptibilities. Iron has a susceptibility of a positive 500, and some other materials are much higher. This means that these materials are strongly attracted even by weak magnetic fields. The permeability of these materials is also very high. With some ferromagnetic alloys, permeabilities of 100,000 are common, with some running much higher than this. The higher the permeability, the stronger the magnet that can be made from the material. A bar magnet of alnico can be made strong enough to lift five times its own weight. Because of their obviously strong attraction to magnetic fields, materials in the ferro-magnetic category are commonly referred to as *magnetic materials.*

circuit terminology

The opposition that a material offers to flux is comparable to resistance. It is called reluctance and is measured in *rels.* The symbol for rel is R. For most calculations $R = l/\mu A$.

Where: R is rels, l is the length of the path in centimeters, μ is permeability, and A is the cross-sectional area of the path in square centimeters.

We have already compared the magnet to a power supply, but instead of providing electromotive force measured in volts, it furnishes *magnetomotive force* (mmf), which is measured in *gilberts.* This gives us a formula for magnetic circuits which is comparable to Ohm's law for electrical circuits; $\Phi = \text{mmf}/R$.

Where: Φ is flux in maxwells, mmf is magnetomotive force in gilberts, and R is rels of reluctance.

One gilbert is the amount of magnetomotive force required to force one maxwell of flux through a reluctance of one rel.

Further comparisons can be drawn between magnetic circuits and electrical circuits.

1. Flux is directly proportional to mmf and inversely proportional to reluctance.
2. Total magnetomotive force is distributed around each closed loop in proportion to the size of the reluctances.
3. The number of maxwells (lines) of flux leaving the north pole is equal to the number entering the south pole.

4. The north pole of the magnet compares to the negative terminal of the battery, and the south pole compares to the positive terminal.

analysis of magnetic circuits

The *only* requirement for the establishing of a complete magnetic circuit is a magnet. The flux always has a complete path because all known materials are magnetic conductors. An ordinary bar magnet suspended in air is a complete magnetic circuit. This circuit has an infinite number of paths for the flux between the north and south poles. The quantity of flux in this circuit is determined by the pole strength of the magnet.

The problem then in magnetic circuits is not to change the total circuit flux, but rather to *concentrate* it into more limited areas. Since the square of the distance between the poles is a primary factor, any reduction of this distance will increase flux concentration. When the bar magnet is bent into a C-shape or a horseshoe shape, the poles can be brought close together. This causes most of the flux to concentrate into the short distance between the poles as shown in Figure 5-17.

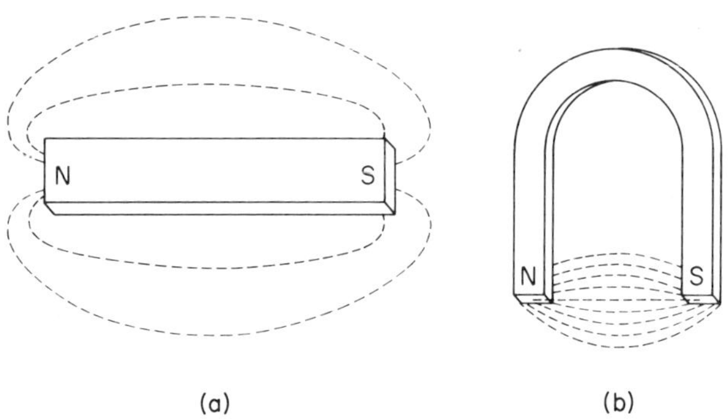

Figure 5-17. Concentration of Flux

The bar magnet has flux dispersed in all directions. The horseshoe creates an easier path for flux by bringing the poles closer together. Magnets can be formed into most any shape to suit the convenience of the designer. The most common shapes are bar, horseshoe, and ring.

Another method of changing flux concentration is to use a different substance between the poles. Since reluctance is the approximate reciprocal of permeability ($R = l/\mu A$), when the material between the poles has a high permeability, the flux has a low reluctance path. Placing a piece of soft iron across the poles of a horseshoe magnet reduces the reluctance of the flux path to a value of less than $\frac{1}{7000}$ when compared to air. Not only does this offer a path of low reluctance, but it also reduces the cross-sectional area of the flux path. The more attractive path through the iron induces more lines to concentrate into this smaller path. This can be seen by comparing Figures 5-17 and 5-18.

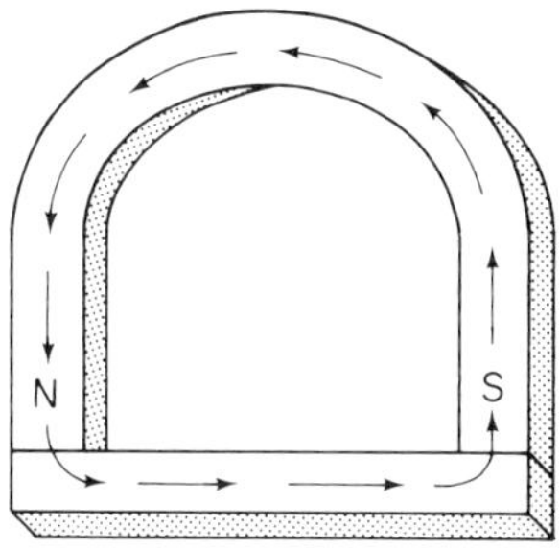

Figure 5-18. Maximum Concentration of Flux

A variation of the complete iron path is a path of part iron and part air. This principle is used in the design of meter movements, motors, and generators. The iron (or other high μ material) is attached to the magnetic poles and shaped to produce a minimum air gap. The iron attachments are called *pole pieces*. A magnetic circuit using pole pieces is shown in Figure 5-19. Notice that the pole pieces have the effect of bringing the poles closer together.

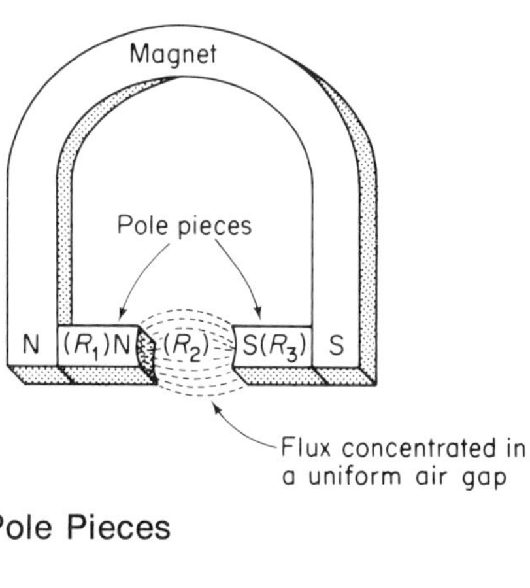

Figure 5-19. Use of Pole Pieces

Actually this is a series circuit consisting of a power supply and three reluctances. The power supply is the magnet and the north pole represents the negative terminal. *Flux* effectively *leaves the north pole,* passes through the north pole piece (R_1), through the air gap (R_2), through the south pole piece (R_3), and *enters the south pole.* Inside the power supply, the flux passes from south to north (+ to −).

Magnetic circuits *are not* limited to series circuits. Parallel circuits and multiple mmf circuits are common enough, and even series-parallel arrangements are possible. Figure 5-20 shows a parallel circuit.

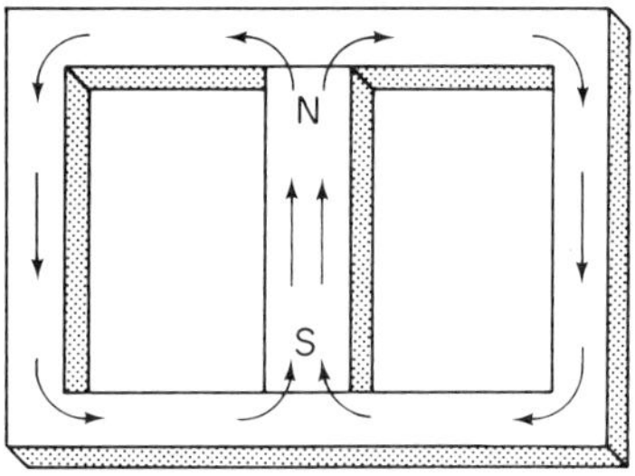

Figure 5-20. Parallel Magnetic Circuit

All magnetic circuits suffer some losses due to *flux leakage.* Even with complete, low reluctance paths, as shown in Figures 5-18 and 5-20, not all flux lines will be confined to the intended path. In electric circuits, we saw that there were no perfect conductors or insulators. Here, it would be better to say that all materials are fairly good conductors. This makes it easier for the flux to leave the confines of the intended path. This flux leakage is considered a loss because it serves no useful purpose.

Apart from being a loss, flux leakage causes other problems. It affects current carrying conductors and other components—*especially instruments.* Since there is no known material that will insulate against these stray lines of flux, *shielding* critical circuits from their influence is a difficult task. The most effective method for such protection is to provide the stray flux with a more attractive path.

Since most flux lines will take the path of least reluctance, an instrument that is enclosed in a shield of low reluctance material has adequate protection. This is illustrated in Figure 5-21.

The stray flux lines, which would otherwise pass through the instrument, first encounter the low reluctance path. This more at-

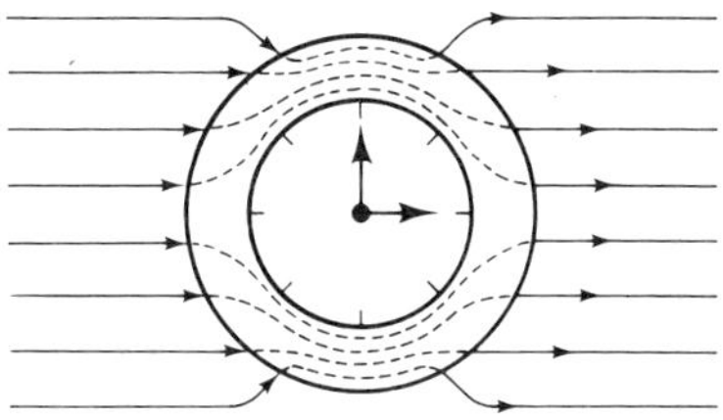

Figure 5-21. Magnetic Shielding

tractive path causes them to bend and detour through the shield. The case on a magnetic resistant watch is an example of a magnetic shield. Don't be misled, however, the watch is by no means magnet-*proof.*

INTERACTION OF MAGNETISM AND ELECTRICITY

It has become well established that magnetism and electricity cannot be separated. As previously stated, current is surrounded by flux, and flux can be used to generate current. Let's examine the magnetic field that is created by current.

flux resulting from current

Currents of any type or magnitude *create magnetic fields* about themselves. Electrons are always in motion, and every moving electron carries with it a magnetic field, which is created by the motion of the electron. This fact was discovered by Hans C. Oersted, a Danish physicist (1777–1851). In 1820, he noticed that a compass which had been placed near an electrical circuit was being affected by the circuit. This discovery was of great importance and was the first proof of *a definite link between magnetism and electricity.*

Further experimentation with the compass established the fact that a current carrying conductor is literally surrounded by a magnetic field. The compass needle aligned itself at right angles to the conductor when current was present and returned to normal when current was stopped. When the direction of current was reversed, the compass needle swung 180° and pointed in the opposite direction. This action is shown in Figure 5-22.

When the field about the conductor was completely charted, the field showed that the current was *encircled* by a continuous field. The circles representing the field are concentric rings emanating

Figure 5-22. Compass Showing the Field About a Conductor

from the center of the conductor and spreading out in all directions. Iron filings on a piece of paper reveal a cross-sectional pattern similar to that of the waves that result when a stone is dropped into a pond. This is shown by the graphic representation in Figure 5-23.

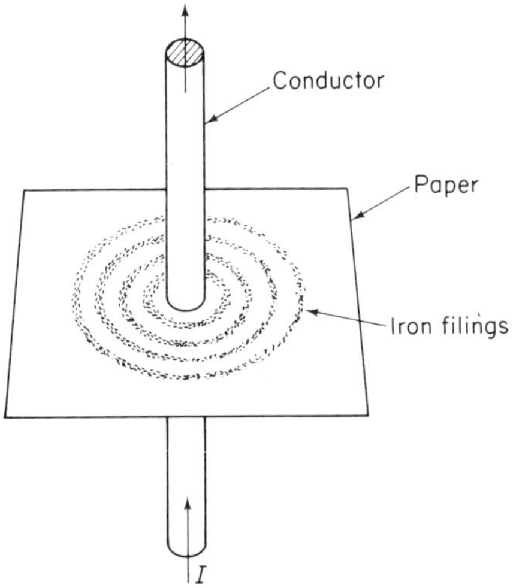

Figure 5-23. Field Pattern About a Conductor

When compasses are placed in the field and spaced through a full 360° around the conductor, a definite field direction is established as shown in Figure 5-24.

Notice the direction of the field with respect to the direction of current. In either diagram, if you grasp the conductor in the left hand with the thumb pointing the direction of current, the fingers will encircle the conductor in the same direction as the flux lines that compose the field. This is known as the *left-hand rule for field direction* about a current carrying conductor. This is handy in many situations where field direction is important. It can also be used to determine the direction of current when the field direction is the known factor. This rule is illustrated in Figure 5-25.

The strength of the field about a conductor can be calculated by $H = 2I/10r$ where: H is field strength in oersteds, I is the current

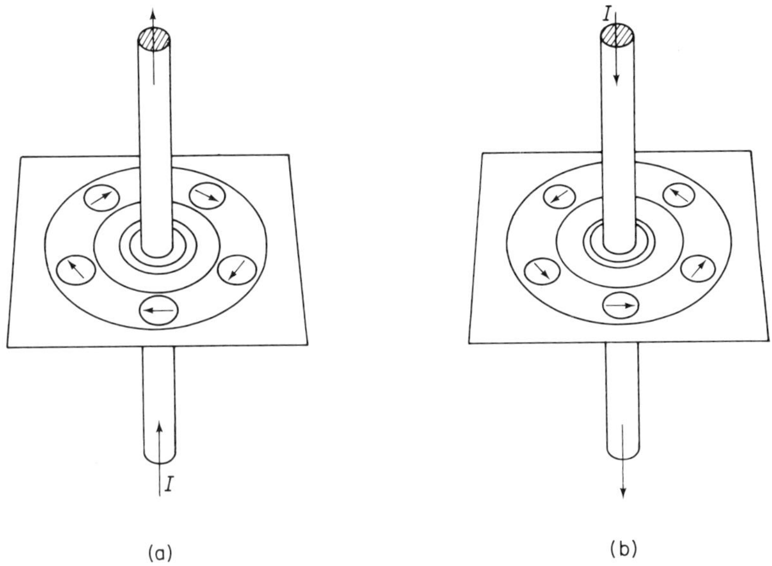

Figure 5-24. Compass Showing Field Direction About a Conductor

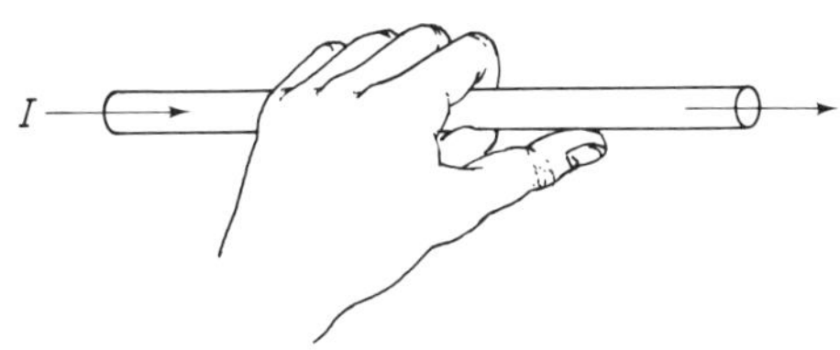

Figure 5-25. Left Hand Rule for Field Direction

in amperes, and r is the distance from the center of the conductor in centimeters.

Since it has already been established that lines of force in the same direction repel and lines in opposite directions attract, it stands to reason that when two current carrying conductors are placed close together, their fields will interact. The type of interaction will be determined by the relative direction of the two currents. The force of the reaction will be proportional to the current amplitudes.

force between conductors

Figure 5-26 shows a cross-sectional view of two current carrying conductors. The *dot* (·) indicates that the current is coming out of the page *toward the viewer.* The *plus* (+) indicates that current

Figure 5-26. Cross Section of Fields About Conductors

is going into the page *away from the viewer.* The two conductors are separated far enough to prevent interaction of the fields. Check the direction of the field in each case by applying the left hand rule.

When the two conductors of Figure 5-26 are moved close together, the fields will exert a physical force upon each other to keep them *separated.* This is shown in Figure 5-27. This is obviously a

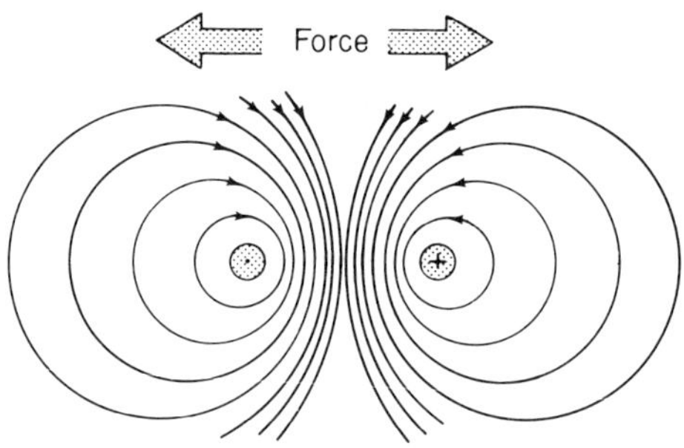

Figure 5-27. Repulsive Force Between Conductors

force of repulsion. The lines of force between the conductors are all in the same direction, and these lines repel each other. Notice that the currents in the two conductors are in opposite directions. Therefore, two conductors carrying *currents in opposite directions repel each other.*

By the same token, two conductors carrying *currents in the same direction should attract each other.* And they do, as you can see from Figure 5-28. In this case, the lines between the conductors are in opposite directions. These lines cancel one another, reducing the field between the conductors. All lines around both conductors are in the same direction. The outer lines combine into a single field and exert a force to bring the two conductors together.

A careful analysis of Figures 5-27 and 5-28 should establish that if either of the conductors were free to move, it would move in the direction of the exerted force. In fact, if a current carrying con-

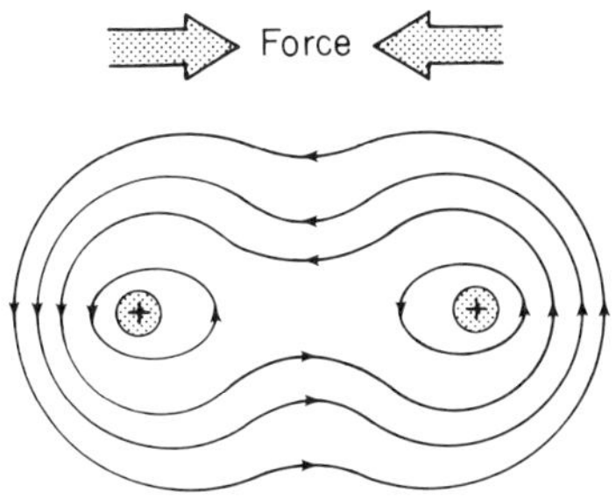

Figure 5-28. Attractive Force Between Conductors

ductor (which is free to move) is placed in any magnetic field, it will move *out of that field.* The direction of motion is determined by the relative directions of the two fields. The speed of motion is proportional to the strength of the two fields. An electric motor is based on this principle as illustrated in Figure 5-29. Lines in the

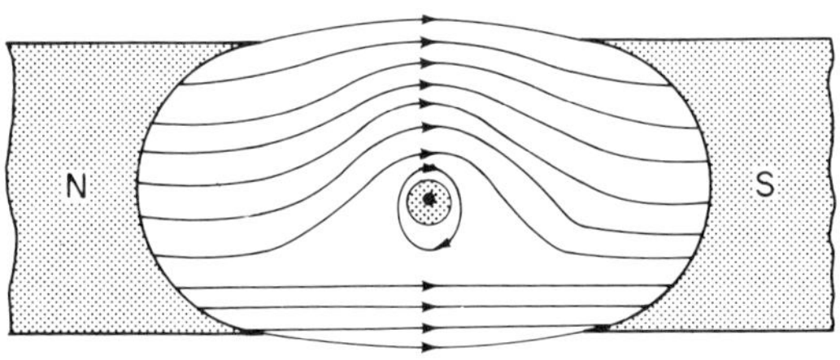

Figure 5-29. Motion of a Conductor Through a Magnetic Field

same direction are warped and exert a force to make the conductor move in a direction that will relieve this pressure. Lines in the opposite direction cancel to relieve the pressure ahead of the moving conductor.

The quantity of force exerted on the conductor is calculated by $F = \beta I l/10$ where F is force in dynes, β is flux density in gauss, I is the current in amperes, l is the length of the conductor in the magnetic field in centimeters, and 10 is a constant.

electromagnets

When a conductor is coiled into a loop, the direction of the lines of force is consistent (all lines in one direction) through the center of the loop. This is illustrated in Figure 5-30.

The field intensity at the center of a loop is $H = 2\pi I/10r$, where

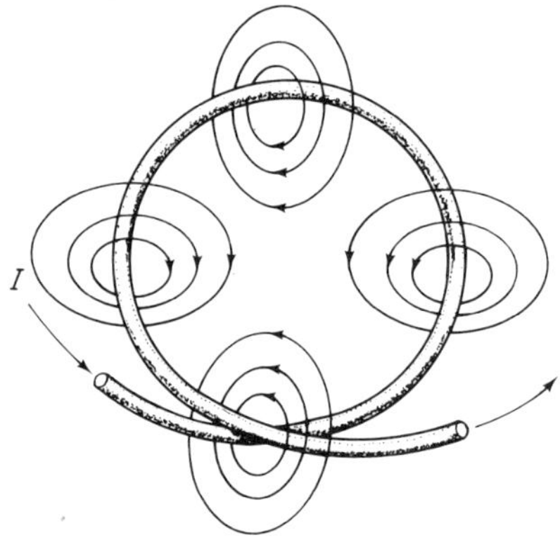

Figure 5-30. Field in a Loop

H is the field strength in oersteds, *I* is the current in amperes, and *r* is the radius of the loop in centimeters. Ten is the same constant we had previously, and the 2π is a constant derived from the fact that the flux is acting through a full 360° circle.

With current as specified, all the flux lines pass through the center of the loop going away from the viewer. They then curve around the outside and come back toward the viewer. When several loops of wire are wound around a cylindrical form the above pattern is extended to include the entire coil. This is illustrated in Figure 5-31.

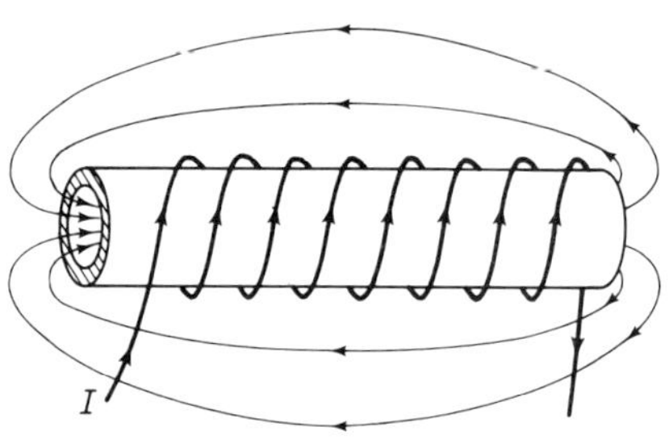

Figure 5-31. Field About a Coil

Here we have the loops of a conductor wound around a form. With the current as indicated, the field direction can be verified by applying the left-hand rule to any part of the conductor on the coil. The lines of force leave the right-hand end of the coil, fan out around the outside, re-enter the left end, and pass through the center of the coil back to the right hand end. Examine this drawing carefully and compare the field to that of a bar magnet. They are not only *similar,* but, in many respects, they are *identical.*

Current through the windings of any coil will create a magnet. These are called electromagnets because their function depends upon electricity. Electromagnets are designed in a wide variety of sizes and shapes, and they serve many useful purposes.

As you have already seen in Figure 5-31, the field about an electromagnet is the same as that of a bar magnet. Therefore, the electromagnet must also have a set of magnetic poles. Since flux leaves the north pole of a magnet and enters the south pole, the magnetic polarity of the coil in Figure 5-31 must be north on the right and south on the left. Electromagnets (or coils manufactured for that purpose) are called *solenoids.* Determining the polarity of a solenoid can be important. There is a left-hand rule for this purpose.

> Grasp the coil in the left hand with fingers pointing the direction of current through the conductor and the thumb extended. The thumb will then point to the north pole.

This rule is illustrated in Figure 5-32.

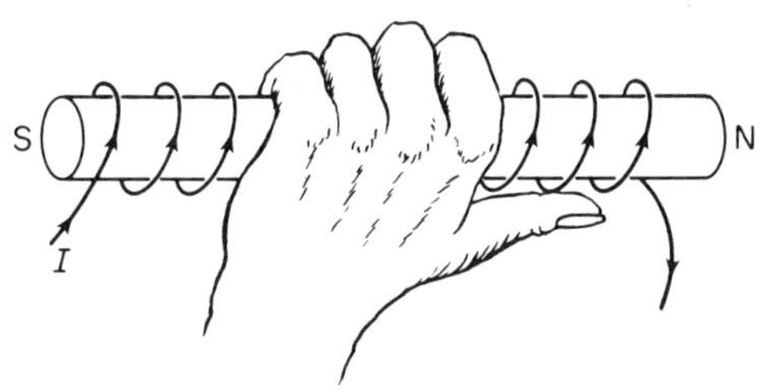

Figure 5-32. Polarity of a Solenoid

Since a solenoid is composed of a number of loops of a conductor, the field strength at the center of the solenoid is the combined field strength of all of the loops. If we were to take the formula for field strength in a loop and multiply it by the number of loops (turns) on the solenoid, we would have the formula for field strength at the center of the solenoid: $H = 2\pi NI/10r$.

The N in the above formula represents the number of turns of conductor on the solenoid. Of course, this is assuming a perfect condition—one where there are no losses. The formula cannot be applied to all coils without regard to their size and shape. It will hold true for a relatively short coil that has a small length compared to its radius.

Electromagnets are used in such delicate work as the precise movement of the needle on a microammeter. They are also employed in such manly tasks as lifting huge loads of metal. Since the magnetic field strength is directly proportional to the number of turns, the solenoid can be constructed to fit almost any task. Another factor which has a direct bearing on the field strength is the amount of current through the coil. Huge magnets, which use only a small current, can be made from delicate wire. Strong solenoids are also constructed from a few turns of heavy wire which carry a strong current.

The amount of current and the number of turns cannot be separated. In fact, the field strength is proportional to the *product* of the number of turns and the current, as the previous formula indicated. Magnetomotive force can be expressed in terms of *ampere-turns.* An ampere-turn is one ampere of current through one turn (loop) of wire. In the metric system of measurement, this is the unit for mmf.

The ideal solenoid would be perfectly insulated, and the turns would be so close together that every line of force would loop every turn of the conductor. This would allow no flux leakage. The term *flux linkage* or *flux-current linkage* is closely related to ampere-turns. One flux linkage is one line of flux linking one turn of wire. Total flux linkage is total flux times the number of turns. The formula for magnetomotive force in gilberts is mmf $= 4\pi NI \times 10^{-1}$ (or $4\pi NI/10$ or $1.26NI$). The total flux is $\Phi =$ mmf/R. The flux density inside the coil then becomes $\beta = \Phi/A$.

We have seen that current causes a magnetic field, but a magnetic field can be used to generate current.

current resulting from flux

Any set of circumstances that causes *flux* to cut *across a conductor will generate a voltage.* This voltage will cause current through the conductor if a complete path exists. This fact was discovered by Joseph Henry, an American physicist (1797–1878). His discovery was made in 1830. Michael Faraday, a British physicist (1791–1867), working independently and without knowledge of Henry's work, made the same discovery in 1831.

The above process is known as electromagnetic *induction,* and since Faraday was first to make the announcement, he received credit for the discovery. Faraday found that when the pole of a permanent magnet is moved in the vicinity of a loop of wire, current is induced into that wire. This is shown in Figure 5-33.

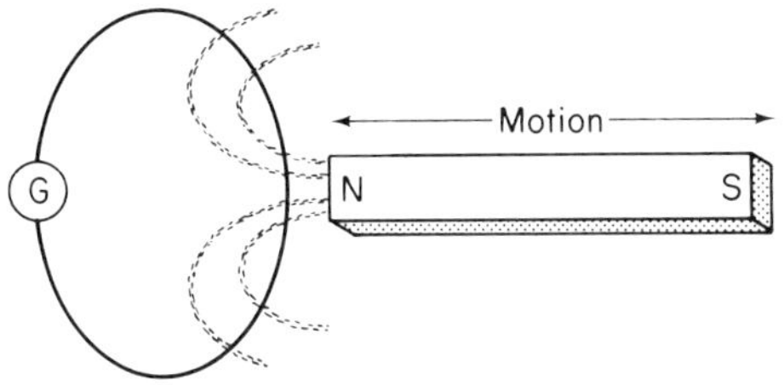

Figure 5-33. Faraday's Experiment with Magnet and Loop

The magnet moving toward the loop caused the meter to indicate a current. When the magnet stopped, the current stopped. As the magnet was withdrawn, the meter showed current in the opposite direction. The relative motion between the field and the conductor induced a voltage, which caused a current in the conductor. Faraday tried several variations of this experiment; one of which made use of a solenoid like the one illustrated in Figure 5-34.

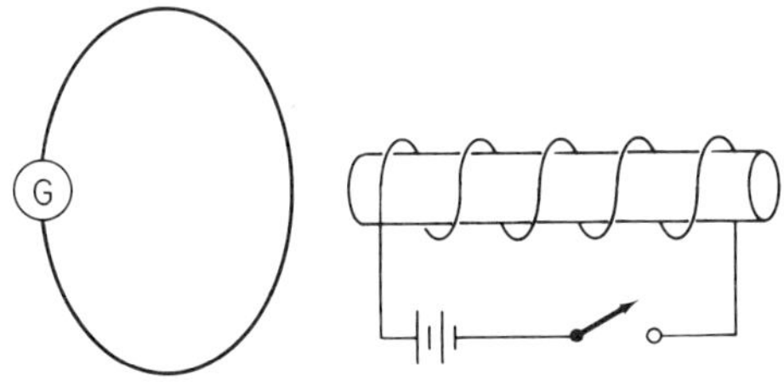

Figure 5-34. Faraday's Experiment with Solenoid

When the switch was closed, the meter showed a current, which quickly died out to zero. As the switch was opened, another surge of current appeared in the opposite direction and died out to zero. The same relative motion between field and conductor is present in both Figures 5-33 and 5-34. In the latter case, when the switch closes, flux builds up to a peak value and cuts across the conductor in the process. When the switch is opened, flux dies out by contracting into the coil. As it contracts, it again cuts across the conductor.

These experiments established the principle that an electromotive force can be induced by *relative motion* between a magnetic field and an electric conductor. This is the principle of the *electrical generator.* Faraday's devices were actually generators in their simplest form. Another variation of the same line of experiments proved that one circuit could induce a voltage into another circuit. This is shown in Figure 5-35.

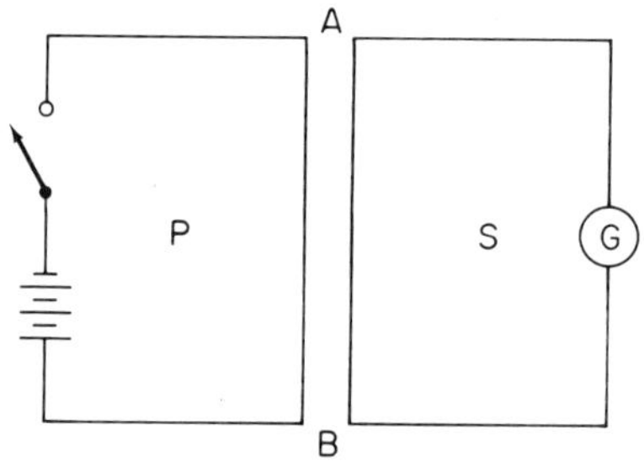

Figure 5-35. Inducing Voltage into a Secondary Circuit

When the switch in the primary circuit was closed, the meter in the secondary showed a surge of current, which quickly returned to zero. When the switch was opened, current surged in the opposite direction and quickly died out. The current in the primary circuit created a magnetic field, which expanded outward from the center of the conductor. This expanding field, where the circuit conductors are parallel (points *A* to *B*), cut across the conductor of the secondary circuit. When current was stopped, the decaying field shrank back into the primary conductor. As the field contracted, it again cut across the secondary conductor. Each time the flux cut across the secondary conductor (once on buildup, and again on decay), a voltage was induced into the secondary circuit, which caused current through the meter.

All electric *transformers* are based on the above principle of current in a primary circuit inducing voltage into a secondary circuit. If Faraday had been content to stop here, his contribution to electrical science would have been great, but he didn't stop. He went on to formulate a law concerning the quantity of induced voltage. His law states that the induced emf is proportional to the rate at which the conductor cuts the lines of force. Three factors, then, are involved with inducing a voltage:

1. Conductor.
2. Magnetic field.
3. Relative motion between the field and the conductor.

But there is only one factor which affects the amplitude of the induced voltage, and that is the rate at which the conductor cuts the lines.

There are several ways to increase the rate of cutting:

1. Increase the speed of relative motion.
2. Increase the number of conductors.
3. Increase the total flux.

Another way to express rate of cutting is flux linkages per unit time. This can be expressed in a formula as $e = N\Phi/t$.

One volt is generated when 100,000,000 flux lines cut a conductor in one second. In order for the induced voltage (e in the above formula) to be in volts, the 100,000,000 must be considered. The formula then becomes $e = N\Phi/t \times 10^8$. Where e is the average induced voltage in volts, t is time in seconds, N is number of conductors (or turns of a conductor), Φ is total flux, and 10^8 is flux linkages per second required for one volt.

Heinrich Friedrich Emil Lenz, a German physicist (1804–1865), expanded Faraday's law to include the *polarity* of the induced voltage. The law states, in effect, that the polarity of an induced emf will be such that it will *oppose the action that induces it.* This means that current in the primary circuit is opposite in direction to the current in the secondary circuit. It also means that as the pole of a magnet approaches the end of a coil, the coil becomes an electromagnet with an opposing polarity. This is illustrated in Figure 5-36.

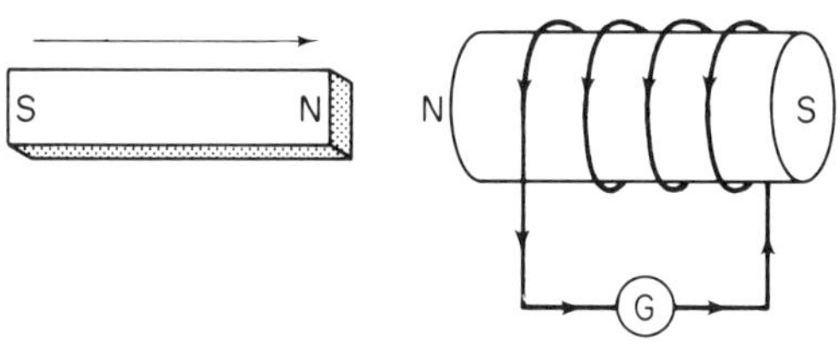

Figure 5-36. Lenz's Law

As the north pole of the magnet is pushed toward the coil, a voltage is induced, which causes current as indicated. Applying the left-hand rule to this coil shows the north pole of the electromagnet to be opposing the motion of the permanent magnet. When the motion ceases, current stops, and there is no electromagnetic polarity. As the north pole of the magnet is withdrawn from the coil, voltage of the opposite polarity is induced. This causes current in the opposite direction and sets up a south pole at the left end of the solenoid. The attraction of this south pole will oppose the withdrawal of the permanent magnet's north pole.

When an isolated north pole is pushed all the way through a coil, the induced voltage changes the magnetic polarity to oppose the motion each step of the way. (See Figure 5-37.)

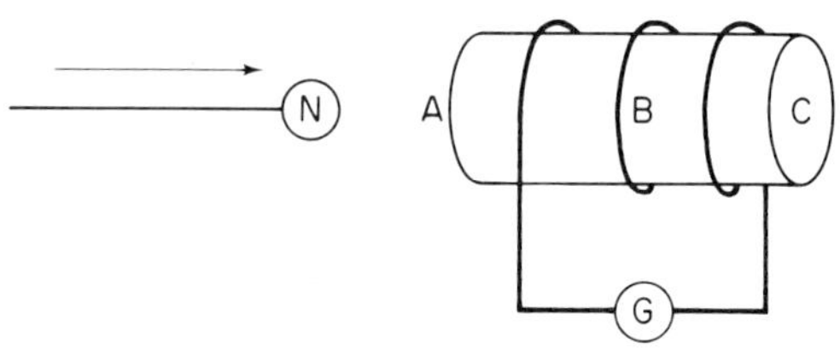

Figure 5-37. Isolated Pole Through a Coil

As the pole approaches, point *A* is north and *C* is south to oppose the approach. When the pole passes point *A,* point *A* becomes a south pole and point *C* a north to oppose further entry of the permanent north pole. When the pole passes point *B,* point *A* becomes a south and point *C* a north to oppose the pole emerging from the coil. When the pole passes point *C,* point *A* becomes north and point *C* becomes south to oppose the pole's motion away from the coil.

So far, we have been discussing *moving fields* and *stationary conductors.* The same effect would be observed with a *moving conductor* in a *stationary field.* This is shown in Figure 5-38. In this

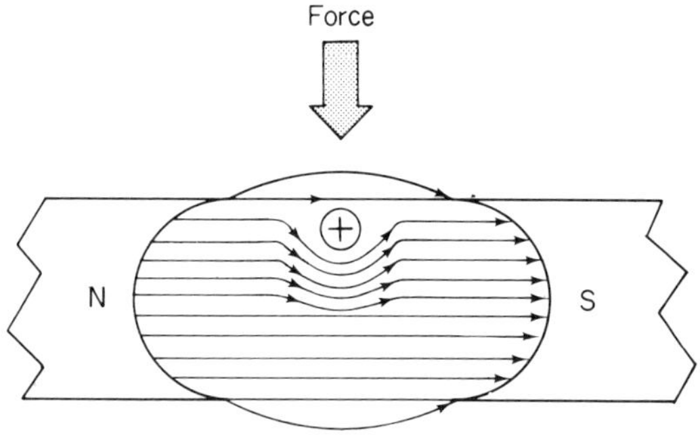

Figure 5-38. Moving Conductor—Stationary Field

drawing a conductor is being pushed through a stationary field. The stationary lines of force are cutting across the moving conductor and inducing a voltage in the conductor.

There is a very simple way to detect the polarity of an induced voltage:

1. Open the left hand with thumb extended.
2. Place the hand into the magnet field in such a manner that the fingers point the direction of flux, and the conductor is moving toward the palm.
3. The thumb is then pointing the direction of the induced current in the conductor.

This concludes Chapter 5. Many important principles have been developed. Your understanding of these principles will have a considerable effect on your understanding of the electric phenomena that you will encounter in later chapters. The following review exercises are designed to reinforce what you have learned here.

REVIEW EXERCISES

1. Knowledge of magnetism is essential to ______________.
2. Materials are commonly referred to as either magnetic or ______________.
3. The two broad classifications for magnets are ____________ and ______________.
4. What are the two types of artificial magnets?
5. The ability of a material to hold magnetism is called ______________.
6. What is residual magnetism?
7. The area surrounding a magnet where its influence can be felt is called a ______________ ______________.
8. The lines of force that compose a magnetic field are called ______________, and they are highly concentrated near the ______________ of a magnet.
9. The ends of a magnet are called ______________. The north seeking end is the ______________ ______________.
10. Describe the force exerted between poles of a magnet.
11. The direction of a magnetic field is the direction that an isolated ______________ pole would move in the field.
12. The flux lines are concentric rings which cannot ______________. They leave the ______________ pole and enter the ______________.
13. Write Coulomb's formula for force between poles.

14. Permeability is the ____________ with which a substance conducts flux.
15. What force is exerted by two unit poles separated by 10 centimeters of a material with a permeability of 2000?
16. The magnetic field grows progressively weaker in proportion to the square of the distance from the ____________.
17. The formula for field strength at any given point is ____________. Field strength is measured in ____________.
18. What is an oersted?
19. Gauss is a unit of measure for ____________. One gauss is one ____________ per square centimeter.
20. A magnet with a cross-sectional area of 20 square centimeters has a flux of 4×10^6 maxwells. What is the flux density?
21. Explain the difference between a law and a theory.
22. The idea that each molecule of a material is a separate magnet is a simplification of ____________ theory of magnetism.
23. When a magnet is broken into small pieces, each piece becomes a ____________.
24. The domain theory is based on electron ____________.
25. An atom becomes magnetized when more than ____________ its electrons spin in one direction.
26. How many magnetized atoms are needed to form a domain?
27. What are the three magnetic categories for materials?
28. A material with permeability of less than one and a negative susceptibility would be classified as ____________.
29. List three outstanding characteristics of ferromagnetic materials.
30. In comparing magnetic and electrical circuits, flux is comparable to ____________, magnetomotive force is compared to ____________, reluctance is similar to ____________, and permeability is about the same as ____________.
31. Write the formula which shows the relationship between flux, mmf, and reluctance.
32. What is the unit of measure for Φ, mmf, and R?

33. What is the principal advantage of bending a magnet into a horseshoe shape?
34. Adding pole pieces to a magnet provides a path of lower ______________ and concentrates the ______________ in a uniform air gap.
35. One of the principal losses in a magnetic circuit is ______________________________.
36. How can a sensitive instrument be protected against stray flux?
37. Magnetism and electricity cannot be separated because every current is surrounded by ______________ and flux can be used to generate ______________.
38. When a current carrying conductor is held in the left hand with the thumb pointing the direction of current, the fingers are in the direction of ______________.
39. When the viewer is looking into the end of a conductor with current going away from him, what is the direction of the field about that conductor?
40. Lines in the same direction ______________ while lines in the opposite direction ______________.
41. What type of force (attraction or repulsion) will be exerted between two parallel conductors with current in:
    (a) The same direction?
    (b) Opposite directions?
42. If the conductor shown in Figure P5-42 were placed in the magnetic field, in which direction would it move with respect to this page?

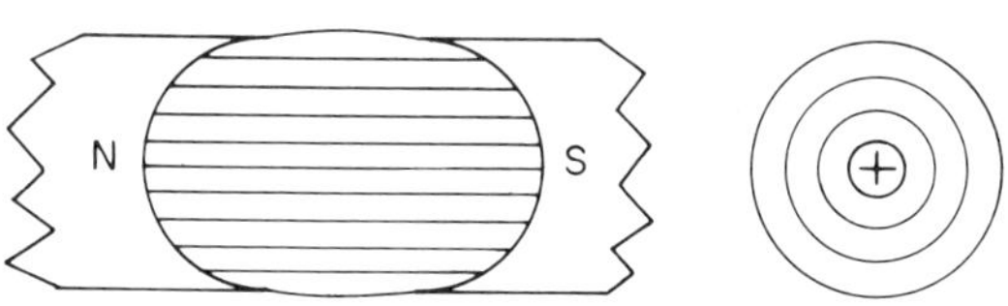

Figure P5-42

43. Write the formula for field intensity at the center of a loop.
44. Which end of the solenoid shown in Figure P5-44 is the north pole?

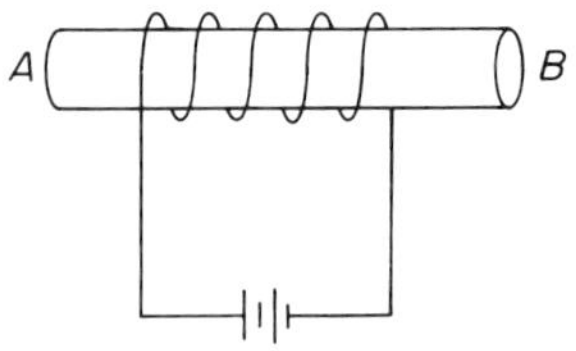

Figure P5-44.

45. The formula, $H = 2\pi NI/10r$, is used to determine field strength at the center of a ____________.
46. $4\pi NI \times 10^{-1}$ expresses mmf in ____________.
47. When relative motion exists between a conductor and a magnetic field, a current will be ____________ into the conductor.
48. What principle is expressed by the statement in item 47?
49. When one circuit induces current to flow in another circuit, what principle is being demonstrated?
50. State Lenz's law of polarities.
51. The formula for average induced voltage is ____________.
52. If the conductor shown in Figure P5-52 were pushed upward through this magnetic field, the induced current would ____________ (come toward-go away from) the viewer.

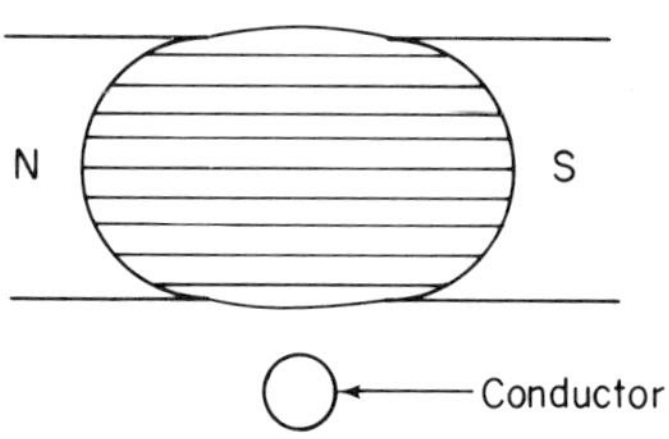

Figure P5-52

6

application of magnetic properties

This chapter title may be a bit on the presumptuous side. Actually an entire text, and many others this size, would be required to treat even superficially *all* the electrical applications of magnetism. What we will do is examine some of the work that is done by combining the forces of electricity and magnetism. Since this survey takes us inside the electric meters, we will take a little side trip and become familiar with the meter circuits.

INDUCTORS

We are using the *broad definition* for inductor—any coil with an electric current through it; or for that matter, any coil of wire whether it has current through it or not. After we examine the characteristics of coils a bit further, we will put some of them to practical use.

inductance

Inductance is the *property* of a coil that causes it to *oppose a change of current* through the windings. It is measured in henries. L is the symbol for inductance; H is the symbol for henry. Inductance is caused by an emf, which the coil induces into itself. Since all induced voltages oppose the force that induces them, this induced voltage opposes every change of the current within the coil. For this reason, it is commonly called a *counter* electromotive force (cemf) or back emf.

The construction of a coil determines its ability to produce cemf. As current passes through each loop, the magnetic field from that loop cuts across several other loops and induces cemf into each of them. The placement of the loops in respect to one another determines the number of loops that are cut by each individual field.

A one-henry coil is described as a coil that will have one volt of cemf with a current change of one ampere per second. Please note that we are speaking of *design.* Once a coil is constructed for a specified number of henries, it will have that amount of inductance even when it is sitting on a shelf. The amount of inductance refers to the coil's ability rather than its action.

The process whereby a coil induces a voltage within itself is *self-induction.* Inductance is a property which causes opposition to a change in current. Once the current reaches a stable value, there is no longer any relative motion between fields and conductors. The conductors are still there and so are the fields; but without relative motion, the lines of force are not cutting across the conductors. With a steady current, then, there is no back emf, and the coil offers very little opposition. Does it still have inductance? Yes, the ability is still there. It will reassert itself the very instant that the current tries to change either its amplitude or direction.

Any coil is properly referred to as an inductor. The schematic symbol for an inductor is  or  . The first symbol indicates an air core inductor, while the second shows that the inductor has an iron core. Inductors are used extensively in all areas of electric circuits. In dc circuits, their chief function is that of an electromagnet. The magnets operate relays, lift weights, move indicators on meters, separate magnetic from nonmagnetic materials, focus beams of electrons, and do many other specialized tasks. The inductor, however, doesn't really come into its own until it is used in ac circuits. We'll have a great deal more to say about inductors later in the text.

There is one more area of inductance that needs some further exploration at this time.

mutual inductance

Mutual inductance is the *ability of one circuit to induce a voltage into another* when the two are electrically isolated. Of course, mutual inductance is generally associated with two coils, but there is some mutual inductance between even two single conductors. It is quite proper to think of mutual inductance as a transfer of energy from one circuit to another through a magnetic field.

As we have already seen from Faraday's experiment, flux will expand from the conductor of one circuit and induce a voltage into another circuit. This becomes much more pronounced when two inductors are used. The coils of wire on the inductors provide a stronger magnetic field in one case and more conductors for that field to cut across in the other.

Mutual inductance is measured in henries (H), but it is symbolized as M to prevent confusion with self-inductance. The amount of mutual inductance that exists between two coils is determined by the inductance of each coil, their position relative to one another, and the type of material which separates them. The type of material between them and the physical position of the coils can be summed up into a single expression called the *coefficient of coupling.* Ideally, all flux lines from one coil would cut all the loops of the other. This would be a perfect coefficient of coupling. The coefficient of coupling is symbolized as K, and, since the perfect coupling is impossible, K is always less than one.

When two coils are arranged in such a manner that a change of one ampere per second in one coil induces one volt into the other, the mutual inductance is one henry. The relationship between mutual inductance, self inductance, and the coefficient of coupling is:

$$M = K\sqrt{L_1 L_2}$$

Where: M is mutual inductance in henries, K is a constant (less than 1), and $L_1 L_2$ are the inductances of the two coils in henries.

Mutual inductance in terms of induced voltage is expressed as

$$M = \frac{e}{\frac{\Delta i}{\Delta t}}$$

Where: e is volts induced into the secondary coil, Δi is a small change of current in the primary coil, and Δt is a small change in time.

The process of mutual induction is illustrated in Figure 6-1.

When the switch closes, current in the primary changes from zero to maximum in a very short time. Flux builds up and expands outward from L_1. This expanding field cuts across the windings of

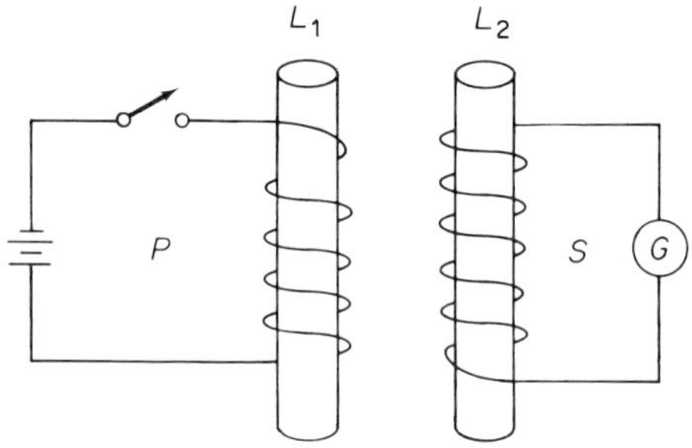

Figure 6-1. Mutual Induction

L_2 and induces a voltage. The meter registers a current in the secondary. When the primary current becomes steady, the magnetic field becomes steady as well, and current in the secondary ceases. When the switch has been opened, the process is reversed. Primary current changes from maximum to zero; the field decays back into L_1. The contracting flux lines again cut across the windings of L_2 and induce a voltage. This voltage is opposite in polarity to the first, and the meter shows current in the opposite direction. This is transformer action. In such cases, the designer strives to accomplish a maximum K. Among the methods used are:

1. Both coils wound on one iron core.
2. One coil wound on the top of the other.
3. Both coils placed close together on parallel axes.

There are other cases when mutual inductance causes undesirable effects. The best way to minimize these effects is to decrease K. This can be done by:

1. Separating coils as far as possible.
2. Placing on perpendicular axes.

Methods used to increase K are called *close coupling;* methods used to decrease K are called *loose coupling.* Some types of coupling are shown in Figures 6-2 and 6-3.

Example 1: Mutual inductance between two coils is 7 henries. Current is changing at the rate of 3 amperes per second. What average voltage is induced into the secondary?

Solution:

$$e = \frac{M\Delta i}{\Delta t} = \frac{7 \times 3}{1} = 21\text{V}$$

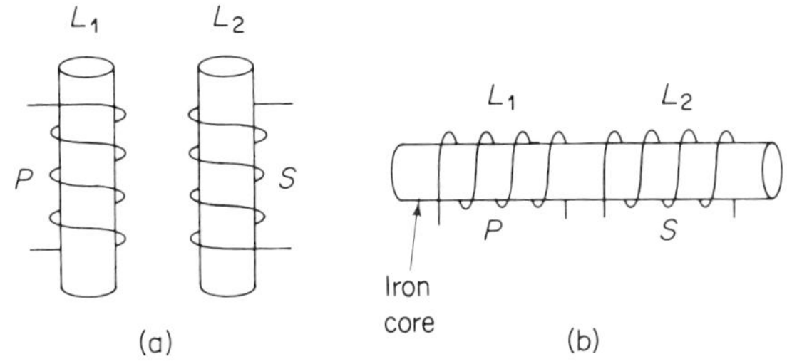

Figure 6-2. Close Coupling (maximum *K*)

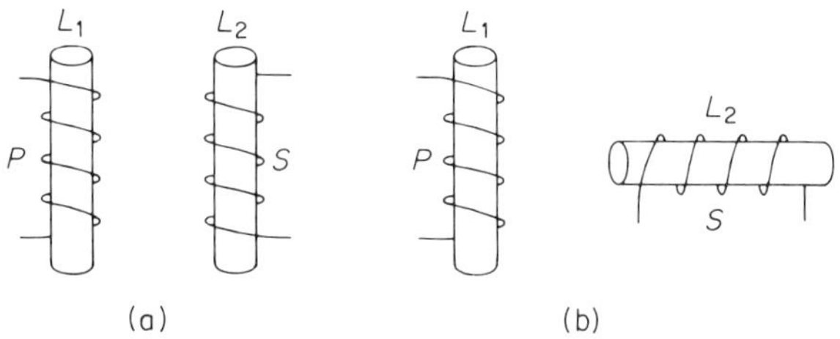

Figure 6-3. Loose Coupling (minimum *K*)

Example 2 (Figure 6-4): What is the mutual inductance between the two coils?

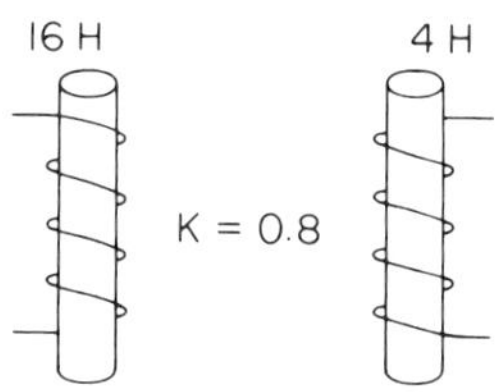

Figure 6-4. Solve for Mutual Inductance

Solution:

$$M = K\sqrt{L_1 L_2} = 0.8\sqrt{16 \times 4} = 0.8 \times 8 = 6.4\text{H}$$

We have used soft iron in several examples as core material for coils. Actually, soft iron is not the best of materials for this purpose. The permeability of the core has a direct bearing on the flux density of a coil.

magnetic curves

When *flux density* is plotted against the *field strength* of a magnetizing force, curves are produced that show the permeabilities of the different materials. The table of sample permeabilities, which we used earlier, was intended only for use as examples. Permeability is much more complicated. Not only do the permeabilities of two pieces of the same material differ, but the permeability of a given piece of material will change according to its own magnetic history.

Figure 6-5 shows magnetic curves for some sample materials.

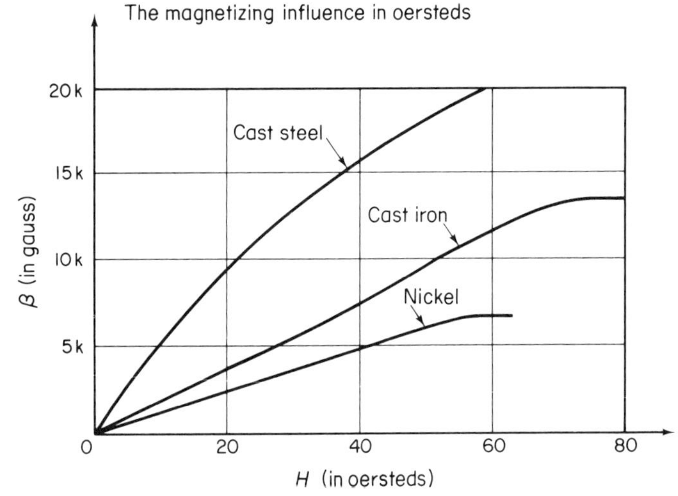

Figure 6-5. *B-H* Curves

Since permeability at any given instant is flux density divided by field strength, the permeability (μ) is actually changing from instant to instant while under the influence of the magnetizing force. The formula is $\mu = \beta/H$, where β is the flux density in gauss and H is the field strength of the magnetizing influence in oersteds.

Cast steel on this chart has a gradually decreasing μ as H increases. When the field strength is 20 oersteds, it has a flux density of nearly 10k gauss. Permeability at that point is $\mu = \beta/H = 10k/20 = 500$. When the magnetizing force reaches 60 oersteds, flux density is 20k gauss. Now $\mu = 20k/60 = 333.3$.

The same trend, to a more marked degree, is seen for cast iron and nickel.

When the material is fully magnetized, the curve levels off. The cast steel runs off the chart, but leveling off can be seen for both the nickel and the iron: the nickel at 60 oersteds and the iron at 80 oersteds. From this point on, flux density remains fairly constant for these materials. This is the point of *magnetic saturation.*

When the magnetizing influence has been removed, each metal retains a certain amount of magnetism depending upon its retentivity. This residual magnetism will have some effect on how this material reacts to the next magnetizing force that it encounters. For instance, before one of these materials could be magnetized in the opposite direction, some force would be required to cancel out this residual magnetism.

Starting with a *new* piece of metal and charting its magnetization from *zero to saturation in both directions* produces a curve similar to that of Figure 6-6. Such a graph is called a *hysteresis*

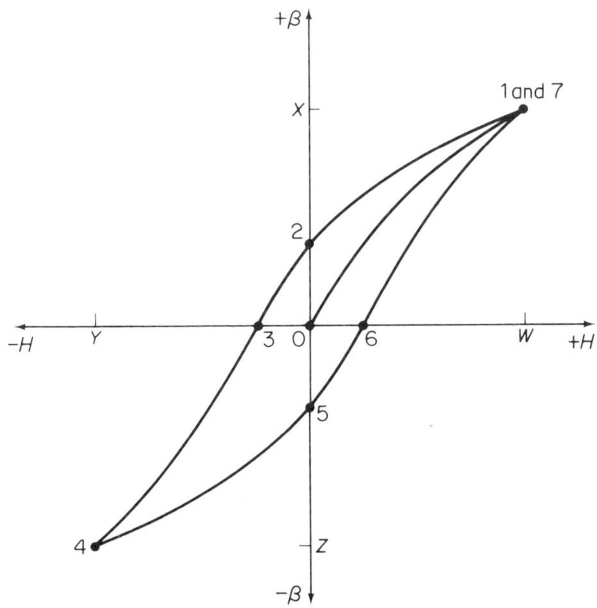

Figure 6-6. Hysteresis Loop

loop, and it is constructed by plotting the various values of flux density against various strengths of a magnetizing force.

The magnetizing force is plotted along the *horizontal* axis with — and + to show direction from zero (which is a point of reference as well as a starting point). Flux density is plotted on the *vertical axis* with — and + to indicate the direction of flux. The numbers (0 through 7) and letters (*W,X,Y,Z*) are points of reference to keep us oriented.

The material has never been magnetized, so it has *no flux* when we start. When the magnetizing force is applied and increases along the $+H$ line, flux density begins to build up in the metal along the curved line running from 0 to 1. When the magnetizing field strength reaches point *W,* the metal is saturated with a flux density of *X.* A further increase of *H* will have no effect on flux density.

Decreasing the strength of *H* from *W* to 0, allows the flux density to decrease along the curved line from 1 to 2. The difference in flux density between points 2 and 0 is the *residual magnetism.* Building up the magnetizing force along the $-H$ line from points 0 to 3 reduces the residual magnetism to 0 as flux density moves from point 2 to 3. The force represented by the distance from 0 to 3 represents a loss. This is the *coercive force* required to cancel the residual magnetism.

Further increase of the magnetizing force from points 3 to *Y* causes the material to become magnetized to saturation in the opposite direction. Flux density moves along the line from points 3 to 4. The flux density at this point is represented by *Z.* Any further increase of $-H$ will have no effect on the flux density. Decreasing the magnetizing force from points *Y* to 0 decreases the flux density along the line from points 4 to 5. The difference between points 5 and 0 is the residual magnetism, again because the magnetizing force is now zero. To overcome this residual magnetism, the magnetizing force must be increased in a positive direction from points 0 to 6. The flux density then dies out along the line from points 5 to 6. The material now has zero flux. The distance from points 0 to 6 is a loss. It is the coercive force used to cancel the residual magnetism. Further increase of $+H$ from points 6 to *W* will increase flux in the metal along the line from points 6 to 7. The metal is now saturated again in the first direction.

The loop is now complete. The width of the loop at the horizontal axis represents the total coercive force used to cancel the residual magnetism.

What is the loop used for? It is used by the designer to choose a suitable material for the core of his coils. If he intends his coil to be a quick acting relay, he must choose a metal with a narrow hysteresis loop. If he wants the coil to store information, he chooses a material with a wide loop and a high value of residual magnetism. If his coils are intended for transformers, he must select a metal with a narrow loop and an extended saturation point.

Now let's analyze the operation of a circuit component that is strictly an electromagnetic device.

magnetic switches

When Samuel Morse was experimenting with the telegraph, one of his chief problems was loss of signal strength when he tried to send signals distances of 20 miles or more. He discovered that a very small current through a multiple winding coil creates a magnet strong enough to close a switch. He set up these magnetic switches at distances of about 20 miles. Station 1 placed current on the line by means of a manual key. This current operated a *magnetic switch* at Station 2 which replenished the current in the line. The stations, from number 3 on up, repeated the action of Station 2 until the end of the line was reached. Thus, magnetic switches, acting in *relay* to pass information from station to station, enabled the establishment of the first practical telegraph. Today, all electromagnetic switches are called *relays.* This name very likely comes from the manner in which Morse used these switches to relay his messages.

The principle of the relay is quite simple. An electric coil energizes (magnetizes) a ferromagnetic core, and the resulting attraction opens or closes a metal switch. This principle is illustrated in Figure 6-7.

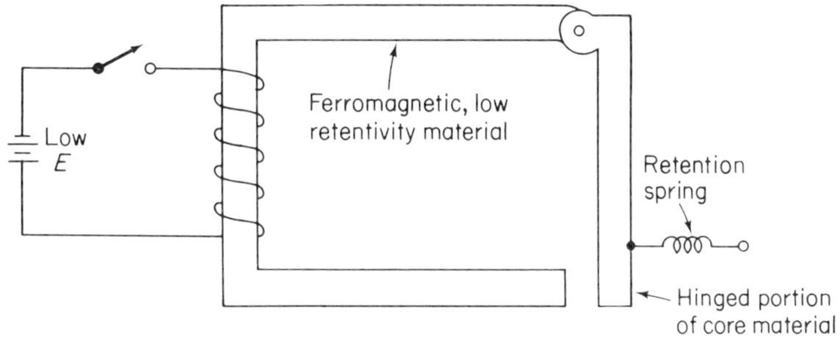

Figure 6-7. Relay Principle

When the manual switch has been closed in the low voltage circuit, a current flows in the coil. This current magnetizes the core, and places a north pole on one side of the air gap and a south pole on the other. The hinged portion of the core material moves to close the air gap. When the low voltage switch has been opened, the core quickly loses its magnetism. The movable arm is no longer attracted, and the retention spring pulls it back to its original position.

This movable arm can be made to serve as many uses as the designer can imagine. Morse used a low voltage circuit some 20 miles long and had his movable arm close the switch in another low voltage circuit. One very popular use is for remote control of a circuit that has a dangerously high voltage. When a switch closes in a 330kV circuit, it is a very good idea to be some distance away from that switch.

The five basic parts of a relay are the coil, the core, the armature, the contacts, and the spring. These parts are shown in Figure 6-8.

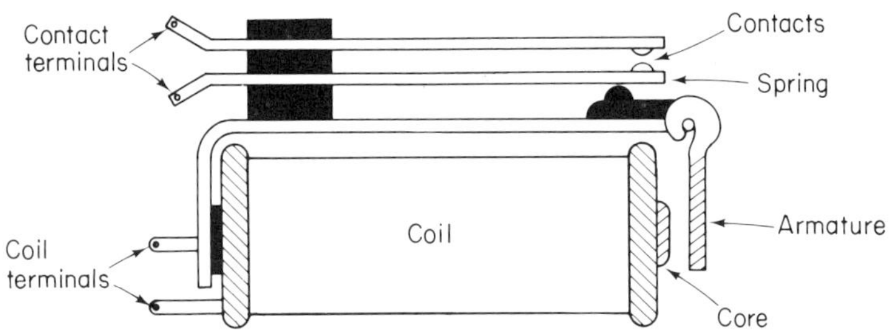

Figure 6-8. Parts of a Relay

The solid black areas in this illustration represent electric insulation material. Current through the coil energizes the iron core which pulls the armature. The pivoted armature pushes a strip of insulation against the spring, which contains one of the contacts. As the spring bends, the contacts close. The circuit being controlled is connected between the *contact* terminals. The low voltage control circuit is connected between the *coil* terminals.

When the current is removed from the coil, the electromagnet de-energizes; the spring tension opens the contacts and forces the armature back to its original position.

Relays may be constructed with any reasonable number of contacts and used to control many circuits at the same time. It is not uncommon to find a relay with eight sets of contacts. A double set of contacts is illustrated in Figure 6-9.

The popular symbol for the relay is a coil and contacts. These appear as in Figure 6-10 (either (a) or (b)).

Frequently, these symbols are placed apart from the schematic diagram of a circuit to facilitate identification of the relay, its contacts, and the normal condition of the contacts. N.O. indicates that

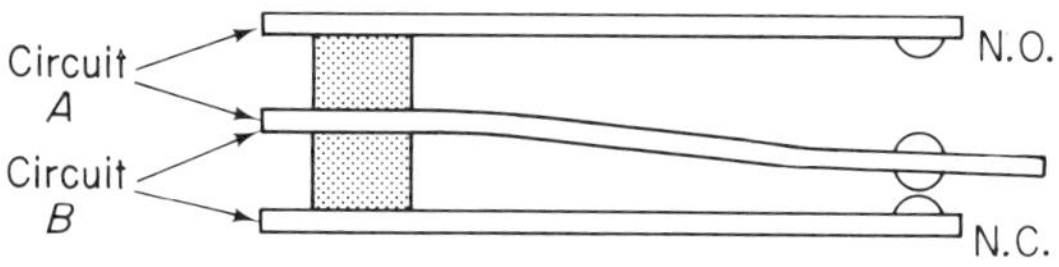

Figure 6-9. Dual Set of Relay Contacts

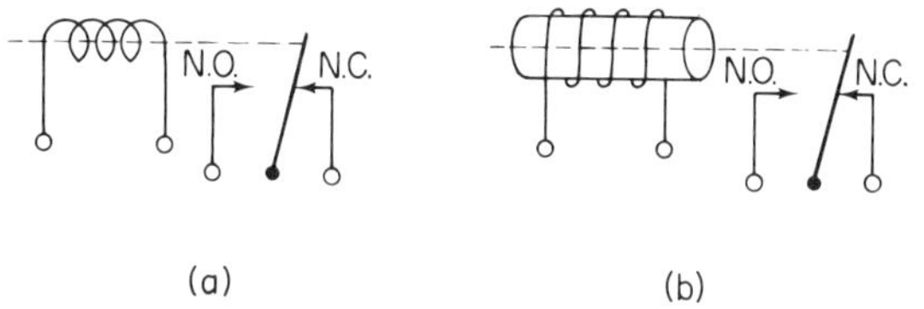

Figure 6-10. Schematic Symbols for Relays

the contacts are normally open, while N.C. shows that they are normally closed.

The contacts themselves, along with the identifying number of the relay, will appear again in the schematic of the circuits which they control.

High speed relays are designed by using a high permeability core material that has minimum retentivity. The high permeability allows fast armature action on energizing; low retentivity allows quick release of the armature on de-energizing.

The coil of a relay may be connected in series with a set of its own normally closed contacts. This would be called an *overload relay.* Its function would be to open the circuit when the current became strong enough to energize the coil.

Relays are frequently connected in a manner that causes energizing current to feed back to the coil through one set of contacts. When this relay energizes, it holds itself energized until all power is removed from the circuit. This particular action accounts for the fact that relays used in this manner are termed *holding relays.*

There is another type of automatic switch which is classified as a relay although it does not operate on the electromagnetic principle. This is the *heat-sensitive* (*thermal*) *switch,* which opens or closes when the metal reaches a specified temperature. Such a switch is shown in Figure 6-11.

The *thermal switch* shown here is connected in series with the circuit. It is intended to heat up when excessive current is drawn. The heat causes the dissimilar metals to expand at different rates. The strips then curve and lift the contacts apart. This stops the

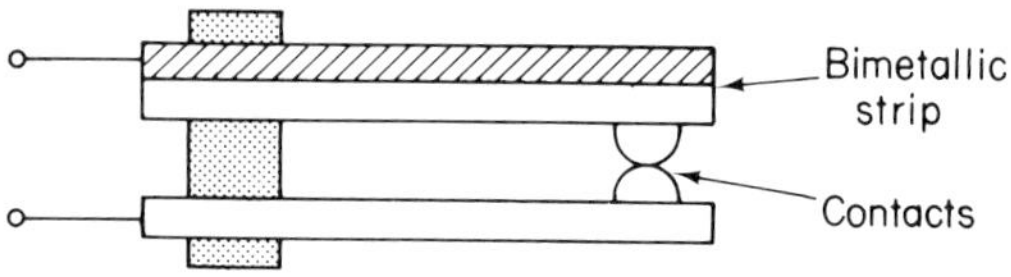

Figure 6-11. Thermal Switch

circuit current and prevents damage from excessive current. These switches may also be constructed with heating elements to enable a very accurate prediction of exactly what level of current will open the switch.

The schematic in Figure 6-12 illustrates the use of a remote control relay utilizing the hold feature.

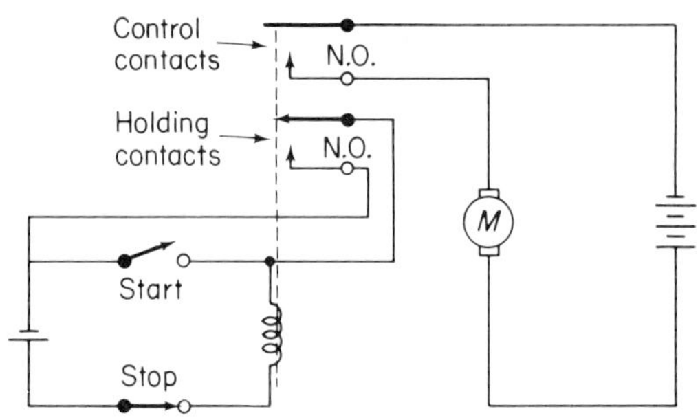

Figure 6-12. Start and Hold Relay Circuit

When the start switch is *closed,* the relay *energizes* and closes both sets of contacts. The control contacts complete the high voltage circuit and start the motor. The holding contacts complete a path for the relay coil, which *bypasses* the start switch. The Start and Stop switches would probably be push buttons, which are spring loaded to the position indicated. Releasing the Start switch has no effect on the relay. In order for the motor to be stopped, the Stop switch has to be *opened.* This allows the relay to *de-energize,* and both sets of contacts *open.* Once it is released, the Stop switch will close. But now, the Start switch is open, leaving the relay in a de-energized condition.

In Chapter 3 the use of dc measuring instruments was discussed. Since then, we have covered enough principles to enable us to explore the inside of these instruments.

METER MOVEMENTS

Since current is invisible, it must be measured by the effects it produces. The *chemical effect* of current in an electroplating process has been used to measure current. The *physiological effect,* the reaction of the human body to current, is not an accurate measurement of current, but it certainly is one method of detecting the presence of current. This method is not recommended because there is always a chance of detecting a lethal quantity of current. The *heating effect* of current has been used to move pointers along a calibrated scale. The most popular meter movements, however, are based on the *electromagnetic effect* of current. The ammeter, the voltmeter, the ohmmeter, and the wattmeter all utilize the effects of current to move the pointer along the scale. The scales, of course, are calibrated and are marked to indicate amperes, volts, ohms, or watts.

the D'Arsonval movement

The *D'Arsonval movement* has been a favorite among designers and manufacturers for a number of years. You are sure to encounter it in many ammeters, voltmeters, and ohmmeters (including the multimeter). Its popularity demands that it be given detailed treatment.

The D'Arsonval movement is constructed on the principle of a *coil pivoting in a permanent magnetic field.* This is illustrated in Figure 6-13.

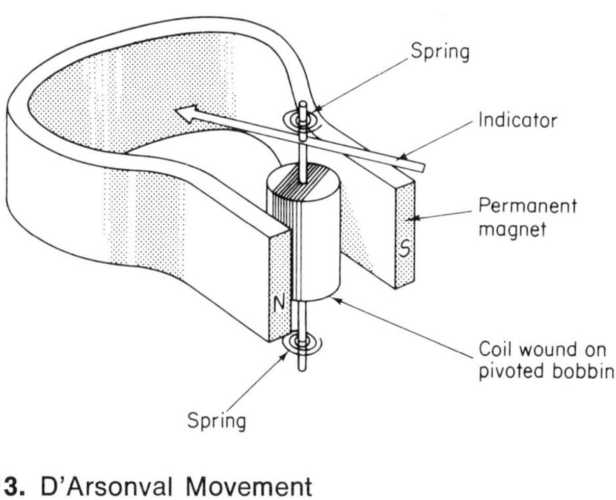

Figure 6-13. D'Arsonval Movement

The vertical shaft is pivoted on each end in jeweled bearings. Current from the circuit being tested enters through one spring and exits through the other, passing through the coil windings in the process. The energized coil is an electromagnet, which interacts with the field of the permanent magnet. A turning force is exerted on the bobbin which swings to neutralize this force. In turning, the bobbin swings the indicator across the dial. When current ceases, the de-energized coil and the indicator are returned to their original positions by the spring tension.

Current through the coil must be limited to one direction. Otherwise, the electromagnet is energized with the wrong polarity and pivots in the wrong direction. This can be seen by careful examination of Figure 6-14.

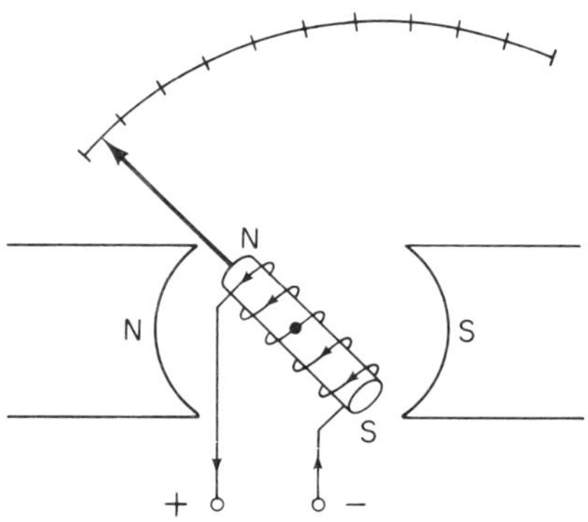

Figure 6-14. Turning Force on Pivoted Coil

With current as shown, the *repulsion* between the two north poles, plus the repulsion between the two south poles, causes the pivoted coil to swing clockwise. Reversing the current would reverse the polarity of the coil. This would result in a rather emphatic counterclockwise swing of the coil. These movements are delicate, and sending current through them in the wrong direction could cause permanent damage.

The turning force on the coil is *proportional* to the current through the coil. Thus, the scale can be linearly calibrated. The internal circuitry of the meter will determine whether the scale is marked to indicate amperes, volts, or ohms.

taut band meter movement

For years, the most reliable meter movement in the industry was the D'Arsonval. Recently, it has been *excelled* by the *taut band movement.* This movement has no bearings that might wear out and features a *constant level of sensitivity* throughout the range of movement. The principles of construction are illustrated in Figure 6-15.

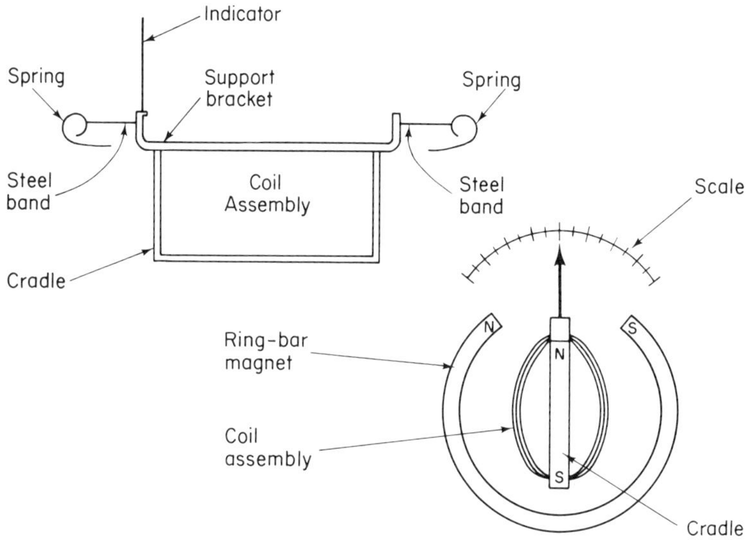

Figure 6-15. Taut Band Meter Movement

The movement operates by electromagnetic deflection similar to the D'Arsonval, but there the similarity ends. The solenoid is mounted in a *cradle* and surrounded by a ring-bar magnet. When the coil is energized, the cradle swings to the side. The needle is mounted on the support bracket and swings with the cradle. The movement is proportional to the amount of current entering the coil. When the coil moves, the support bracket exerts a *twisting force on the taut steel bands,* which support the entire assembly. These twisted bands supply the torque to return the meter to zero when the current is removed.

electrodynamometer

This movement of the *electrodynamometer* works on the same principles as that of the D'Arsonval except for the fact that the former uses two more coils to replace the permanent magnet. This is the *most popular* movement used for *wattmeters* even though it is not limited to that function. A schematic representation of the dynamometer is shown in Figure 6-16.

It is not apparent from the schematic diagram, but the coils would have to be wound in a manner that produces *opposing poles*

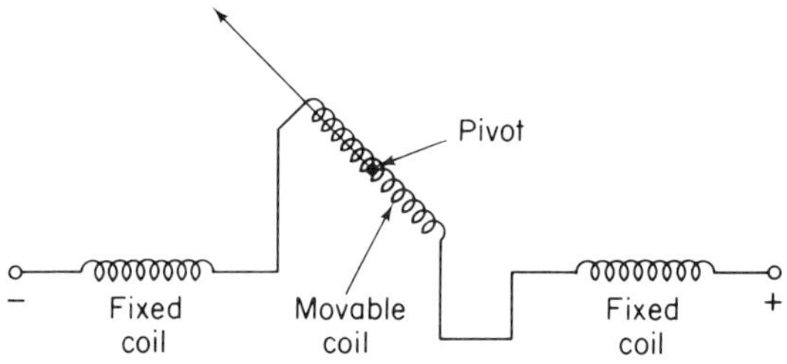

Figure 6-16. Electrodynamometer

between the movable coil and each of the fixed coils; that is, north to north, and south to south. Otherwise, the movable coil would simply rotate to the horizontal position and remain there. This polarity arrangement is illustrated in Figure 6-17.

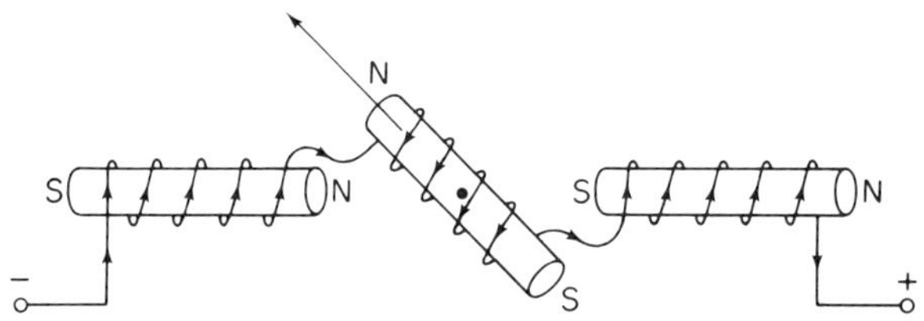

Figure 6-17. Opposing Poles

The two north poles and the two south poles repel each other. This action supplies a turning force for the movable coil, which rotates and sweeps the indicator across the dial.

meter damping

The movement of an electrical indicator must be *damped;* otherwise, the indicator would move rapidly toward its mark, overshoot, and *swing to and fro* for a considerable period of time. Damping is accomplished either mechanically or electrically. Mechanical damping is shown in Figure 6-18.

The movable vane swings with the indicator. This *compresses the air* ahead of the vane and slows the movement of the indicator. The movement can only be as fast as the air pressure can equalize through the small air vent. The movement is slightly retarded in either direction, but when the indicator has reached its mark, it holds steady with very little tendency to oscillate.

The electrical method of damping came about naturally with

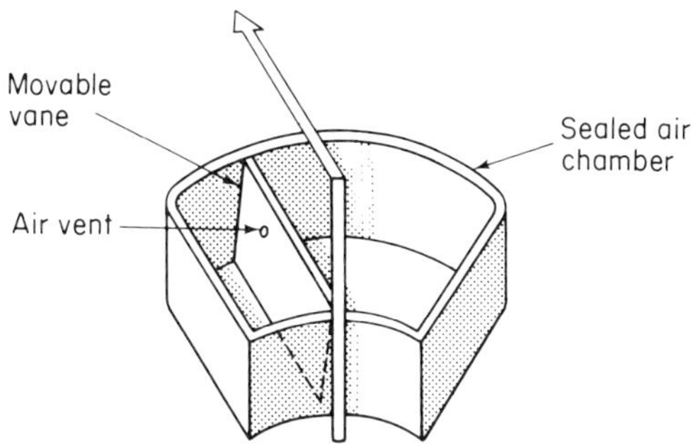

Figure 6-18. Mechanical Damping

the trend toward electromagnetic meter movements. The movable coil needs a frame anyway, so it is wound on an aluminum bobbin. As this bobbin turns in the magnetic field, a current is induced into it. According to Lenz's law, the direction of this current is such that it will oppose the motion that induced it. Remember the action of the coil as the isolated pole was pushed through it. It opposed the motion every step of the way. The action is the same with electrical damping. Every movement induces a current, which sets up a field to oppose that movement. The result is a steady indication without the need to wait for the indicator to settle down.

Now that meter movements are no longer a mystery, let's examine the internal circuits that work with these delicate movements.

METER CIRCUITS

The circuits that we will deal with here are dc circuits, pure and simple. Their only mark of distinction is their close tolerances. Plus or minus one percent can be tolerated in some cases, but 20 percent is out of the question. In this section, we will analyze the circuits for the ammeter, voltmeter, and ohmmeter. After that, we'll combine the three to form the circuits for a multimeter.

In this discussion, we will use the term *ammeter* to mean any meter that is calibrated to read *current of any quantity,* and *voltmeter* will mean a device that measures *any magnitude of voltage.* Technically, a meter that reads in milliamperes is a *milliammeter.* If it reads in microamperes, it is a *microammeter.* There are also *microvoltmeters* and *millivoltmeters.* These specific names are not important to this discussion and may add unnecessary confusion.

meter sensitivity

The basic *sensitivity* of most common moving-coil meters is expressed in *millivolts.* It is a statement of the number of millivolts required across the coil in order to produce a full scale deflection. Standard sensitivities are 50mV, 100mV, and 200mV. Millivolts of sensitivity is a term that is gradually dying out due to lack of use. Currently, it is more popular to refer to the movement according to its *maximum current.* For instance, if a given meter has a coil resistance of 100 Ω and a sensitivity of 100mV, it can handle currents from 0 to 1mA. Voltage is the sensitivity standard. Dc resistance of the coil is seldom mentioned; but current handling capability is another way to express the same idea.

ammeter

A 0 to 100mA meter movement may be used to measure *any reasonable quantity of current.* The procedure is to place a *shunt* in parallel with the meter coil to bypass a specified percentage of the current. For instance, in order to measure 0 to 1A on a 0 to 100mA movement, the shunt would need to bypass 90 percent of the current. Therefore, the shunt would be one ninth the size of the coil resistance. If coil resistance (R_m) were 100Ω, the shunt resistance would be $\frac{100}{9}$ or 11.1 ohms. This ratio is usually easy to work out mentally, and it can be checked quickly by applying Ohm's law.

Example 1 (Figure 6-19): Putting the above problem into circuit form we have the circuit in Figure 6-19.

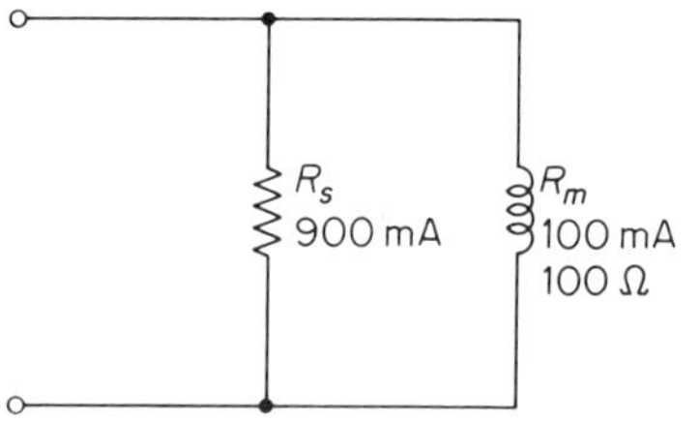

Figure 6-19. Calculating Resistance of Shunt (R_s)

Terminal voltage (E_a for this circuit) is

$$I_mR_m = 100\text{mA} \times 100\Omega = 10\text{V}$$

Shunt Resistance is

$$\frac{E_a}{I_s} = \frac{10\text{V}}{900\text{mA}} = 11.1\Omega$$

Example 2 (Figure 6-20): What size shunt would be required to extend the range of a 0 to 5mA meter with a 90-Ω coil to enable it to measure 0 to 50mA?

Mentally:

The shunt must handle 90% of the current. Ratio of R_s to R_m must be 1 to 9. $R_m = 90\Omega$; R_s must be 10Ω. Check by Ohm's law:

$$E_a = I_m R_m = 5\text{mA} \times 90\Omega = 450\text{mV}$$

$$R_s = \frac{E_a}{I_s} = \frac{450\text{mV}}{45\text{mA}} = 10\Omega$$

Since we are dealing with parallel circuits, the product of IR for each branch must be the same because the applied voltage appears across each branch. In terms of IR, this is $I_s R_s = I_m R_m$ and $R_s = I_m R_m / I_s$. Using values from Figure 6-20 in this formula will produce the same results as before.

$$R_s = \frac{I_m R_m}{I_s} = \frac{5\text{mA} \times 90\Omega}{45\text{mA}} = \frac{5 \times 90}{45} = 10\Omega$$

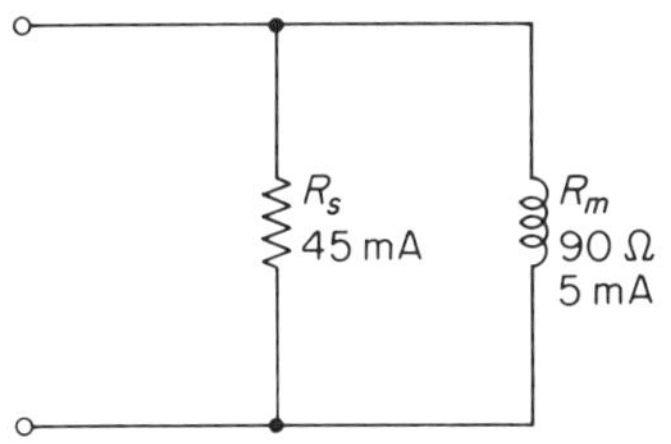

Figure 6-20. Checking Size of Shunt

In the above examples, we are dealing in values that allow a tangible resistor to be used as a shunt. But in many cases the shunt is not only a straight piece of conductor, but also a conductor of very low resistance. Frequently, a shunt will be a copper block or a strap with a large cross-sectional area to minimize the resistance while building in very strict tolerances.

Example 3: What size shunt would be required to measure 10A on a 0 to 100μA meter when the coil resistance is 100Ω?

$$R_s = \frac{I_m R_m}{I_s}$$

$$I_s = 10\text{A} - 100\mu\text{A} = 10 - 0.0001 = 9.9999\text{A}$$

$$R_s = \frac{(100 \times 10^{-6})(100)}{9.9999} = \frac{0.01}{9.9999} = 0.001\Omega$$

Extending the range of this same meter to measure 50A would reduce the shunt size to approximately 0.0002Ω.

Many ammeters are designed to measure *more than one range* of current. This requires more than one shunt with a means of selecting the shunt to match the range. One method of accomplishing this is to install a separate shunt for each range and a rotary selector switch to select the desired range along with a shunt to match. This type of device is shown in Figure 6-21.

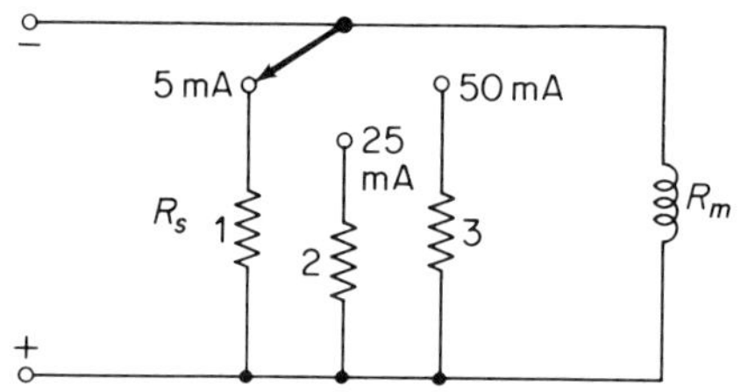

Figure 6-21. Multiple Range with Selector Switch

Selecting the proper shunts for this arrangement is done in the same manner as above. This is not the most popular arrangement, probably because selector switches are expensive as well as being a source of trouble.

Most multiple range ammeters use a *tapped resistor* for all the shunts as shown in Figure 6-22.

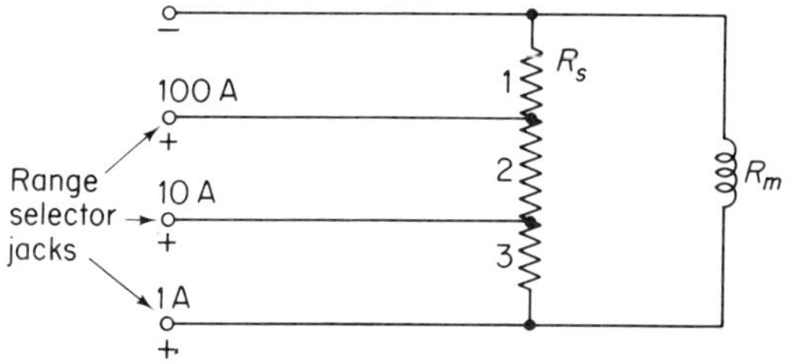

Figure 6-22. Multiple Range with Tapped Shunt

The range is selected by plugging the positive test lead into the proper range jack. The tapped resistor shunt is called the *universal shunt.* The arrangement in Figure 6-21 is sometimes *improperly* called the universal shunt.

Example 4 (Figures 6-23 and 6-24): Calculate the value of each shunt resistor in this meter circuit.

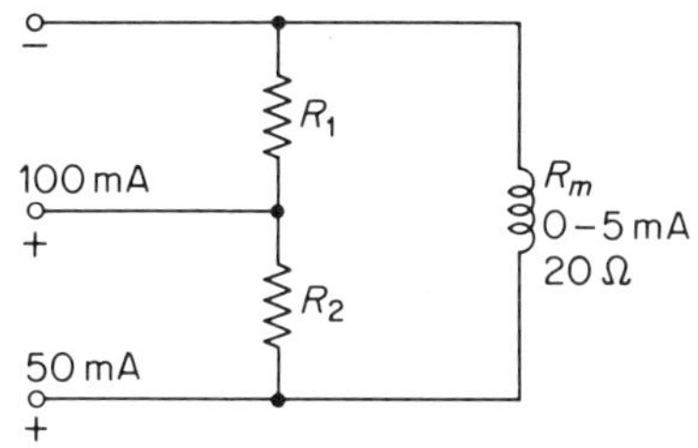

Figure 6-23. Calculate the Universal Shunt

Solution:

Find total R_s:

$$R_s = \frac{I_m R_m}{I_s}$$

$$I_s = 50\text{mA} - 5\text{mA} = 45\text{mA}$$

$$R_s = \frac{5\text{mA} \times 20\Omega}{45\text{mA}} = \frac{5 \times 20}{45} = 2.2\Omega$$

This is the resistance of sections 1 and 2 of R_s in series. This full shunt (2.2Ω) will be used when measuring current on the 50mA range. Going to the 100mA range places section 2 of R_s in series with the coil, and section 1 becomes the shunt as illustrated in Figure 6-24.

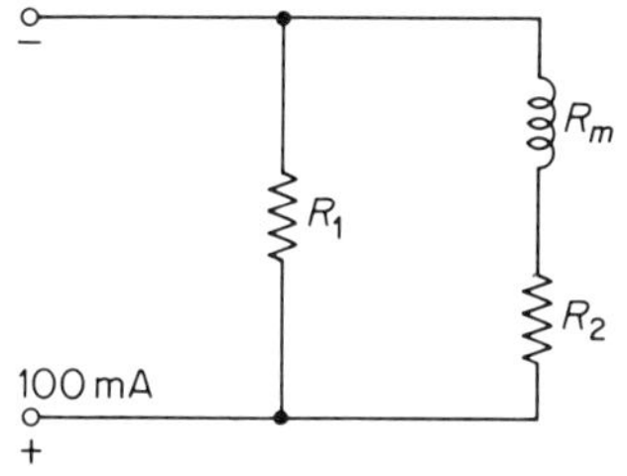

Figure 6-24. Part of Shunt in Series with Coil

Summation:

We know that:

1. $R_1 + R_2 = 2.2\Omega$.
2. $R_m = 20\Omega$.
3. $I_m = 5\text{mA}$.
4. $I_s = 95\text{mA}$.

5. Voltage is the same across each branch.
6. Resistance of the coil branch is $R_m + R_2$ (which is $20\Omega + R_2$).

Solution:

Set the voltages equal:

$$95\text{mA}(2.2\Omega - R_2) = 5\text{mA}(20\Omega + R_2)$$

Simplify and solve for R_2:

$$0.209 - 0.095R_2 = 0.1 + 0.005R_2$$

$$0.1R_2 = 0.109$$

$$R_2 = \frac{0.109}{0.1} = 1.09\Omega$$

Solve for R_1:

$$R_1 = 2.2\Omega - 1.09\Omega = 1.11\Omega$$

Calibrating the ammeter is a simple, although rather delicate, process. The ammeter is connected into a *test circuit* in series with another meter, which is *known to be accurate*. The shunts in the meter being calibrated are then *adjusted until the meter readings match on all ranges.*

voltmeter

Voltmeters draw some current from the circuit in order to activate the meter movement. This usually is a negligible amount, and it is properly ignored in ordinary circuits. However, it must be taken into consideration in meter design.

The *sensitivity of the voltmeter* is expressed in *ohms per volt.* Generally speaking, the larger the value of ohms per volt, the better the meter.

Since the meter movement is the same as the ammeter movement, our problem now is how to apply a larger voltage across the meter without driving too much current through the coil. The answer is to connect a suitable size *resistor in series* with the coil to limit the current. Since this resistor extends the range of voltage that can be measured, it is called a *multiplier.* For instance, when a 0–5mA movement of 50Ω is used to measure 0 to 50V, the meter circuit would be that of Figure 6-25.

I_mR_m = voltage across the movement for a full scale deflection. The multiplier must limit meter current to that amount specified for a full-scale deflection. In this case, the meter can stand 5mA, and

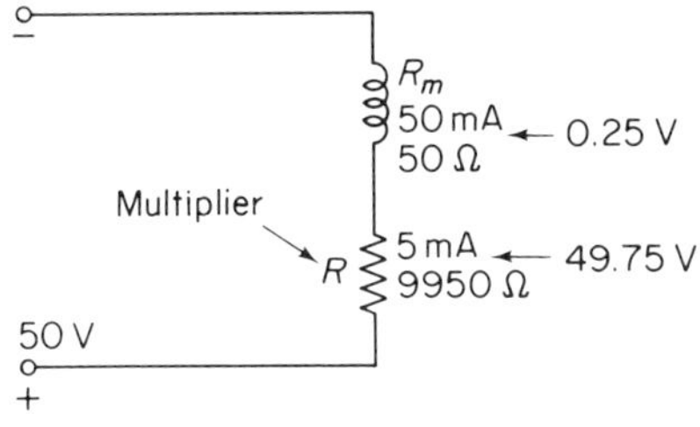

Figure 6-25. Use of Multiplier

it has a resistance of 50Ω. This gives a voltage drop across the coil of 0.25V. The remainder of the 50V (49.75V) must be dropped across the multiplier. Since the same current goes through both the coil and the resistor, $R = 49.75\text{V}/5\text{mA} = 49.75/0.005 = 9950\Omega$.

The sensitivity is total resistance divided by full scale voltage. Thus, the meter in Figure 6-25 has a sensitivity of

$$\frac{9950\Omega + 50\Omega}{50\text{V}} = \frac{10{,}000\Omega}{50\text{V}} = 200\Omega/\text{V} \text{ (200 ohms per volt).}$$

Example 1: What size multiplier will extend the range of a 0–1mA, 10-Ω movement to measure 0 to 500V?

Solution:

$$R_t = \frac{500\text{V}}{1\text{mA}} = 500\text{k}\Omega$$

$$R_{\text{mult}} = 500\text{k}\Omega - 10\Omega = 499{,}990\Omega$$

Example 2: What is the sensitivity of this voltmeter?

Solution:

$$\text{Sensitivity} = \frac{500\text{k}\Omega}{500\text{V}} = 1\text{k}\Omega/\text{V}$$

A good voltmeter may have a sensitivity of from 20,000 to 50,000 ohms per volt.

For multiple range voltmeters, a range selector switch may be used to select a different multiplier for each range. This is shown in Figure 6-26. The proper size for R_1, R_2, and R_3 would be calculated in the same manner as above in three different operations.

Another means of converting a voltmeter to a multiple range meter is to use a tapped multiplier as shown in Figure 6-27.

Here, the range is selected by plugging the positive meter lead

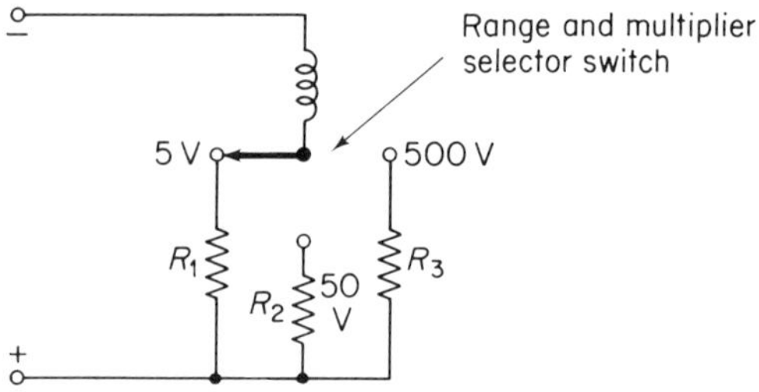

Figure 6-26. Selecting Different Multipliers

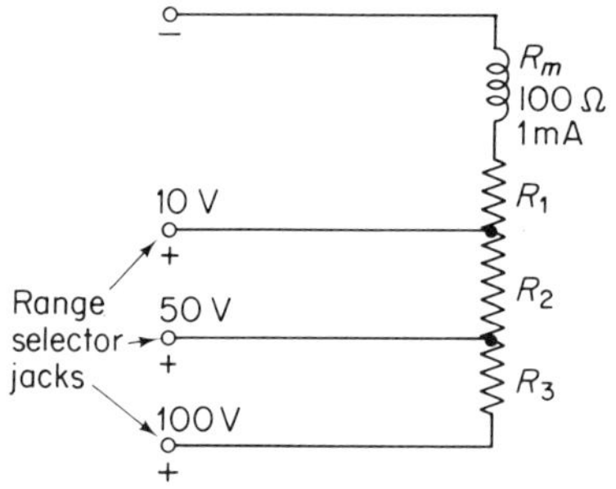

Figure 6-27. Tapped Multiplier

into a particular selector jack. The higher the voltage range, the more of the multiplier used.

Example 3 (Figure 6-27): Calculate the size of R_1, R_2, and R_3 in the figure.

Solution:

On the 100V range:

$$R_t = \frac{100\text{V}}{1\text{mA}} = 100\text{k}\Omega = R_1 + R_2 + R_3 + R_m$$

On the 50V range:

$$R_t = \frac{50\text{V}}{1\text{mA}} = 50\text{k}\Omega = R_1 + R_2 + R_m$$

On the 10V range:

$$R_t = \frac{10\text{V}}{1\text{mA}} = 10\text{k}\Omega = R_1 + R_m$$

$$R_1 = 10\text{k}\Omega - 100\Omega = 9.9\text{k}\Omega$$

$$R_2 = 50\text{k}\Omega - 10\text{k}\Omega = 40\text{k}\Omega$$

$$R_3 = 100\text{k}\Omega - 50\text{k}\Omega = 50\text{k}\Omega$$

Wattage ratings for multiplier resistors are of little concern because the current is always very small. However, the *tolerance* on the resistor sizes *is fairly critical.* They should be wire wound resistors with no more than one percent tolerance.

The voltmeter is calibrated in much the same manner as the ammeter. It is *connected in parallel* with a variable voltage source and *another voltmeter of known accuracy.* The *multipliers* in the meter being calibrated *are then adjusted until the meter readings match on all ranges.*

ohmmeter

The series ohmmeter is an *inverse* scale meter, and it is the type most common in the field of electricity. The circuit consists of a standard meter movement in series with a battery, a fixed resistor, and a variable resistor. This is illustrated in Figure 6-28.

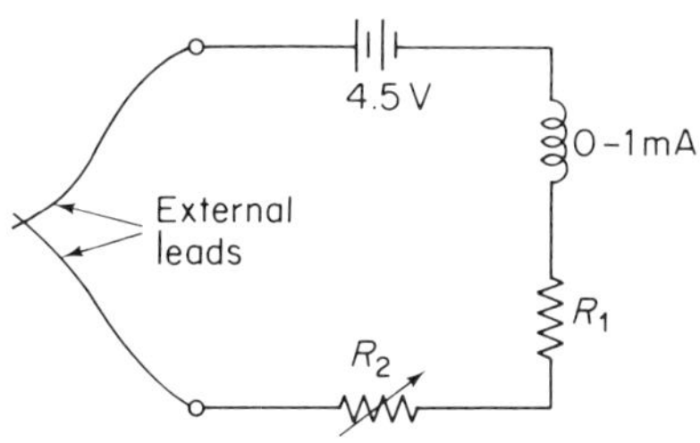

Figure 6-28. Series Ohmmeter Circuit

The battery supplies the current to activate the meter movement. R_1 is a current limiting resistor, and R_2 is the zero adjustment. With the leads as shown, the circuit is complete, and there is current in the circuit. Under these conditions R_2 is adjusted until a zero reading is obtained. Keep in mind that this will be a full scale deflection of the indicator. The meter scale is calibrated in ohms, and this is in respect to the total internal circuit resistance.

The *center point* on the scale is a value *equal* to the total *internal resistance.* The reasoning behind this is as follows:

1. Full scale current exists when the leads are shorted together.
2. Doubling the resistance halves the current.
3. When the leads are across a resistor equal to the total internal resistance, the resistance is doubled, and half-scale current is present.

The same process is repeated for all other values on the scale. Here is one more example:

1. A resistance twice the size of the internal resistance would triple the circuit resistance.
2. Tripling the resistance allows enough current for one-third of full-scale deflection.
3. The point located one-third up the scale from the left is labeled at twice the value of the internal resistance.

This is illustrated by the schematic in Figure 6-29.

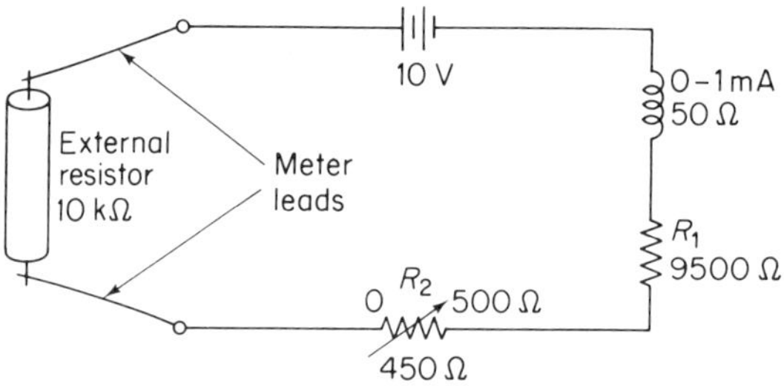

Figure 6-29. Measuring an External Resistor

Before measuring the resistor, we adjusted R_2 for full-scale current (1mA) with the leads shorted together. We have assumed that this setting is 450 ohms. At this time the internal resistance is $50\Omega + 9500\Omega + 450\Omega = 10k\Omega$.

This 10kΩ allows full scale current of one mA. When the leads are separated and placed across another 10kΩ resistor, the same 10V is now dropped across a total of 20kΩ of resistance. Meter current is now $10V/20k\Omega = 0.5mA$. 0.5mA is the half-scale current. The center of the scale on this meter is marked as 10kΩ.

$$R_x = R_i \left(\frac{I_1 - I_2}{I_2}\right)$$

Where: R_x is the resistor to be measured, R_i is the total internal resistance, I_1 is full-scale current, and I_2 is circuit current with the external resistor connected.

As already stated, the series ohmmeter is the most common in electricity. However, it is difficult to design a single ohmmeter that will produce accurate measurements for both very low and very high resistance values. The series design is intended for average work, and special designs are required for extremes in either direction. If greater accuracy is required on very low values, a shunt meter could be used. This is illustrated in Figure 6-30.

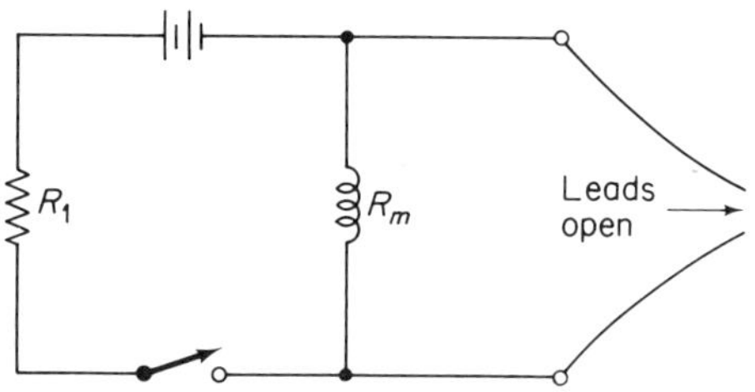

Figure 6-30. Shunted Ohmmeter

In this type of arrangement the switch is open when the meter is not in use to avoid useless drain on the battery. With the switch closed, the values are such that full-scale current is present with the external leads open. When the leads are shorted together, current bypasses the meter movement and gives a zero deflection. Zero for this meter appears on the left side, making the scale direct instead of reverse.

Placing the leads across an external resistor provides another path in parallel with the meter movement. Actually, we now have a series-parallel circuit similar to that in Figure 6-31.

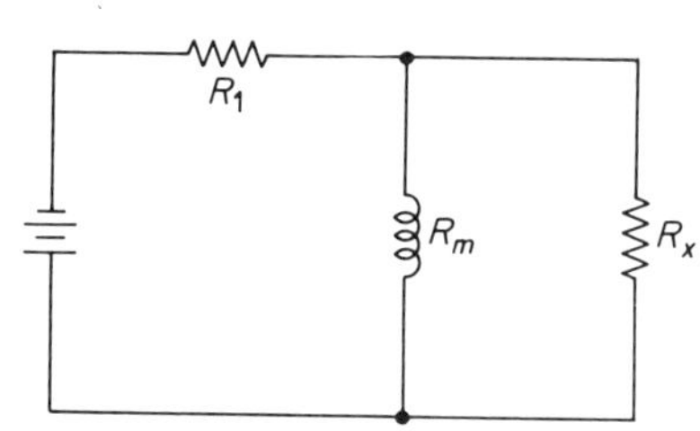

Figure 6-31. Shunt Ohmmeter with External Resistor

The size of R_1 is so large by comparison to the resistance of the coil that it allows full scale current to flow regardless of the value of R_m. When R_x is added, the full scale current is divided through two branches. Any change in the size of R_x results in a new division of current between the coil and the external resistor.

As is apparent in Figure 6-31, the internal resistor is not a shunt, it is a dropping resistor. Its value is high enough to guarantee that no more than full-scale current will flow under any circumstances. The *shunt,* then, *is the resistor being measured* (R_x).

Now that we have seen movements and sample circuits for ammeters, voltmeters, and ohmmeters, let's combine the three circuits.

multimeter

The *multimeter* is a device designed to gain maximum utilization of a meter movement. Since the movement is the most expensive part of the meter, using one movement for several ranges of current, voltage and resistance greatly reduces the cost. A sample multimeter circuit is shown in Figure 6-32.

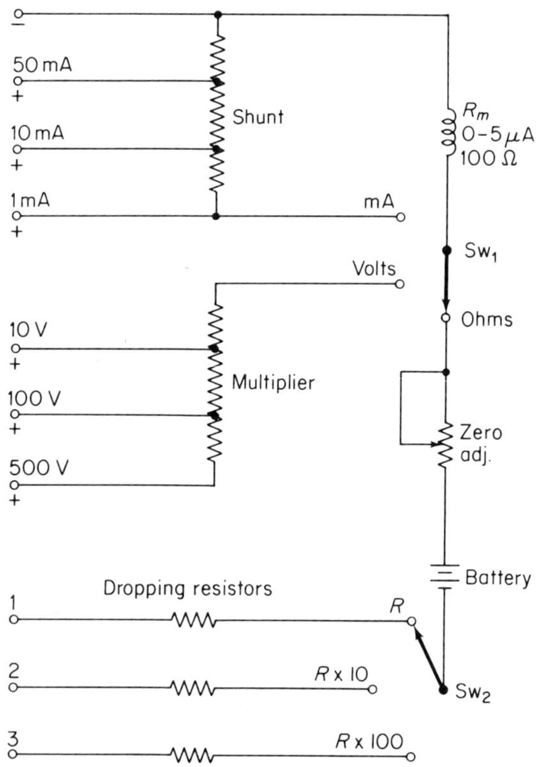

Figure 6-32. Multimeter Circuit

With switches 1 and 2 as shown in the diagram, the meter is ready to measure relatively small resistance values. Switch 1 on the ohms position has selected the operation of an ohmmeter. Switch 2 has selected the low resistance range. The readings will be taken directly from the meter scale. Sometimes a multimeter will have more than one battery, in which case the batteries would be between switch 2 and the dropping resistors.

The meter leads would be connected with one lead in the negative (or common) jack and the other in ohmmeter jack 1. For a higher resistance, switch 2 would be moved to $R \times 10$ or $R \times 100$, and the lead would be connected to the corresponding jack.

For a *voltage* measurement, switch 1 would be moved to the volts position, and the positive meter lead would be moved from the ohmmeter jack to a voltage jack. In this case, the jack selected determines the range of voltage that can be measured. The voltage range could be selected by a rotary switch if the designer so desired.

For a *current* reading, switch 1 is moved to the mA position, and the positive lead to a current jack. Again, the range is selected by choosing the jack for the positive lead. These ranges may also be made a selector switch action, as we have seen earlier when dealing with the ammeter.

Looking at the multimeter circuits *one section* at a time, you will recognize the simpler circuits of the *individual* meters.

This concludes Chapter 6. The following review exercises are designed to reinforce learning on the key points.

REVIEW EXERCISES

1. The property of a coil which causes it to oppose a change in current is called __________.
2. The letter L is the symbol for __________, and it is measured in __________.
3. Write the formula for mutual inductance.
4. Describe the condition that would exist between two inductors if the coefficient of coupling were one.
5. A narrow hysteresis loop indicates that the material has a low __________.
6. The width of a hysteresis loop represents the amount of __________ force required to overcome the __________ magnetism.
7. The term relay is a name for a __________ switch.
8. Name the five vital parts of a relay.

9. The D'Arsonval meter movement is based on the turning force between a ______________ and the field from a ______________ magnet.
10. The electrodynamometer movement has ______________ fixed coil(s) and ______________ movable coil(s).
11. When the electrodynamometer is used in a wattmeter, the turning force is proportional to ______________ ______________.
12. What are the two best methods of meter damping?
13. The D'Arsonval meter uses ______________ damping.
14. What is the correct expression for meter sensitivity?
15. What size shunt would be required with a 0–50mA, 50-Ω ohm movement in order to measure up to 10A?
16. A 20-Ω shunt is used with a 0–5mA, 90-Ω movement. How much current will this meter measure?
17. Multipliers are resistors in ______________ with a meter movement to ______________ the range of voltage it will measure.
18. What is the maximum voltage the meter shown in Figure P6-18 will measure?

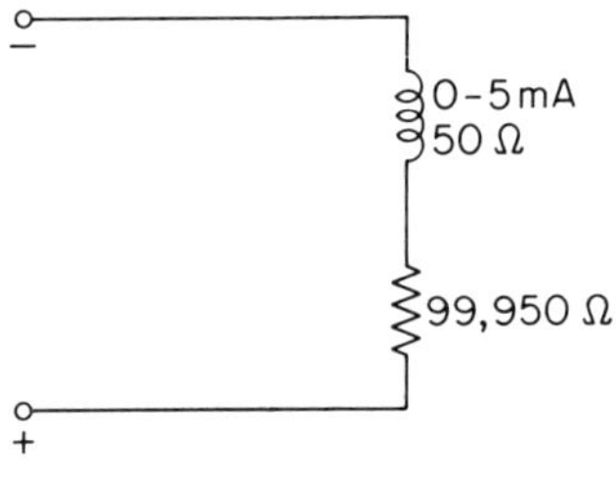

Figure P6-18.

19. What is the sensitivity of the meter in item 18?
20. The scale on a series ohmmeter is in comparison to the ______________ resistance.
21. The shunt for a shunt ohmmeter is the ______________ (internal or external) resistor.
22. The multimeter accomplishes maximum utilization of a single ______________.

7

energy conversion

Alternating current (ac) is current that *changes direction periodically.* The electrons leave one terminal of the power supply, *flow out* along the conductor, *stop,* and then *flow back* toward the same terminal. A voltage that causes current to behave in this fashion reverses its polarity periodically. This is properly called an *alternating voltage.* The power supply that provides the alternating voltage actually *reverses the polarity* of its terminals according to a fixed periodic pattern. A given terminal will be negative for a specific period of time and drive electrons out through the circuit. Then, the same terminal becomes positive and attracts electrons back from the circuit. Obviously, this voltage source cannot be a battery. It must consist of some types of *rotating machinery.*

This difference is the first of many differences between alternating current (ac) and direct current (dc). Some of the laws that you have learned can still be applied with certain reservations, but many characteristics will be *completely new.* The knowledge of dc that you have acquired will serve you well, but less than 10% of the power in the United States is supplied from dc sources. Any serious student of electricity or electronics must take time to *master* the principles of ac. The first step toward the understanding of alternating current is a study of the voltage source.

CONVERTING MECHANICAL ENERGY TO ELECTRICAL ENERGY

Since energy can neither be created nor destroyed, any process of producing voltage must be a *conversion* from one form of energy to another.

principles of ac generation

There are several names for the machines that convert mechanical energy into electrical energy. The *dynamo* is the source of huge amounts of power; the *magneto* supplies minute power outputs; and in between, we have *alternators and generators.* All of these work on the same principle, the principle demonstrated by Faraday when he discovered that *relative motion between a magnetic field and a conductor in that field* would induce a current into the conductor. It makes no real difference whether the conductor is stationary and the field moving, or the field is stationary and the conductor moving. The important factor is the *relative motion* in a manner that will cause flux to cut across the conductor.

During our study of magnetism, we discussed the results of forcing a conductor through a magnetic field. At that time, a left-hand rule was suggested for determining the direction of current through the conductor. We said:

> Place the open left hand with thumb extended into the magnetic field. When the hand is placed in a manner that points the fingers in the direction of flux with the moving conductor being arrested by the palm, the thumb will point the direction of current in the conductor.

According to the left-hand rule, which direction is the induced current in Figure 7-1(a) and (b)?

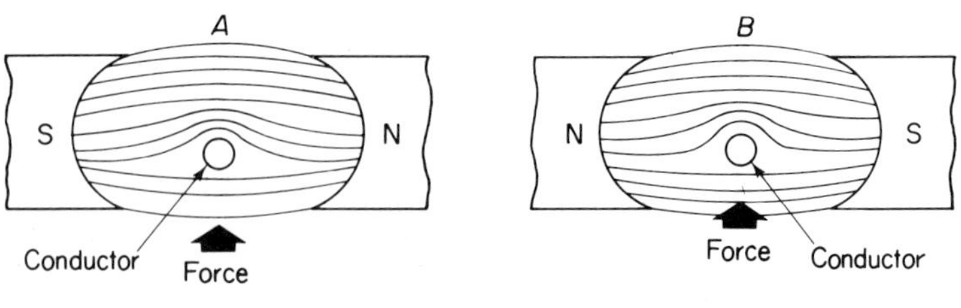

Figure 7-1. Determine Direction of Current

If you decided that the current in (a) is moving away from the viewer, current (b) is moving toward the viewer, you were right, and you were applying the rule correctly. If you missed on either count, review the rule and practice its application. It will be used extensively in this chapter.

Compare Figure 7-1(a) and (b), and pick out as many significant differences as you can between the two drawings. You are right, if you said the only significant difference is the direction of the magnetic field. In (b) the field is reversed with respect to (a). But notice that this reversal of field direction also reversed the current.

What would happen to the direction of current if the direction of conductor movement were reversed? Figure 7-2 is the same as

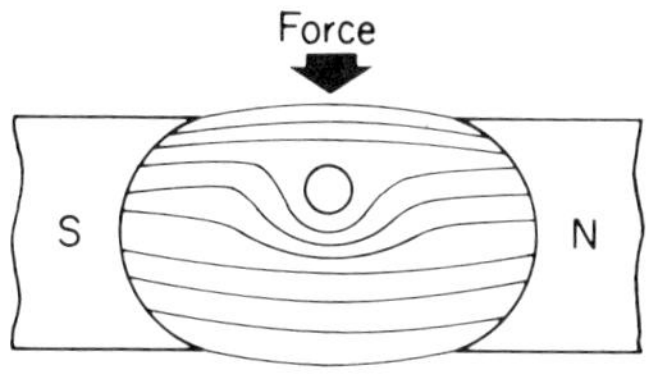

Figure 7-2. Result of Reversing Motion

Figure 7-1(a) except that the conductor in the former is moving in the opposite direction. Apply the left-hand rule, then compare the direction of current in these two figures.

You should have concluded that reversing the direction of movement, in the same field, reverses the direction of current. Actually, the situation is this:

1. Reversing *either* the direction of the field or the movement reverses the direction of current.
2. Reversing the direction of *both* field and movement *will not* change the direction of current.

basic characteristics of alternating voltage and current

If a vector of any length is pivoted at the point of origin and rotated through 360°, the tip of the vector will describe a full circle with a radius equal to the length of the vector. If the vector is rotated at a constant rate, say one revolution per second, it will repeat the same circle 60 times each minute. This is an example of a *periodic function,* and it is illustrated in Figure 7-3.

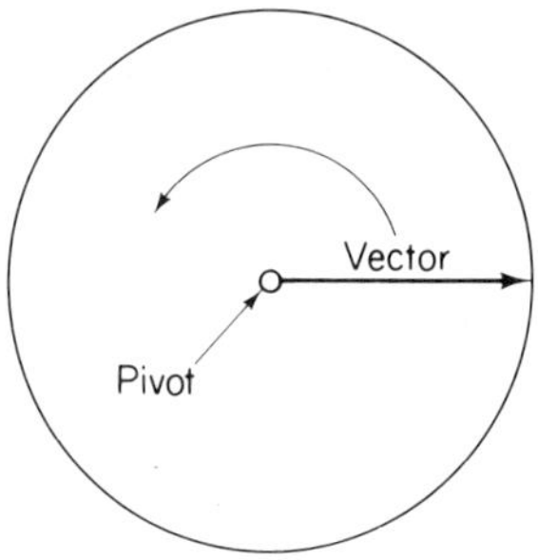

Figure 7-3. Function of a Rotating Vector

Any function which repeats a full set of events in a periodic fashion is called a *periodic function.* The second hand on your watch is a rotating vector. The tip of the second hand describes a 360° circle every minute. That is a periodic function.

Alternating voltage and current are examples of periodic functions. In fact, we can use the rotating vector to draw a picture of either alternating voltage or current. The vector will start at a certain point, rotate in a given direction, and maintain a specified rate of movement. Let's divide the circle along the horizontal axis, and designate the top half as positive and the bottom half as negative. Designate the right side of the horizontal as the starting point and rotate the vector counterclockwise at the rate of one revolution per second. The vector will then be in the positive half for 180° and in the negative half for 180°. This description is illustrated in Figure 7-4.

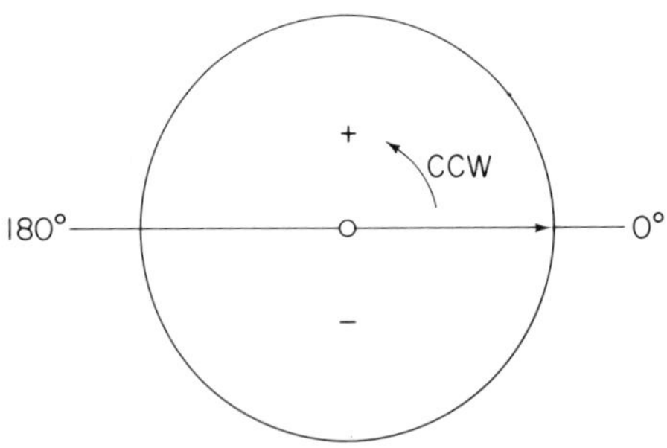

Figure 7-4. Positioning a Vector

Now, if we give the vector a specified length, we can determine the vertical displacement of the vector tip for any point in its 360° rotation. A time-ampltitude graph, where vertical displacement is plotted against time, will give a perfect picture of either alternating voltage or current. This is illustrated in Figure 7-5.

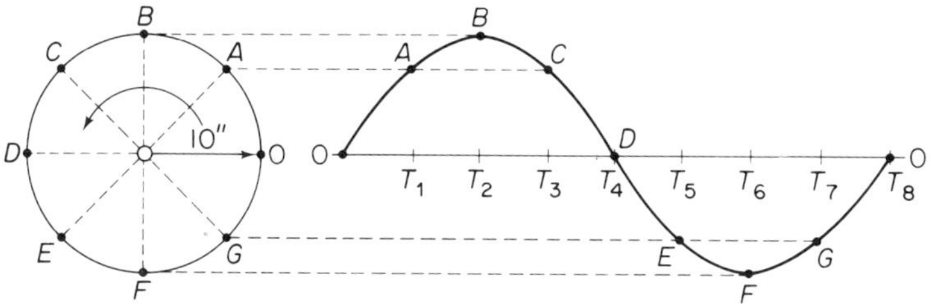

Figure 7-5. Time-Amplitude Graph of Vector Rotation

Vertical *displacement* is the vertical distance from the tip of the vector to the horizontal axis. We have elected to use a 10-inch vector, and time and vertical displacement have been plotted at eight different points around the circle. The circumference of the circle represents one second. Therefore, the time between each two consecutive points is one eighth of a second, and the angular distance is 45°. The zero reference and time base for our graph is just the circumference of the circle laid out in a straight line.

After one eighth of a second, the vector tip is at 45° from the starting position. The vertical displacement at that time is 7.07 inches. This is plotted at the position T_1 on the graph. The vector is 90° from the start at T_2, and its vertical displacement is 10 inches. At T_3 the vector is 135° from the start with a vertical displacement of 7.07 inches. At 180° (T_4) the vector is on the horizontal line with no vertical displacement.

At T_5 the vector is in the negative portion of its rotation and 45° from the horizontal. This position is 225° from the start. The vertical displacement is a -7.07 inches. At T_6 the vector has traveled 270° and has a -10-inch vertical displacement. At T_7 the vector has traveled 315° and has a vertical displacement of -7.07 inches. At T_8 the vector has made the full circle and is back at zero with no vertical displacement. Projecting the described points onto the time-amplitude graph and connecting them with a smooth curve produces the graph in Figure 7-5.

The *length* of the vector is equal to the *radius* of the circle. If we let r represent the vector length, we can write a formula for vertical displacement.

$$\text{Vertical displacement} = r \sin \theta$$

Where: r is the length of the vector and θ is the angle between its position and the horizontal axis.

On the graph, this vertical displacement converts directly to amplitude. The *curve* on the graph is called a *sine wave* because its amplitude at any point is equal to r sine θ.

Alternating voltages and currents are described by just this type of sine wave. The voltage changes amplitude constantly and changes polarity periodically. The *current changes amplitude constantly and changes direction periodically.* The time base can be extended in both directions so that we can plot as many repetitions of the sine wave as we care to see. One sine wave contains a full set of events. The positive portion is the positive alternation; the negative portion is the negative alternation; and the two alternations form one period. The *frequency* is the number of periods generated per unit time. In the above case, this was one period per second, which is defined as one hertz; $f = 1/t$, where f is frequency in hertz, and t is time in seconds for one period.

Now, let's examine a machine that produces such a voltage.

the basic ac generator

Instead of rotating a vector, if we rotate a loop of wire in a magnetic field, we can produce a sine wave of voltage. Take a section of a conductor and bend it into a U-shape. Place this loop between the poles of a magnet so that you are looking into both ends of the conductor as shown in Figure 7-6.

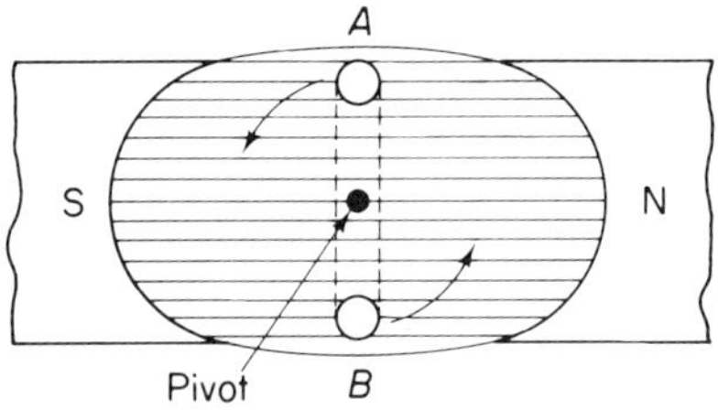

Figure 7-6. One Loop in a Magnetic Field

When the loop starts to turn in the direction indicated, current will go in at point *B,* circle around the back of the loop, and come out at *A.* Check this with the left-hand rule. If we were to think of *A* and *B* as the terminals of a power supply, which would be the positive terminal and which would be the negative terminal? Point *B* is positive, and point *A* is negative. Why not the other way around? —because we are now *inside* the power supply where the *electrons flow into the positive terminal,* through the power supply, and exit at the negative terminal.

Any convenient method from a hand crank to a water fall may be used to rotate the loop. As the loop turns through the first 90°, the induced voltage builds up to a maximum value. During the second 90°, the induced voltage dies out to zero. The ends of the loop have now exchanged positions and are as shown in Figure 7-7.

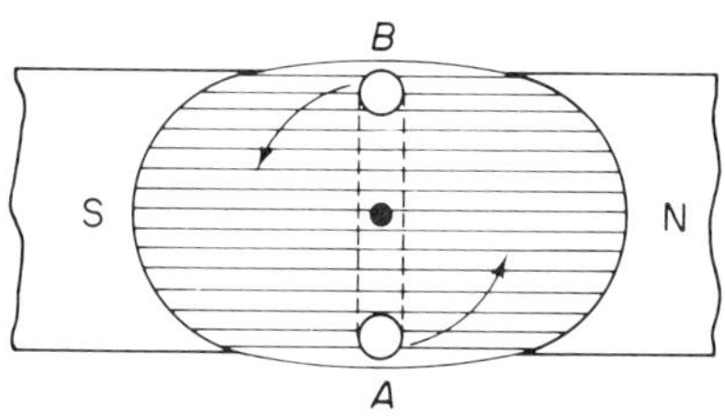

Figure 7-7. Position After 180° Rotation

No lines are being cut; no voltage is being induced. When this point is passed, each conductor cuts into the flux field opposite to its previous direction of cutting. Voltage *polarity reverses; A* becomes positive, and *B* becomes negative. Current enters at *A* and comes out at *B.* Check this by applying the left-hand rule. During the next 90° of rotation, the voltage builds up to a maximum *negative* value (assuming that the first value was positive). During the final 90° of the 360° rotation, the negative voltage falls from maximum to zero. This puts the loop back into its starting position as shown in Figure 7-6.

The one rotation of such a loop generates one complete period of alternating voltage. Therefore, a rotating loop in a magnetic field must be an ac generator. If we can devise a way to apply this voltage to an external circuit, we will have the complete basic generator. Since the loop is rotating, we will need a device that can rotate with the loop and still be electrically connected to an external circuit. This is accomplished by *slip rings* and *brushes* as illustrated in Figure 7-8.

The slip rings are two cylindrical pieces of metal, and one is fastened to, and rotates with, each end of the loop. The brushes are sliding contacts which press against the rotating slip rings. The brushes are commonly made of carbon. They are carbon blocks, hollow ground for a good electrical contact with the ring. The

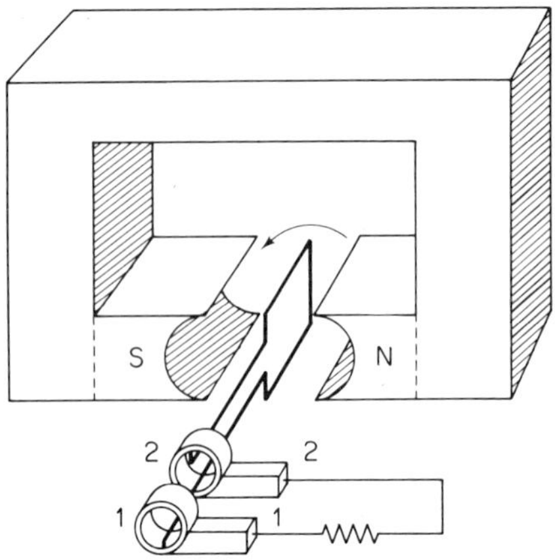

Figure 7-8. Use of Slip Rings and Brushes

brushes are spring loaded to assure a good snug fit. One brush rides on each slip ring, and the two brushes represent the terminals of the power supply. The external circuit is connected between these two brushes in the same manner that a dc circuit is connected between the terminals of a battery.

The drawing in Figure 7-8 shows the loop in the zero position. There is no polarity at the brushes at this time. As the loop rotates 180° counterclockwise, brush 1 becomes positive, and brush 2 becomes negative. Electrons flow out of brush 2, through the circuit, and into brush 1. During this time, the direction of internal current is from slip ring 1, through the loop and back to slip ring 2.

When the rotation passes the 180° point, the polarity of the brushes reverses. Brush 2 becomes positive, and brush 1 negative. The electrons now leave brush 1, flow through the external circuit and into brush 2. Internally, current direction has also reversed. It is now from slip ring 2, through the loop, and out of slip ring 1.

Since positive and negative voltages are relative terms, we need a point of reference before we can say which alternation of the output is positive and which is negative. If we add a ground to our external circuit and speak of polarity with respect to ground, it will simplify our terms of reference. This has been done in Figure 7-9.

Now, when electrons flow through the load resistor *toward ground,* the voltage across the load is *negative.* When electrons flow *from ground* through the load resistor, the voltage across the load is *positive.*

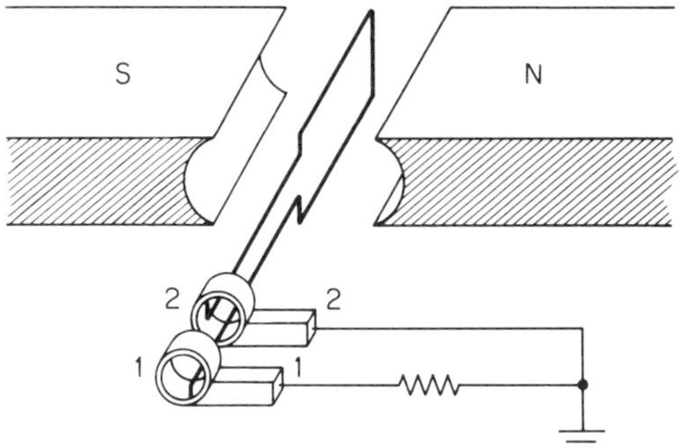

Figure 7-9. Adding Ground for Voltage Reference

Let's take the basic generator and analyze the action through each 45° step of the loop rotation. At the same time, we will plot on a time-amplitude graph the voltage that appears across the load resistor.

analysis of generator action

The starting position is shown in Figure 7-10. The loop is in the *zero* position and is stationary. The letters *A, B, C,* and *D* have been added to the corners of the loop for purposes of reference.

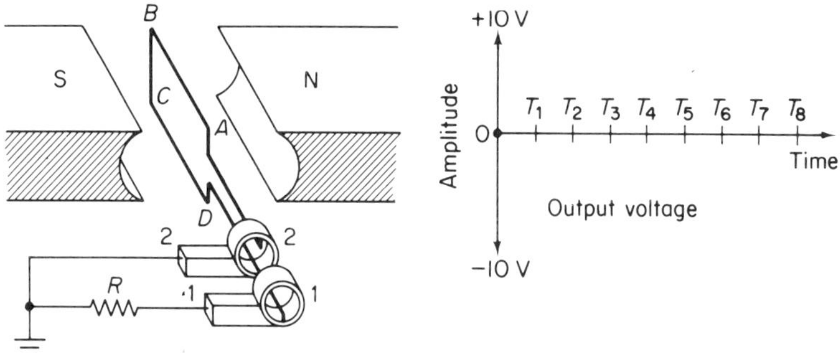

Figure 7-10. Loop in Starting (0°) Position

Conductor *AB* is the top of the loop, and conductor *CD* is the bottom of the loop. Positive and negative amplitudes will be plotted vertically, and time will run along the horizontal line from 0 at the left to time 8 (T_8) on the right. In the zero position, there is no relative motion and no induced voltage.

Assuming that some values will give more concrete details, let's rotate the loop at one revolution per second. This gives us a timing interval between each two consecutive numbered points of one eighth of a second. Elapsed time at each check point will be:

$T_1 = \frac{1}{8}$ second
$T_2 = \frac{1}{4}$ second
$T_3 = \frac{3}{8}$ second
$T_4 = \frac{1}{2}$ second
$T_5 = \frac{5}{8}$ second
$T_6 = \frac{3}{4}$ second
$T_7 = \frac{7}{8}$ second
$T_8 = 1$ second

Assume that the *maximum value* of induced voltage will be 10 volts. This will enable specific amplitudes to be determined at each time point.

Start the loop rotating in a counterclockwise direction at the one revolution per second rate, and it will turn through 45° each one eighth of a second. We'll analyze the action and plot the voltage across resistor *R* at the end of each 45° turn of the loop. Figure 7-11 shows the loop position and the graph at the first 45° position.

Figure 7-11. Loop at 45°

When the loop started to turn, conductor *AB* cut into the flux from the *top* while conductor *CD* cut into it from the *bottom.* Application of the left-hand rule to either section of the conductor establishes the following facts:

1. *Internally,* the direction of current is from slip ring 1 to point *D,* and from there, to *C,* to *B,* to *A,* to slip ring 2.

2. *Externally,* the direction is from brush 2 to ground, and from there to brush 1.
3. Voltage across *R* is *positive* with respect to ground.
4. Brush 2 is the *negative terminal,* and brush 1 is the *positive* terminal.

The amplitude of the induced voltage at any point is calculated by using the formula: $e_i = e_m \sin\theta$, where e_i is the instantaneous voltage, e_m is the maximum voltage, and θ is the angle between the flux and the conductor. In this case, our maximum voltage is 10 volts, and the loop has turned 45°. The sine of 45° is 0.707. Therefore, the voltage at this time is: $e_i = e_m \sin\theta = 10\text{V} \times 0.707 = 7.07\text{V}$. This 7.07V occurs at T_1; one eighth of a second has elapsed. The point where +7.07V intersects T_1 has been marked on the graph and connected to 0 with a smooth curve. Any number of points could be plotted along this line to show the amplitude of e_i for any position of the loop between 0° and 45°. Figure 7-12 shows the conditions after the next 45° rotation.

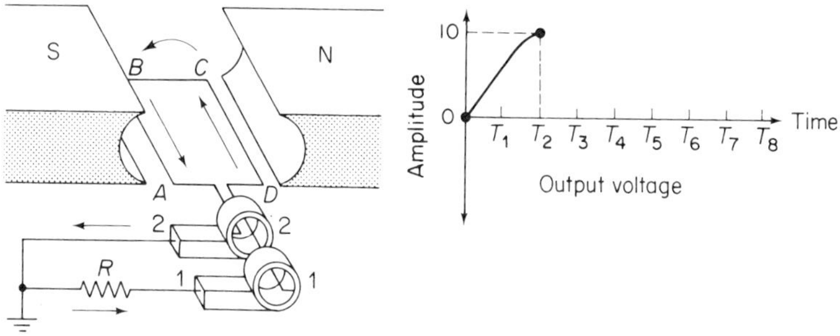

Figure 7-12. Loop at 90°

During the *second* timing interval (T_1 to T_2), the loop rotated from 45° to 90°. Voltage polarity remained the same, while amplitude continued to increase. At 90°, the conductors (*AB* and *CD*) are both cutting flux at a 90° angle. The sine of 90° is 1. Therefore, the induced voltage at T_2 is: $e_i = 10\text{V} \times 1 = 10\text{V}$. The intersection of 10V and T_2 has been marked on the graph and connected to the end of the previous curve. Elapsed time is one fourth of a second; the loop has turned 90°; and the induced voltage has increased from 0 to positive 10 volts. The current direction is the same as before: from slip ring 1 to point *D,* to *C,* to *B,* to *A,* to slip ring 2. Externally, current travels from brush 2 to ground, then, through *R* to brush 1.

This polarity for voltage and direction for current will hold until 180° of rotation has been completed.

During the first 90° of rotation of the loop, voltage and current gradually increase until they *reached a peak* at the 90° position. Figure 7-13 shows the conditions after the third 45° turn.

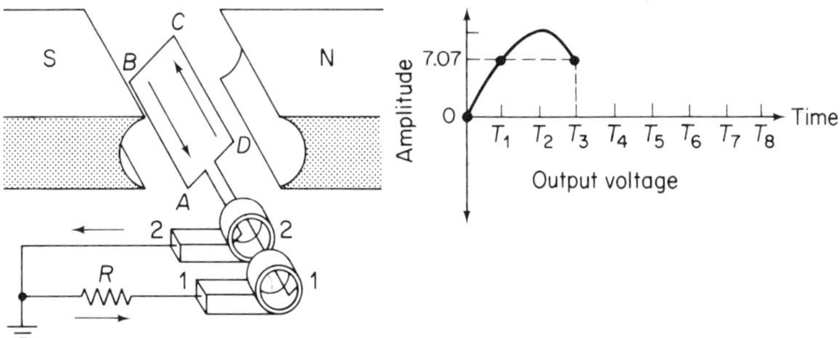

Figure 7-13. Loop at 135°

The loop is now at the 135° position; elapsed time is three eighths of a second; and the angle between conductor and flux is 45°. The amplitude of induced voltage at this time is: $e_i = 10V \times 0.707 = 7.07V$. T_3 and 7.07 has been plotted and connected to the previous plot. Voltage polarity is still the same (brush 1 positive; brush 2 negative). The amplitude of the voltage has steadily decreased since the loop passed the 90° position. This decrease will continue for the next one eighth of a second as the loop rotates from 135° to 180°. Figure 7-14 shows the conditions after the fourth 45° turn.

Conductor *CD* is now at the top center, and *AB* is at the bottom center. Both sections are moving parallel to the flux. No flux lines

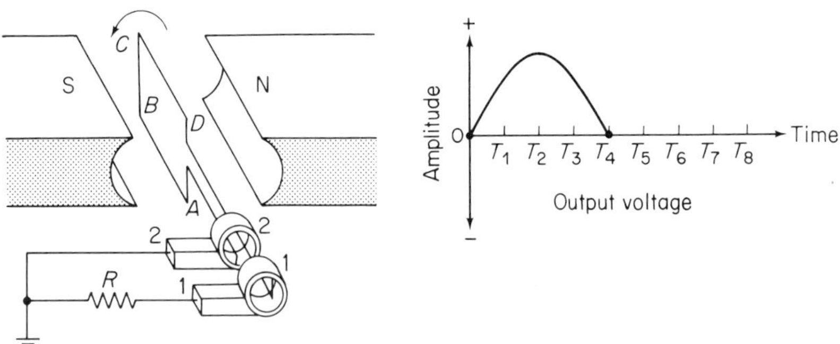

Figure 7-14. Loop at 180°

are being cut, and there is no induced voltage. To put it another way, the angle of cutting is 0°, and the sine of 0° is 0. (Verify by checking tables of trigonometric functions provided in the appendix.) Therefore, $e_i = 10V \times 0 = 0$.

The elapsed time is now one half of a second. The point, 0 and T_4, has been marked and connected to the previous entry on the graph. The graph now shows a complete positive alternation of the output voltage. During this alternation, the voltage built up from zero to maximum and fell to zero; the voltage maintained a constant polarity, and current held a constant direction.

Notice that the voltage on the graph has some amplitude right up to T_4, which is the 180° position of the loop. Figure 7-15 shows the conditions after the fifth 45° turn.

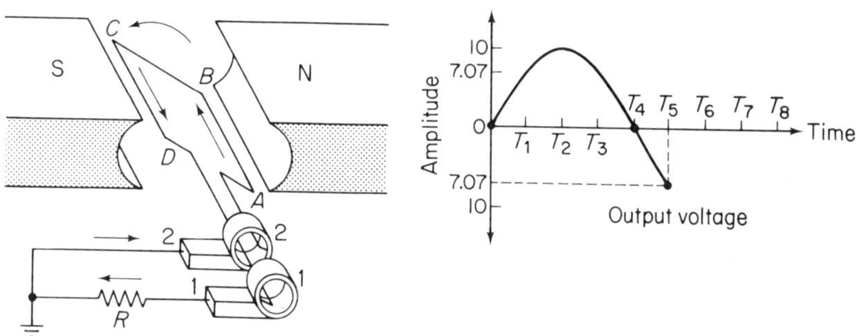

Figure 7-15. Loop at 225°

As the loop rotated past the 180° position, conductor *CD* turned into the flux from the top while *AB* came in from the bottom. Voltage *polarity reversed,* making brush 2 positive and brush 1 negative. Current reversed direction, both internally and externally. Internally, current path is now from slip ring 2 to point *A,* from there to *B,* to *C,* to slip ring 1. Through the resistor, the current travels from brush 1 to ground. This produces a *negative voltage across R* with respect to ground. This polarity will hold for the next 180° of rotation.

At the 225° position, the conductor is cutting the flux at a negative 45° angle. The sine of this angle is —0.707. Therefore, the voltage amplitude is: $e_i = 10V \times -0.707 = -7.07V$. The elapsed time is five eighths of a second, which brings us to T_5. The point, —7.07V and T_5, has been marked on the graph and connected to the previous entries. The next check point is T_6, and Figure 7-16 shows the conditions at that time.

During the 45° turn from 225° to 270°, the voltage polarity re-

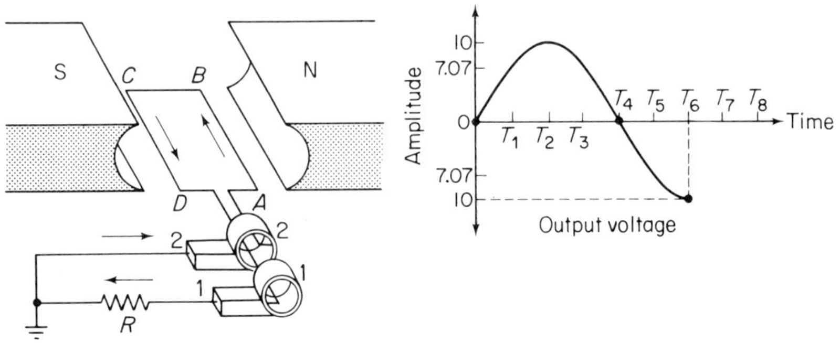

Figure 7-16. Loop at 270°

mained negative, and current direction held constant. Both current and voltage increased amplitude during this time. At this position the cutting angle is —90° which has a sine of —1. The voltage amplitude is: $e_i = 10V \times -1 = -10V$. The elapsed time is three fourths of a second which brings us to T_6. The point, —10V and T_6, has been marked on the graph and connected to the previous entries. The next check point is T_7, and the conditions at that time are shown in Figure 7-17.

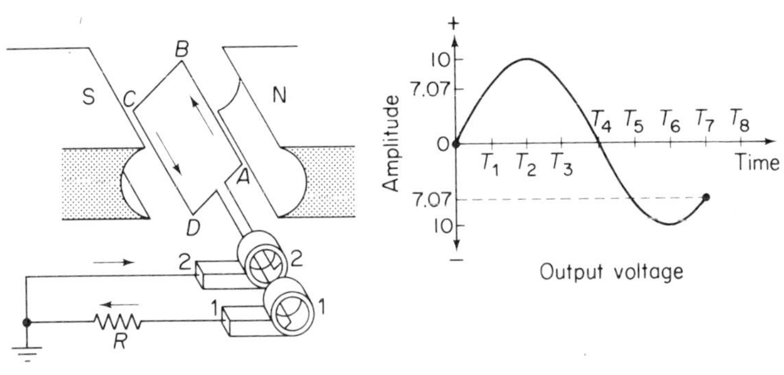

Figure 7-17. Loop at 315°

The loop has now rotated through 315°. During the last 45°, the angle of cutting has been decreasing. At this time, the angle is —45°, which gives us a voltage amplitude of —7.07 volts. The elapsed time is seven eighths of a second, which brings us to T_7. The point, —7.07V and T_7, has been marked on the graph and connected to the previous entries. The graph now shows that the negative voltage has reached a negative peak at 270° position of the loop. During the 45° period from 270° to 315°, this voltage has been steadily decreasing, but its polarity is still negative. The final check point is T_8, and the conditions at that time are shown in Figure 7-18.

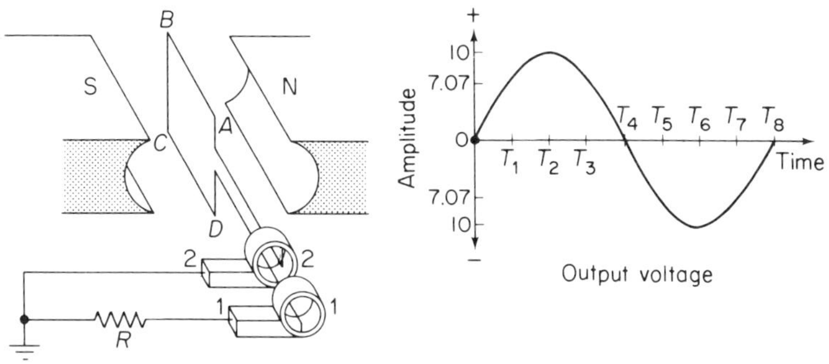

Figure 7-18. Loop at 360°

The loop has rotated a *full circle* (360°) and is now back in the *starting* position. During the final 45° turn, the angle of cutting decreased from 45° to 0°, and the voltage decreased from −7.07 volts to 0 volts. The elapsed time is one second, which brings us to T_8. The point, 0 and T_8, has been marked on the graph and connected to the previous entries.

The graph now shows a *complete sine wave,* which is a history of a full set of events. This is a sine wave of an alternating voltage. *It changes amplitude constantly and reverses polarity every 180°.* This is the output voltage of an ac generator. When the external load is resistive, as has been the case here, the current can be represented by exactly the same sine wave. The only difference would be in the unit of measure.

Some type of generator built on these principles is the voltage source for most alternating voltages. But do not assume that all direct voltage is supplied by batteries.

PRINCIPLES OF DC GENERATION

Technically speaking, direct voltage is not generated. All generators *internally* produce alternating voltage. Yet, there are generators that supply direct voltage. The difference then between an ac and a dc generator must be in the method of applying the output voltage to the external circuit.

All generators operate on the principle of rotating a loop in a magnetic field. As you have already seen, such an action reverses the polarity for each 180° of loop rotation. If the output of the generator were taken through a device that counteracted the polarity change, the output would be dc. Actually, this counteracting device is the principal difference between an ac generator and a dc generator.

basic dc generator

The output of an ac generator is taken through *slip rings and brushes.* With the dc generator, the brushes are still used, but a device known as a *commutator* replaces the slip rings. The slip ring is a complete circle of metal, and a ring is attached to each end of the loop. The commutation is composed of two segments of metal. They are insulated from each other, and each segment is connected to one end of the loop. This is illustrated in Figure 7-19.

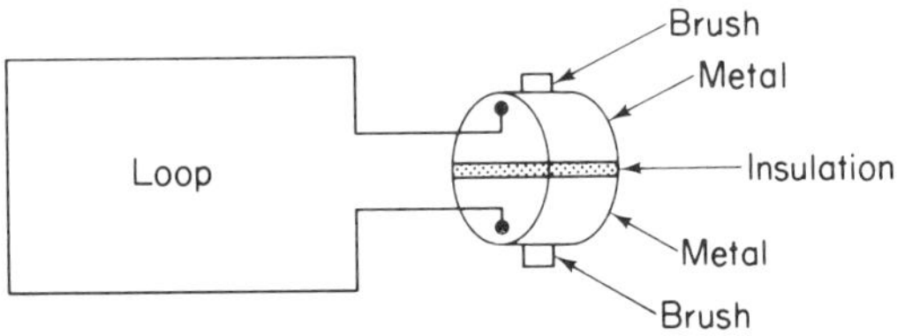

Figure 7-19. Commutator for dc Generator

The commutator rotates with the loop, and the stationary brushes ride against the commutator. The brushes are so positioned that they ride across the insulated area at the same instant that the loop passes the point of zero voltage. The *brush polarities never change;* the positive one remains positive; the negative one remains negative. When the internal polarities of the generator are about to switch (after every 180°), the segments of the commutator change brushes. This is illustrated in Figure 7-20.

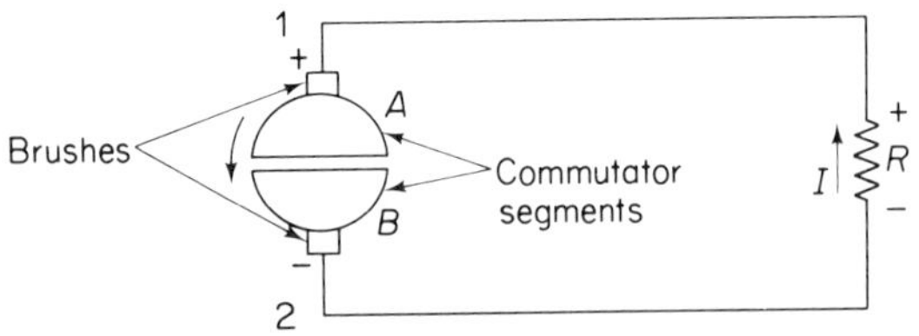

Figure 7-20. Constant Brush Polarities

The rotating commutator segments (*A* and *B*) are *alternately* in the positive and negative positions. Presently, segment *A* is positive, and segment *B* is negative. The polarities of the stationary brushes will remain as indicated and keep the current in a constant direction in the external circuit. This is accomplished as follows:

1. When the commutator turns another 45°, induced voltage will be zero, and the insulated gap will be crossing under both brushes.
2. Slightly beyond this zero point, segment *B* makes contact with brush 1, and segment *A* makes contact with brush 2.
3. At the same time, segment *B* becomes positive, and segment *A* becomes negative.

Figure 7-21 shows both the internal and external wave shapes for a single-loop dc generator along with the loop positions. The letters (*a, b, c, d,* and *e*) on the voltage wave shapes correspond to loop and commutator positions in the accompanying drawings.

The *internal current* reverses direction every 180° of rotation as do the internal polarities of the induced voltage. At the same time that the internal polarities reverse, the segments of the commutator swap the brushes that they feed. The output from this basic (one-loop) generator would be dc in the sense that polarities and current direction are constant. However, the voltage and current will be *constantly changing in amplitude.* This is called *pulsating dc.*

We mentioned earlier that a force would be exerted on a current carrying conductor in a magnetic field. We called this "the motor principle."

CONVERTING ELECTRICAL ENERGY TO MECHANICAL ENERGY

Mechanical energy must be obtained by converting some other type of energy. In this case, electrical energy will be converted because the electric current provides the turning force. The device that performs the conversion is a *motor.*

principles of alternating current motors

Mechanically, there is little difference between an ac *motor* and an ac *generator.* The generator is turned by some mechanical

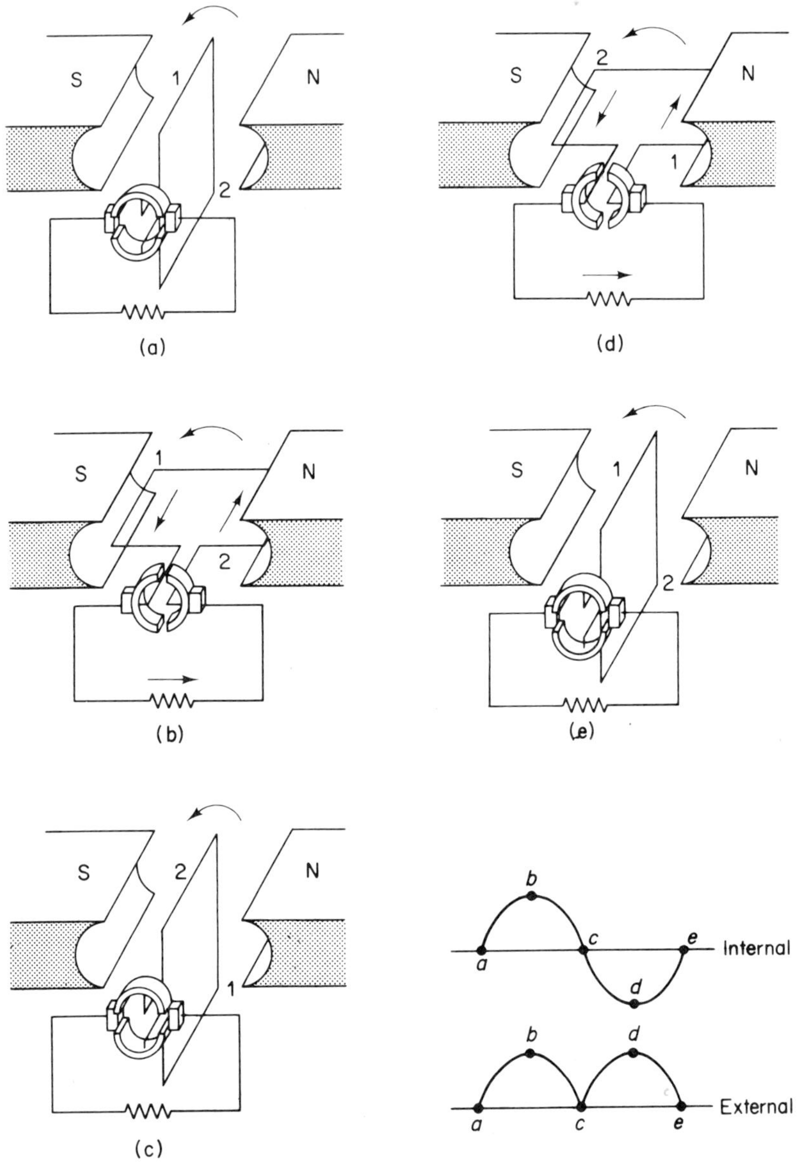

Figure 7-21. Wave Shapes and Loop Positions

means and produces alternating voltage. The motor is activated by an alternating current and produces a mechanical turning force (*torque*). Therefore, the motor is used to convert electrical energy to mechanical energy.

the basic alternating current motor

In its simplest form the motor consists of a conductor loop in a magnetic field as shown in Figure 7-22. With current and flux as

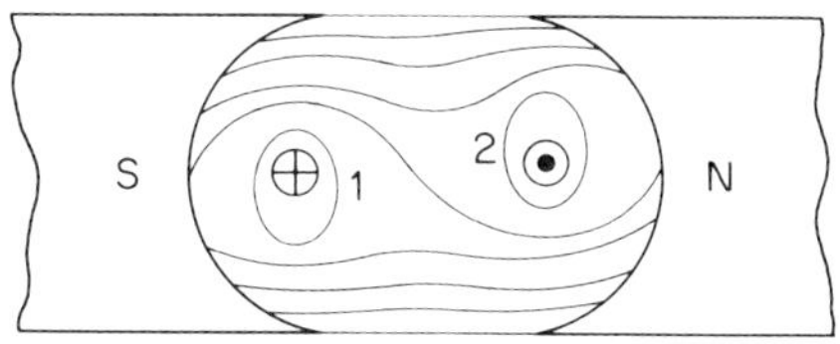

Figure 7-22. Motor Principle

indicated, pressure has increased at the top and decreased at the bottom of conductor 1. At the same time, the opposite is true of conductor 2; it has more pressure at the bottom and less at the top. If the loop is pivoted and free to rotate, conductor 1 will move down, while conductor 2 moves up. This produces a counterclockwise rotation of the loop.

If you have trouble visualizing the direction of rotation, here is another left-hand rule that should help. Place the open left hand in the field with the conductor on the palm; the fingers in the direction of flux; and the thumb in the direction of current. The conductor will then *move away* from the palm. Test this rule on both sections of the conductor in Figure 7-22 and verify the counterclockwise rotation.

In order to construct a basic ac motor, we shall need a voltage source, brushes, slip rings, a magnetic field, a loop of wire, and some means of extracting the torque that will be produced. These parts are assembled as shown in Figure 7-23.

If this appears to be a copy of the basic generator, it is—almost. The external circuit is now an imput power supply, and a shaft has been added to the loop. The shaft fastened on the wire loop, as it is shown here, is not a very practical arrangement. We need something of a far more rigid construction. The important principle to be established is that *the current causes the loop to rotate,* and the rotation of the loop imparts a torque to the shaft.

analysis of torque

If we can cause the loop to rotate 360°, the same actions will be repetitive for as long as power is applied. This action is a periodic function, and all periodic functions are repetitive. Let's

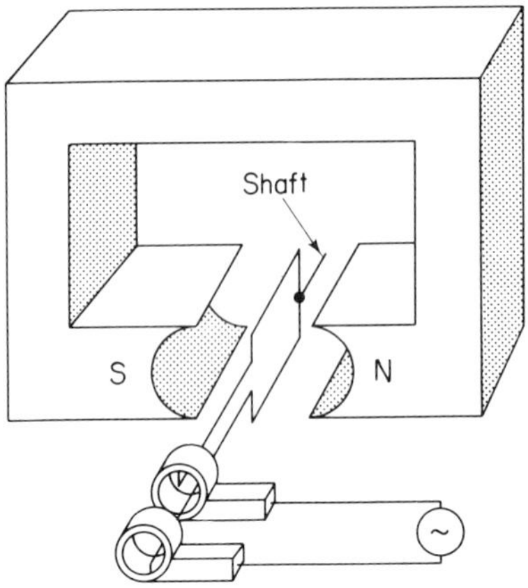

Figure 7-23. Basic Alternating Current Motor

apply a voltage and analyze the conditions at the end of each 45° turn of the loop as we did with the generator. We *cannot* start at the zero position because there is no field interaction at that instant. The loop is moving parallel to the flux of the permanent magnet, and the zero current has allowed the field about the conductor to decay. We'll start under the conditions depicted in Figure 7-24.

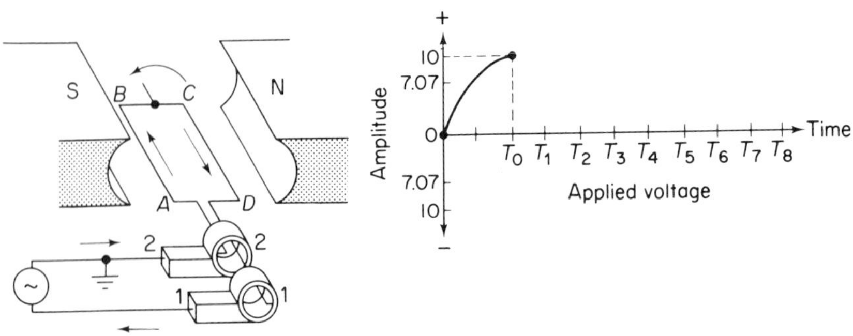

Figure 7-24. Loop at 90° Start Position

We are assuming that the applied voltage has a 10-volt peak and a frequency of one hertz. The loop is in the position where the angle between conductors and flux is 90°. Voltage, with respect to ground, is at a peak positive value of 10 volts. Maximum current is present in the loops in the direction of the arrows. The graph designates this position as T_0 and breaks up the next second into eight equal parts. The loop is rotating counterclockwise.

Maximum turning force is being exerted on the loop at this time. The flux density is maximum, the current is maximum, and the angle is maximum.

Apply the left-hand rule to both conductor *AB* and *CD*, and verify the direction of rotation. Trace the current both internally and externally, and verify voltage polarities with respect to ground.

Now we examine the conditions after the next 45° turn as depicted in Figure 7-25. During the past eighth of a second (T_1 to

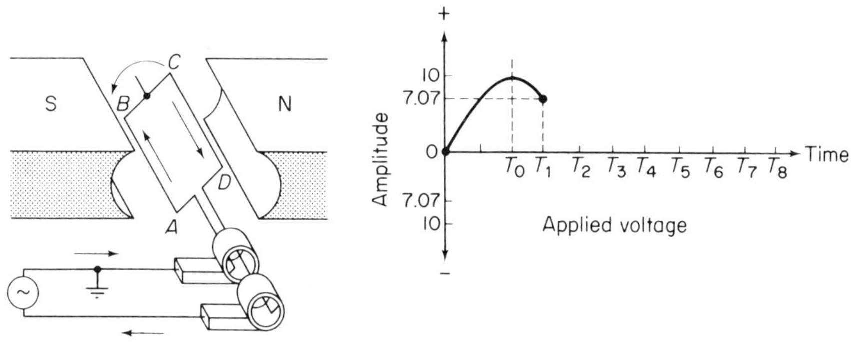

Figure 7-25. Loop 45° from Start

T_0), the loop has turned counterclockwise 45°, which moved it from 90° to 135°. The applied voltage passed the peak and dropped to a positive 7.07 volts. The angle between flux and conductors has changed from 90° to 45°. The point, T_1 and 7.07V, has been plotted on the graph and connected back to the starting voltage, which was present at T_0.

Figure 7-26 shows the conditions after the second 45° turn. At

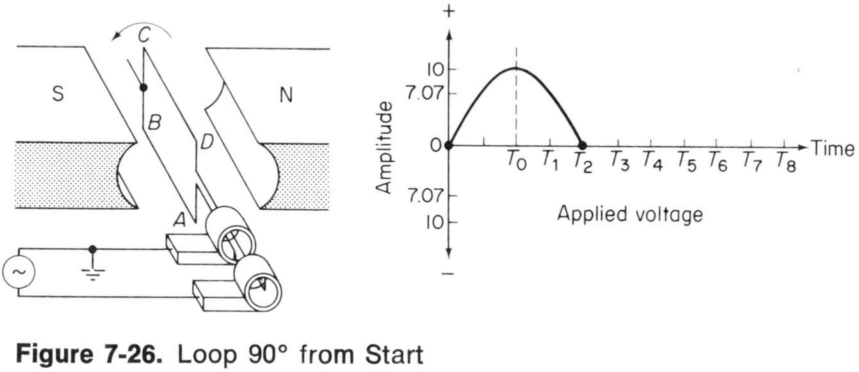

Figure 7-26. Loop 90° from Start

the 90° (from start) position the conductors are actually 180° from the 0° position. Therefore, the conductors are moving at a zero angle to the flux. The applied voltage has dropped to zero. If there had been a dc voltage applied in this manner, the loop would stop in this position. However, it is an alternating voltage. The *inertia* of the loop carries it past the zero angle, and by that time, the applied voltage is starting to build up in a negative direction. Figure 7-27 shows the conditions after the third 45° turn.

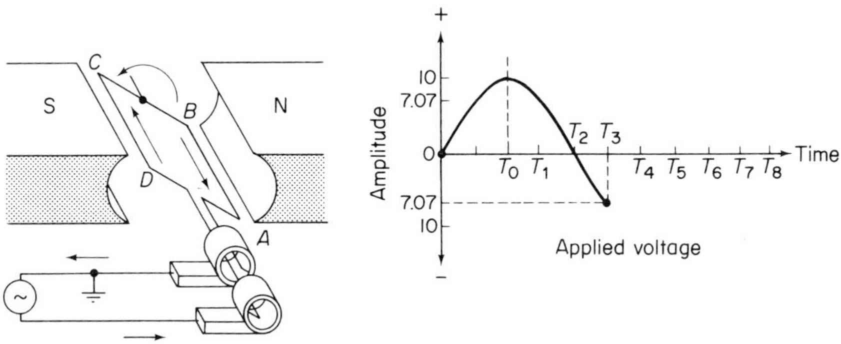

Figure 7-27. Loop 135° from Start

The angle between conductors and flux is now a negative 45° with respect to our previous position. The applied voltage has built up to a −7.07 volts with respect to ground, and the current has changed direction. The torque is still counterclockwise and can be verified by applying the left-hand rule.

Figure 7-28 shows the conditions after the fourth 45° turn. The loop has now turned 180° from our 90° start position. The applied voltage has changed from peak positive to zero, to peak negative.

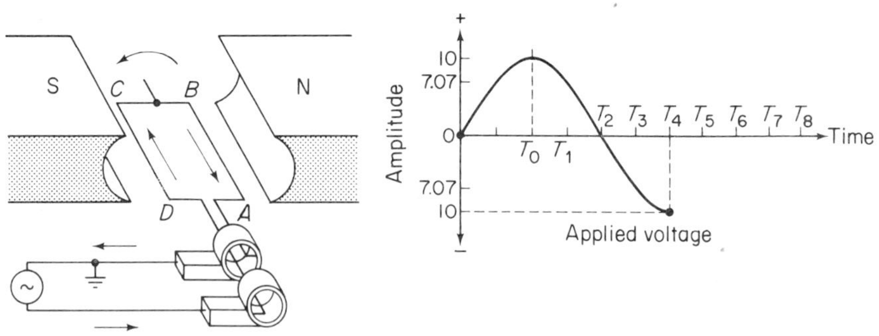

Figure 7-28. Loop 180° from Start

The angle between conductors and flux has changed from 90°, to 45°, to zero, to —45°, and is now at —90°. The torque is back to maximum and is still in a counterclockwise direction.

Figure 7-29 shows the conditions after the fifth 45° turn. During

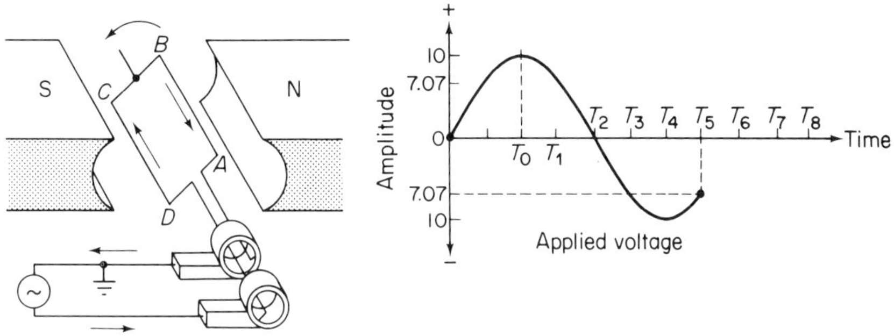

Figure 7-29. Loop 225° from Start

this last eighth of a second (T_4 to T_5), the applied voltage has dropped from —10 volts to —7.07 volts. The loop turned from 180° to 225°, and the angle between conductors and flux is now —45°. Torque is still applied in a counterclockwise direction. The conditions after the sixth 45° turn are depicted in Figure 7-30.

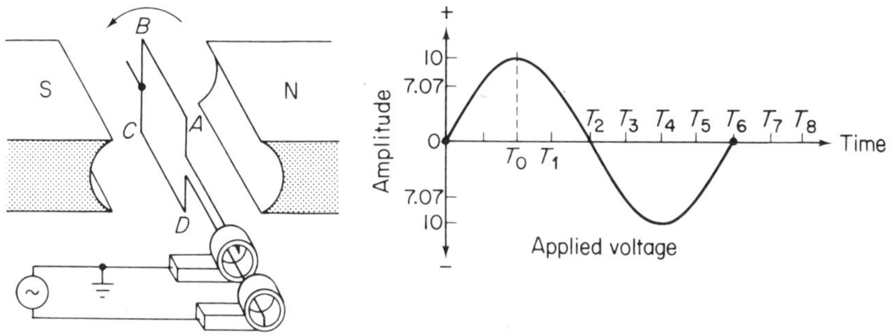

Figure 7-30. Loop 270° from Start

The elapsed time is now three quarters of a second, and the loop has turned 270° from the 90° start position. The applied voltage has dropped to zero. The conductors are traveling parallel to the flux to give a cutting angle of zero. The inertia of the loop will carry it through this area.

T_7 is the next check point, and the conditions at that time are depicted in Figure 7-31. Voltage has crossed the zero reference and built up to a positive 7.07 volts. It has pulled the loop with it to this position, 315° from the 90° start position. Current has changed direction through the loop for the second time. Application of the

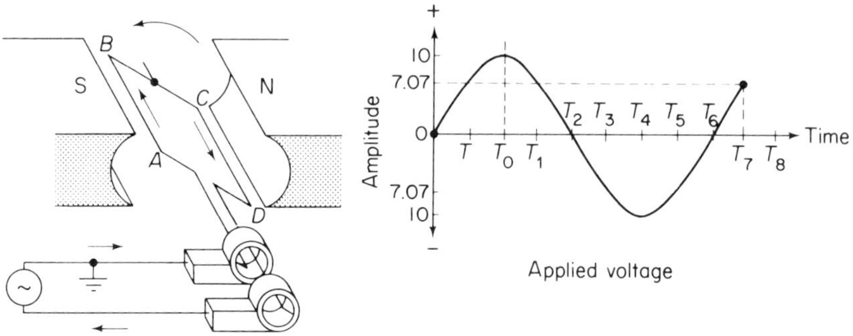

Figure 7-31. Loop 315° from Start

left-hand rule will show that torque is still in a counterclockwise direction. The angle between conductors and flux has built up from zero to 45° and is increasing.

The final check point is T_8. The conditions at that time are depicted in Figure 7-32. The elapsed time is one full second. The ap-

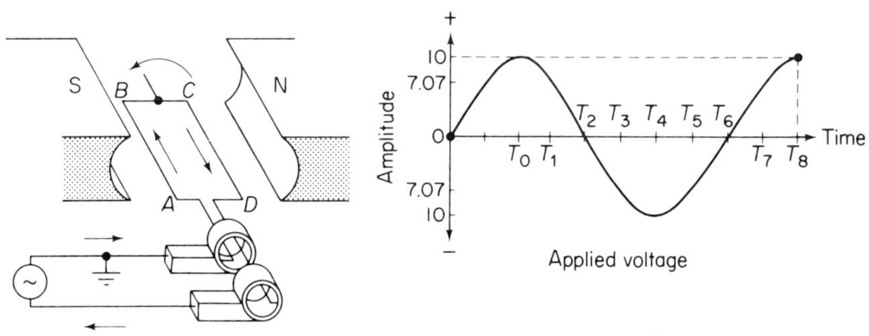

Figure 7-32. Loop 360° from Start

plied voltage has passed through one complete period from peak positive to peak negative and back to peak positive. The loop has been rotated counterclockwise through a full 360° and is back in the start position. As long as the conditions remain the same, the loop will continue to rotate in the same direction. The loop makes one complete rotation for each period of the applied voltage.

We stated that this motor would not function with direct voltage. Yet, there are dc motors.

PRINCIPLES OF DIRECT CURRENT MOTORS

As we showed in the last section, when the loop reaches a position where there is a zero angle between conductors and flux, a change of current direction is required in order to keep the torque applied. If a motor is to be powered by a dc source, there must be a means of switching the polarity of the applied voltage at these critical points.

basic direct current motor

The means of switching the polarity in the motor is the same device used to prevent switching in the generator. That's right. We are back to the *commutator* again. We'll use the single-loop dc motor, and analyze the actions every 90° turn of the loop for one complete rotation. We'll start in the 90° position as shown in Figure 7-33.

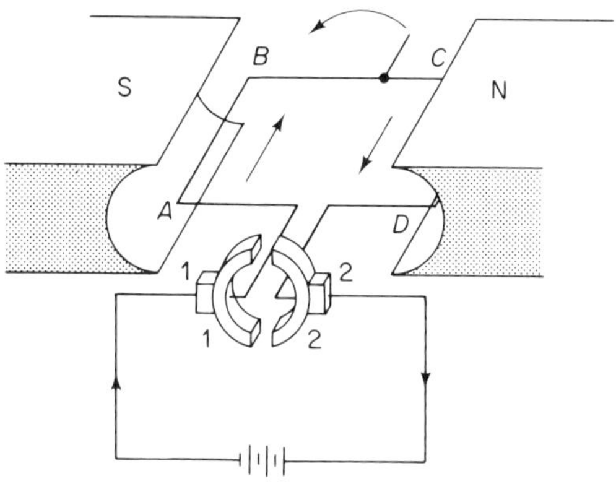

Figure 7-33. Loop at 90° (start) Position

When direct voltage is applied as indicated, current direction will be from the negative terminal of the battery to commutator segment 1, through the loop from *A* to *B,* to *C,* to *D,* to commutator segment 2, and back to the positive battery terminal. Application of the left-hand rule will show the direction of rotation (and torque) to be counterclockwise. Torque is maximum at this position. Torque will remain in this direction and will decrease in force through the next 90° of rotation.

Figure 7-34 shows the conditions after the next 90° turn. The

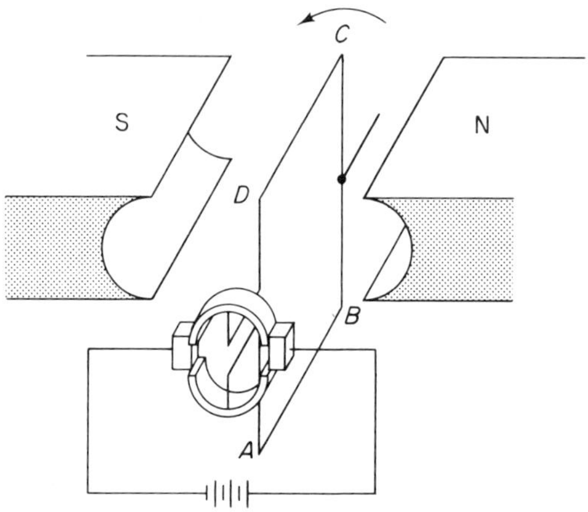

Figure 7-34. Loop 90° from Start

loop has traveled 90° from the 90° start position and is now at the 180° point. This, as well as the zero position, is a point where current in the loop must be reversed in order for torque to be maintained. As the loop passes this point, there is no current because insulated sections of the commutator are passing under both brushes. This is a *dead center position.* The inertia will carry the loop beyond this point. When the loop turns just a little off the dead center mark, the brushes will contact the opposite segments of the commutator. Current will reverse in the loop, and counterclockwise torque will again be applied.

Figure 7-35 shows the conditions after the next 90° turn. This drawing shows the current direction, which was initiated immediately after passing the point of dead center. Notice that the *commutator segments have swapped brushes.* Current has changed direction in the loop, but torque is still in a counterclockwise direction. Torque has been increasing during the past 90° (180° to 270°). This current direction will hold for another 90°, but torque will be decreasing all the time.

Figure 7-36 shows the conditions after the third 90° turn. This is *zero degrees* and another dead center position. The brushes are riding the insulation between commutator segments, and there is

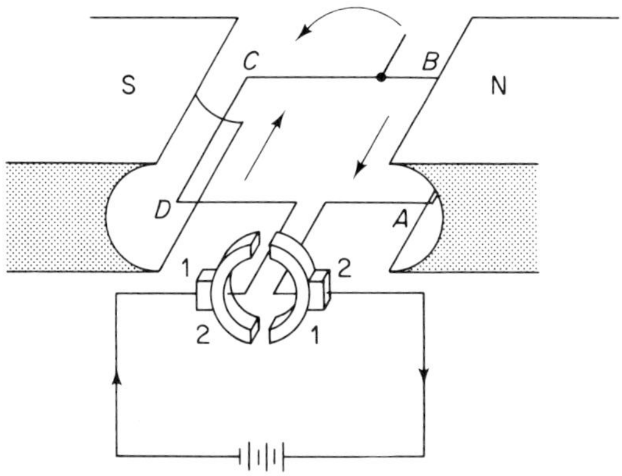

Figure 7-35. Loop 180° from Start

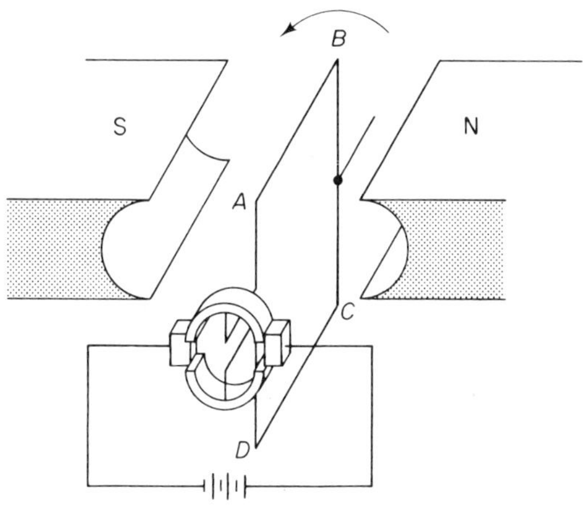

Figure 7-36. Loop 270° from Start

no current. Inertia will carry the loop through dead center, and the commutator will again reverse current direction in the loop.

Figure 7-37 shows the loop back in the starting position. From zero dead center to this 90° position, current was in the direction indicated here. The torque started small and built up to a maximum at this point.

Review Figures 7-33 through 7-36 and notice that at zero and 180° (dead center positions) there is no current. When these points have passed, *internal current reverses* and again applies torque in the same direction. At these two positions, there can be no torque

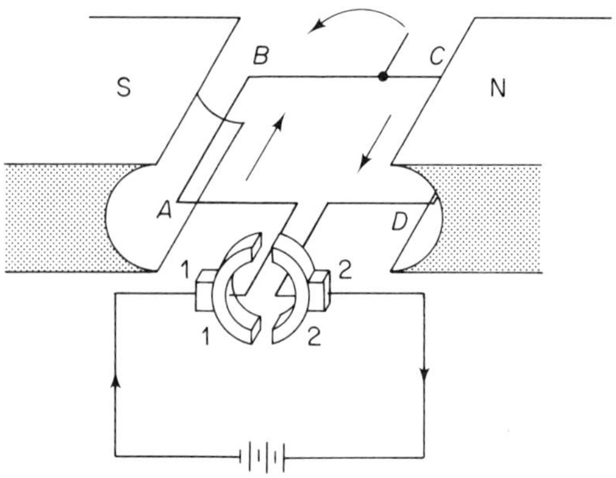

Figure 7-37. Loop Back in Start Position

in any case because of the zero angle between conductors and flux. This being the case, the momentary interruption of current in the loop has no effect on the turning force. By the time that inertia has carried the loop past dead center, the path is again complete, and current is again present in the loop.

In summary, it can be said that *all generators produce ac,* and *all motors operate on ac.* In the case of the dc generator, a commutator changes the output from ac to dc. In the dc motor, the commutator changes the input from dc to ac.

This concludes Chapter 7. Although an attempt has been made to keep all work electrically oriented, there must have been times when the principles seemed rather abstract. If so, this was a false impression. You will see practical applications very soon. The following review exercises should help to reinforce what you have learned here.

REVIEW EXERCISES

1. Alternating current reverses ____________ periodically and changes ____________ constantly.
2. An alternating voltage changes amplitude ____________ and reverses polarity ____________.
3. What is the principal difference between the dynamo and the magneto?
4. When a conductor moves through a magnetic field, a voltage is induced. This is a statement of what principle?

5. What happens to current direction in a conductor when relative motion and field direction are both reversed?
6. What must be done in order to reverse the direction of induced current?
7. Any function which is repeated regularly is a ________ function.
8. The vertical displacement of the tip of a rotating vector that starts at zero describes a ________ ________.
9. The instantaneous amplitude of a sine wave is obtained by ________ ________ multiplied by sine θ.
10. One rotation of a generator loop starting at zero produces one ________ of an ac ________ wave.
11. What is the purpose of slip rings?
12. At what degree points does the ac sine wave cross the zero reference line?
13. Maximum amplitude of ac is being generated as the loop passes through degree points ________ and ________.
14. What is the formula for instantaneous amplitude of induced voltage?
15. The maximum voltage from a rotating loop is five volts. What is the amplitude of the output at the 315° point?
16. Internally, all generators produce ________ voltage.
17. The outstanding difference between the dc and ac generator is the means of taking the output. The dc generator uses ________ while the ac generator uses ________ ________.
18. As the dc generator loop passes through 0° and 180°, the brushes are riding on the ________ segments of the ________.
19. The output of a single-loop dc generator is a ________ dc. The amplitude ________, but the ________ is constant.
20. The electric motor is based on the principle that a force is exerted on a current carrying ________ in a ________ ________.
21. A device which converts electrical energy to mechanical energy is called a ________.

22. The current through the motor loop causes the loop to ____________, and the turning force is imparted to the ____________.
23. The turning force of a motor is called ____________.
24. Maximum turning force is exerted on a loop as it passes through ____________ and ____________ degrees.
25. A single-loop dc motor won't start from the zero degree position. Why does it keep turning after it has started?
26. The loop of an ac motor makes one complete revolution for each ____________ of the applied voltage.
27. In a dc motor the ____________ is converted to ____________ as it enters the loop.
28. What function does the commutator serve in a dc motor?
29. 0° and 180° positions of a motor loop are called ____________ ____________ positions because no ____________ is developed at these times.
30. Why is torque unaffected by the interruption of current when the brushes are crossing the insulated segment of the commutator?

8

periodic functions

Now that we know the general characteristics of alternating current (ac) and how ac can be produced, we need to analyze it in greater detail. We have already described it as a *periodic function.* We represented it as a generated sine wave by rotating a vector; this shows that it is a *vectorial function.* As the vector rotated, we referred to the sine of the angle to determine the amplitude of the sine wave; this indicates an *angular function.* When we combine these characteristics, we recognize that ac has *magnitude, direction,* and *angular velocity* (or amplitude, polarity, and frequency). Let's analyze each of these characteristics a bit further.

GENERATION OF ANGLES

Mathematically speaking, it makes no difference whether we are dealing with the rotating vector or the rotating loop of a generator. Both generate a *sinusoidal* wave shape, which is repeated in a periodic fashion. We must learn to cope with this constantly changing wave shape.

When a vector is pivoted at the origin and rotated, it creates an angle between its final position and its original position. If the vector is in constant motion, we could think in terms of the angle between its *start position* and its *instantaneous position.* We will refer to the starting position as the *initial side* of the angle, and the final position as the *terminal side.*

Our previous illustrations have used counterclockwise rotations that generate *positive* angles; clockwise rotations generate *negative* angles. Therefore, angles may be either positive or negative, and may be of any size. Our sample generator covered one revolution per second. In one second, it generated an angle of 360°. In one minute, it would have generated an angle of 60 times 360°, which is 21,600°. Figure 8-1 illustrates a rotating vector generating both positive and negative angles.

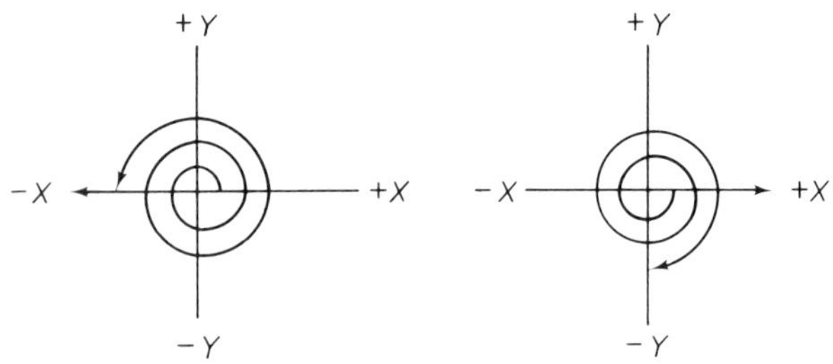

(a) 900° Positive angle (b) 810° Negative angle

Figure 8-1. Generation of Angles

angular measurement

There are several systems for measuring angular distance. The one we have been using (degrees) is called the *sexagesimal system.* A circle is divided into 360 equal parts called *degrees.* The degree is divided into 60 equal parts, which are called *minutes.* The minute is divided into 60 equal parts known as *seconds.* The term sexagesimal is derived from the Latin "sexagesimus," which means sixty.

The sexagesimal system is frequently used in electricity. It is a convenient system and is flexible enough to fit most situations. Most electrical texts use *decimal parts* of a degree rather than minutes. This divides the degree into *ten parts* instead of sixty. It is a bit more convenient, and it is accurate enough. An angle of 36 degrees and 24 minutes (36°24′) becomes 36.4° with the decimal system.

The *right angle* is a standard unit of angular measurement. This represents a segment of the circle equal to 90°. It is a rather large unit, but there are many relationships in electricity that can be expressed as 90° or some multiple of 90°.

Another popular system of angular measurement is the *circular system.* This is sometimes called the *natural* or the *radian* system of measurement. It is based on that *arc* of the circle that is *equal to the radius* of the circle. The circumference of a circle is equivalent to 6.28 times the radius: $C = 2\pi r$, where C is circumference, r is the radius, and 2π is the ratio of C to r; and $\pi = 3.14$. This can also be expressed as: $C/r = 2\pi$.

When two lines are extended from the center of a circle and intercept an arc on the circumference equal to the radius, they form an angle of approximately 57.3°. This section of a circle is called a *radian.* (See Figure 8-2.)

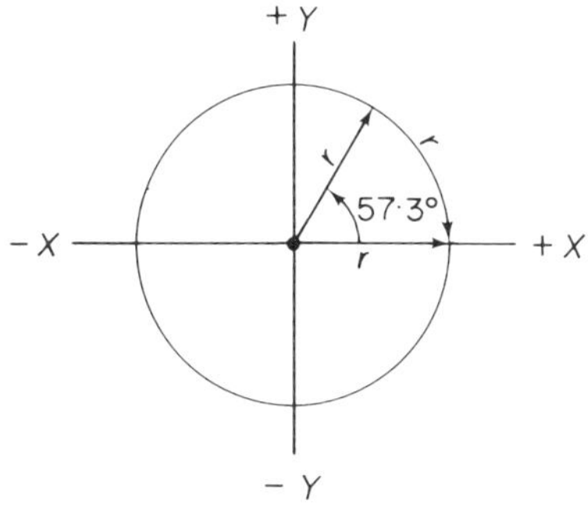

Figure 8-2. One Radian

The circle can be divided into 6.28 radians, and $6.28 = 2\pi$. Therefore, 2π radians = 360°. Figure 8-3 shows a circle divided into radians. Here are expressions of the degree–radian relationship:

$$1 \text{ radian} = 57.2959^\circ \quad (57.3^\circ)$$

$$1 \text{ radian} = \frac{180^\circ}{\pi}$$

$$\pi \text{ radians} = 180^\circ$$

$$1 \text{ degree} = \frac{2\pi \text{ radians}}{360} = \frac{\pi \text{ radians}}{180} = 0.01745 \text{ radians}$$

Many formulas in the fields of *electricity* and electronics are based on the *circular* system of measurement. We have already encountered a few of these in magnetism. Any formula that contains the constant of 3.14 (which is represented by the Greek letter pi—π) is based on this system. Figure 8-4 shows a sine wave in terms of π radians.

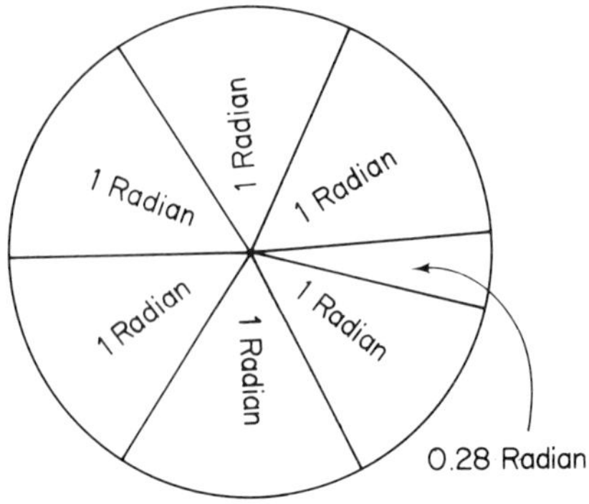

Figure 8-3. 2π Radians = 360°

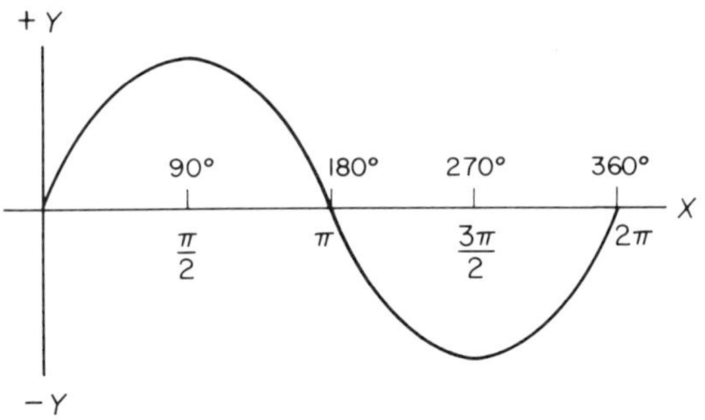

Figure 8-4. Relationship of π Radians and the Sine Wave

angular velocity

When an object moves in a given direction, its rate of travel is its *linear velocity.* When the rate of travel is uniform, the unit of velocity expresses the distance traveled per unit time. For instance, four miles per hour means that four miles are traveled each hour.

Angular velocity is the rate at which an angle is generated. Going back to our sample generator again, we see that it generated a 360° angle in one second. Therefore, if rotation is uniform, *degrees per second* is an accurate description of its angular velocity. Similar to linear velocity, the angle generated per unit time is the unit of angular velocity.

Angular velocity is frequently expressed as *revolutions* per second or revolutions per minute. The tachometer on the sports car measures the angular velocity of the crankshaft in revolutions per minute. The second hand on your watch makes one revolution per minute. When expressing faster rates of angular velocity, one may use revolutions per second.

$2\pi f$ is the number of radians generated per unit time. The angular velocity in *radians per second* is represented by the Greek letter omega (ω), and $\omega = 2\pi f$, where ω is angular velocity in radians per second, 2π represents the radians in a circle, and f is the frequency in hertz.

If the loop of a generator is rotating at 30Hz, its angular velocity is 188.4 radians per second:

$$\omega = 2\pi f = 2\pi \times 30 = 6.28 \times 30 = 188.4 \text{ radians per second}$$

The total angle generated can be expressed in terms of radians, angular velocity, and time: $\theta = \omega t$, where θ is the total angle in radians, ω is angular velocity in radians per second ($2\pi f$), and t is time in seconds.

When the loop in a motor has an angular velocity of 200 radians per second, it generates an angle of 20 radians in one tenth of a second.

$$\theta = \omega t = 200 \times 0.1 = 20 \text{ radians}$$

In degrees, this is $20 \times 57.3°$ or 1146°.

There is an inverse ratio between frequency and time: the higher the frequency, the shorter the time per period. $f = 1/t$, and $t = 1/f$, where f is frequency in hertz, and t is time in seconds for one period.

The frequency of an alternating voltage furnishes a means of *classification. Power frequencies* (frequencies of power supplies) are generally 60 or 400 Hz in the United States. 25 Hz was used at one time, and some European countries still use 50 Hz. Frequencies from 20 to 20,000 Hz are *audio frequencies.* This is the range of frequencies to which the human ear may respond, but both the bottom and top limits are out of the range of hearing for most of us. Above this range are the *radio waves*. We could proceed up the frequency spectrum to the *light wave frequencies.*

The physical *length* of one period of a sine wave is the *wave length.* The wave length can be measured between corresponding points on any two consecutive periods. In other words, the distance between any two points that are 360° apart on a wave shape are said to be one wave length apart. Wave length is generally expressed in meters and represented by the Greek letter lambda (λ).

Electromagnetic waves also have a velocity and may exist in free space. Their *velocity* is approximately the same as the *speed of light,* which is 186,000 miles per second. Converting this to meters gives a velocity of 3×10^8 meters per second.

$$f = \frac{V}{\lambda}, \text{ and } \lambda = \frac{V}{f}$$

Where: f is frequency in Hz, V is velocity of 3×10^8 meters per second, and λ is wave length in meters.

An electromagnetic wave with a frequency of 500 kHz has a wave length of 600 meters:

$$\lambda = V/f = \frac{3 \times 10^8}{500 \text{ kHz}} = \frac{3 \times 10^8}{5 \times 10^5} = 0.6 \times 10^3 = 600 \text{ meters}$$

amplitude

When we plotted the sine wave with the rotating vector, we stated that the amplitude was equal to the vector length multiplied by the sine of the angle. Plotted on the X and Y coordinates this becomes:

$$\text{amplitude} = r \sin \theta$$

$$Y = r \sin \theta = r \sin 2\pi ft = r \sin \omega t$$

This is illustrated in Figure 8-5.

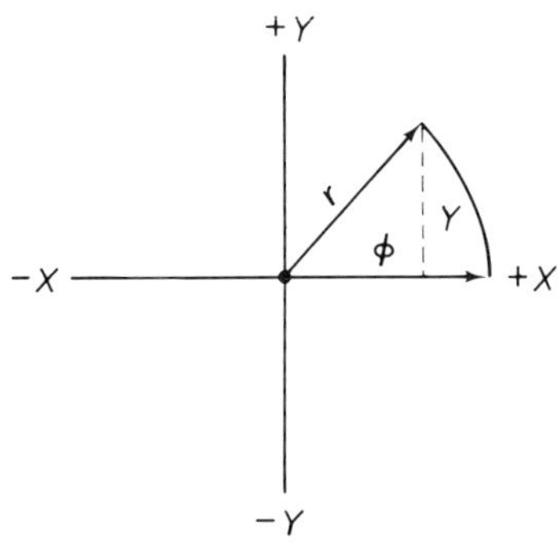

Figure 8-5. Plotting Vertical Amplitude

The previous formula describes the motion of any vector of length r which rotates about a point with uniform direction and velocity starting from a horizontal position at time zero. In other words, the above describes any sinusoidal motion.

If rotation were to start from a vertical position at time zero, the wave shape would be the same, but with a 90° displacement at any given time. This would be called a *cosine wave,* and Figure 8-6 compares it to the sine wave.

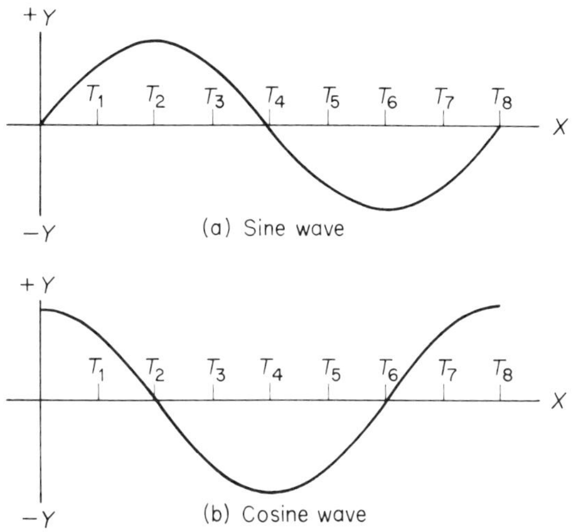

Figure 8-6. Comparison of Sine and Cosine Waves

Since the sine wave was plotted from the sine of ωt, it stands to reason that the cosine wave will be plotted from the cosine of ωt. Therefore, the amplitude of the cosine wave is determined by $Y = r \cos \theta = r \cos \omega t$. Obviously, the only difference between the above waves is their 90° displacement with respect to one another.

Earlier, we spoke of peak amplitude and instantaneous amplitude of a sine wave. There are several other values of ac amplitude that we should be familiar with. The *average amplitude* over a complete period is zero. The positive alternation exactly cancels the negative alternation. But the average value during a given alternation can be useful; it is the sum of a number of instantaneous values divided by the number of values considered. A very accurate average value could be obtained by taking the instantaneous value at each degree point for the 180° alternation, adding the values, and dividing the sum by 180. This is partially shown in Figure 8-7.

Here we have used a peak value of ten and taken four instantaneous values. According to this, the average value is:

$$E_{avg} = \frac{0 + 7.07 + 10 + 7.07}{4} = 6.04V$$

This would mean that for any given alternation, the average value would be: $E_{avg} = 0.604 \times E_{peak}$.

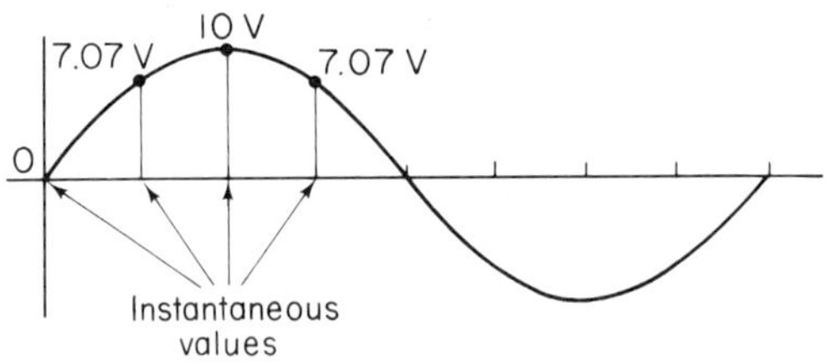

Figure 8-7. Determining Average Value

Since we only considered four instantaneous values, our figures could not have been very accurate. The accuracy would continue to improve as more and more instantaneous values are included in the calculation. The acceptable accuracy is: $E_{avg} = 0.637 \times E_{peak}$. This means that a sine wave with a 100-volt peak (maximum) value would have an average value during each alternation of: 0.637×100 which is 63.7 volts.

Another ac value frequently encountered is *effective value.* Since the voltage is constantly changing in amplitude, the circuit does not receive full benefit from the peak value of the applied voltage. The effective value is determined by the heating effect of ac as compared to that of dc. For example, a sine wave of current that produces the same heating effect as 10 amperes of dc has an effective value of 10 amperes. For both current and voltage, this value is computed by averaging several equally spaced instantaneous values and extracting the square root of the average value. The value is frequently referred to as *root-mean-square* because of this method of computation. The exact effective value is expressed as: $\text{eff} = \text{peak}/\sqrt{2}$.

For practical applications, a slightly less accurate figure is acceptable: eff = 0.707 peak. Thus, an alternating voltage with a peak value of 100 volts has an effective value of 70.7 volts.

$$E_{eff} = 0.707 \times E_{peak} = 0.707 \times 100 = 70.7 \text{ volts}$$

By the same token, an alternating current with a peak value of 10 amperes has an effective value of 7.07 amperes.

$$I_{eff} = 0.707 \times I_{peak} = 0.707 \times 10 = 7.07 \text{ amperes}$$

This amount of alternating current will produce the same heating effect as 7.07 amperes of direct current.

The effective value of alternating voltages and currents is encountered so frequently that if *any other* value is intended, it must be so *specified.* Thus, 120V ac means *120V effective;* 10 amperes ac means *10 amperes effective.* In any ac notation, when value is not specified, it is interpreted to be effective value.

Still another ac value that is sometimes used is *peak to peak.* This is the total vertical displacement from peak positive to peak negative, or: peak to peak = 2 peak.

The peak value of a sine wave is the same as its *maximum* value. In this book, when dealing with ac values, peak and maximum will be used interchangeably. Figure 8-8 illustrates these ac values.

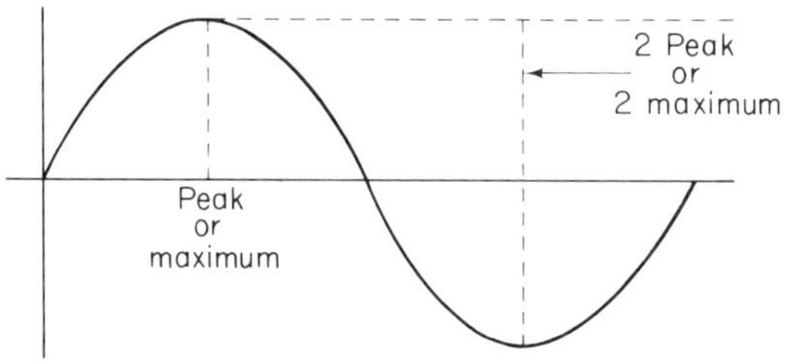

Figure 8-8. Peak and Peak to Peak Values

phase

In Figure 8-6, we saw two wave shapes of the same frequency with a 90° angular displacement with respect to one another. The number of degrees of angular displacement between two wave shapes is called the *phase relationship* of the two waves. If we plot the two waves just mentioned on a common time base, they appear as in Figure 8-9.

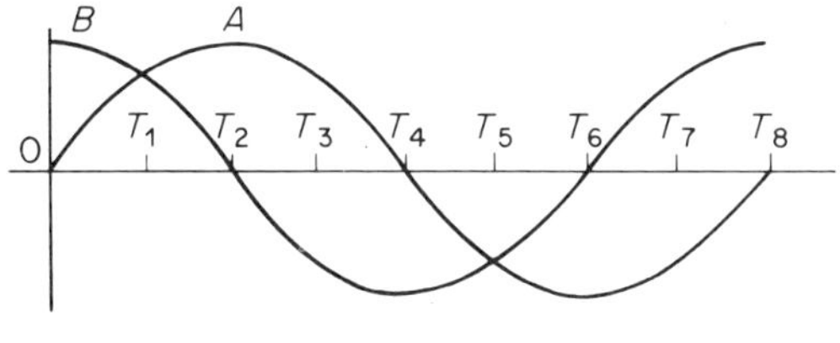

Figure 8-9. Phase Relationship

Reading the wave shapes from left to right, with the increase of time we can see that wave shape *B* passes any selected point 90° *ahead* of wave shape *A.* At time zero, *B* is maximum (90°), while *A* is zero (0°). At time T_1, *B* has passed its peak and decreased in value while advancing to 135°. During this same time (0 to T_1), *A* has increased in value and moved from 0° to 45°. At T_2, *B* is passing through zero, while *A* is passing its positive peak. This same relationship continues on and on. We say that these waves are 90° *out of phase* with one another. In fact, these two wave shapes could be produced by two loops on a generator as shown in Figure 8-10.

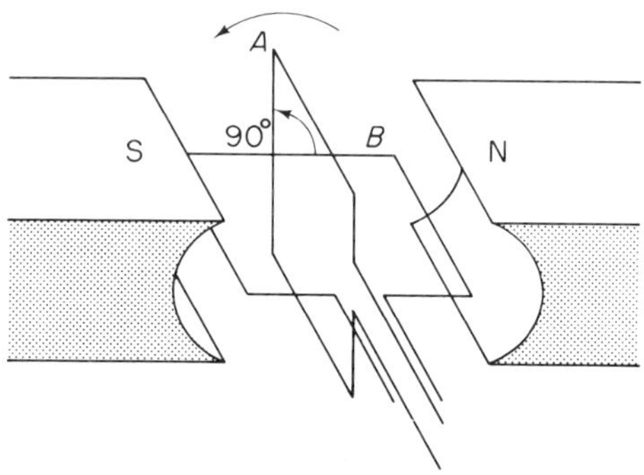

Figure 8-10. Generation of Wave Shapes that are 90° Out of Phase

Assuming that the generator loops are in the start position, a 360° rotation would produce both wave shapes of Figure 8-9, wave shape *A* from loop *A*, and wave shape *B* from loop *B*.

Notice that loops *A* and *B* are fixed with respect to one another. Therefore, the wave shapes will maintain the same fixed phase relationship. Since loop *B* is always ahead of loop *A* by 90°, we say that wave shape *B* is *leading* wave shape *A* by a 90° phase angle. Another way to express the same idea is to say that *A* is *lagging B* by 90°. Both expressions are common, and both accurately describe the phase relationship of the two wave shapes.

The 90° phase difference used above was selected for illustration of phase differences because it is frequently encountered in voltage-current calculations. Actually, two wave shapes of the same frequency may be out of phase with one another by any angle up to 180°. The 180° phase difference is shown in Figure 8-11. At any

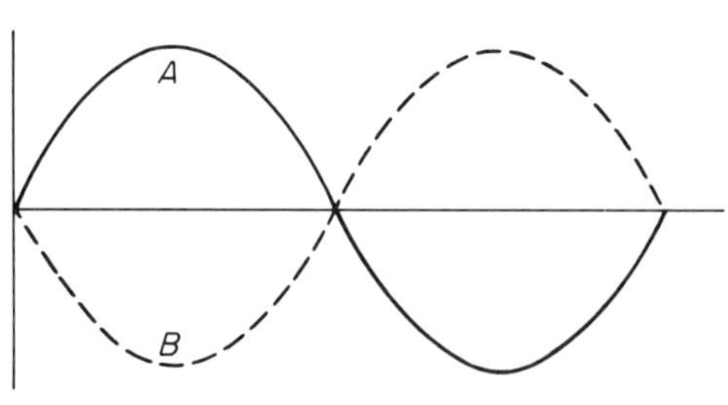

Figure 8-11. 180° Phase Difference

point in time, there is exactly a 180° difference between the angular position of the two wave shapes. It is impossible to say which is leading and which is lagging in such a case. Moving either wave shape from this point in either direction brings the wave shapes closer together in phase, while placing one in a leading and the other in a lagging position.

Since ac is in continuous motion and every motion produces a new angle, we need a refresher exercise on the functions of angles.

ANGULAR FUNCTIONS

Most ac circuit problems involve the solution of one or more triangles. Being adept at these solutions is so important that we must take time and space for a short review.

types of angles

An *angle* is formed when two straight lines intersect. If equal angles are formed by the intersection, all four angles are *right angles.* A right angle is a 90° angle. An angle that is less than 90° is an *acute angle.* An angle greater than 90° is an *obtuse angle.* Two angles whose sum is 90° are *complementary angles.* Two angles whose sum is 180° are *supplementary angles.* Figure 8-12 illustrates the types of angles just described.

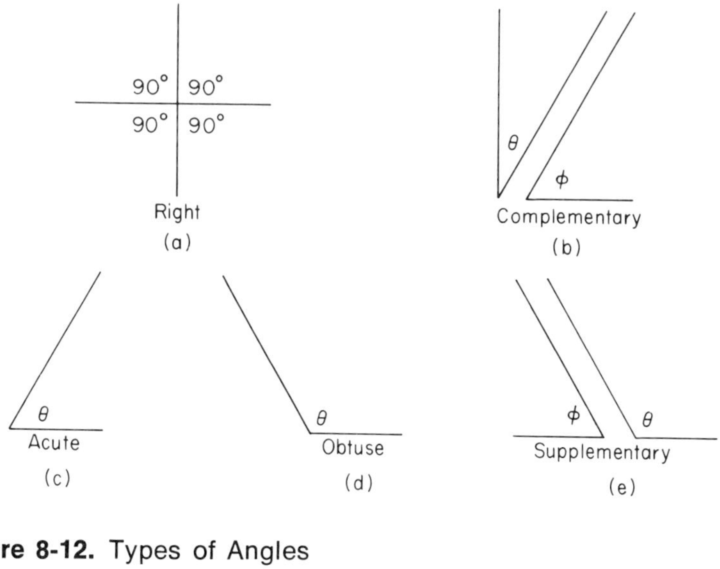

Figure 8-12. Types of Angles

As previously stated, an angle formed by a vector that is rotating clockwise is a negative angle. The counterclockwise rotation of a vector generates a positive angle. This is illustrated in Figure 8-13.

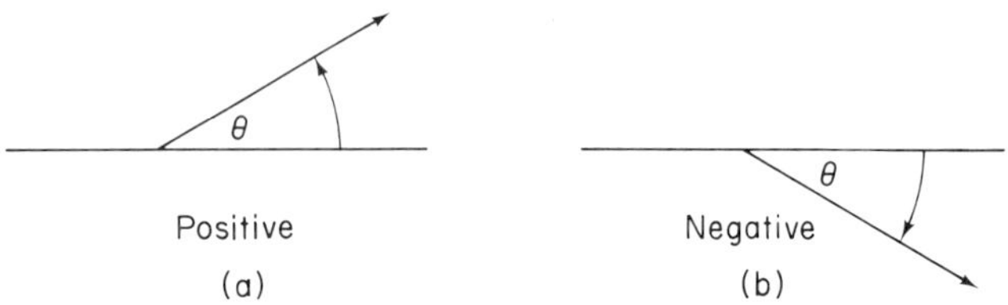

Figure 8-13. Positive and Negative Angles

The pivot point of the rotating vector is the *vertex* of the angle generated. The initial side of the angle is the original position of the vector. The terminal side is the vector's final position. This is shown in Figure 8-14.

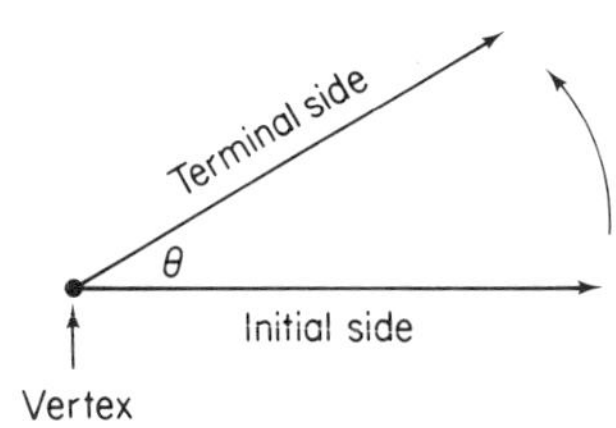

Figure 8-14. Parts of an Angle

triangles

When a vertical line is dropped from the tip of the rotating vector to the horizontal axis, a triangle is formed. Since one of the angles in this triangle is 90°, the triangle is a right triangle. The total of all angles in a right triangle must be 180°. Therefore, this triangle contains a right angle and two complementary angles. One of these acute (complementary) angles is the angular distance traveled by the vector. The complement of this angle is the difference in degrees between it and 90°. These points are illustrated in Figure 8-15.

Vector Z has rotated counterclockwise and created a positive angle θ, which is 60°. Vertical line X forms a right angle with starting position R. Angles θ and Φ are complements of one another;

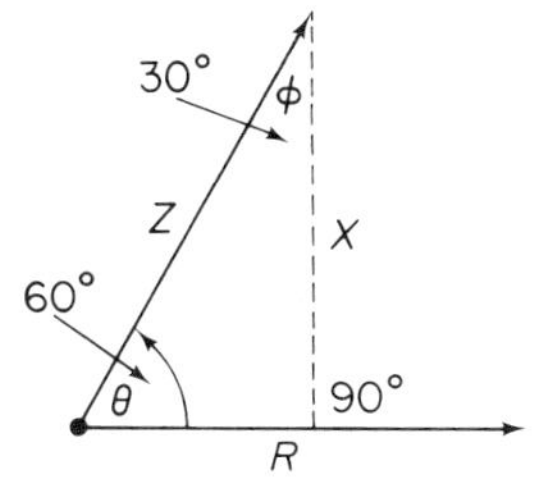

Figure 8-15. Right Triangle

their sum must be 90°. Therefore, $\Phi = 90° - \theta$. Since θ is 60°, Φ must be 30°.

The *three sides* of the triangle are named *in respect to the acute angle that we are dealing with.* The long side is adjacent to both angles and is always called the *hypotenuse.* The side which intersects the hypotenuse to form the acute angle being considered is the *adjacent side* for that angle. The remaining side is the *opposite side.* Figure 8-16 names the sides with respect to both acute angles.

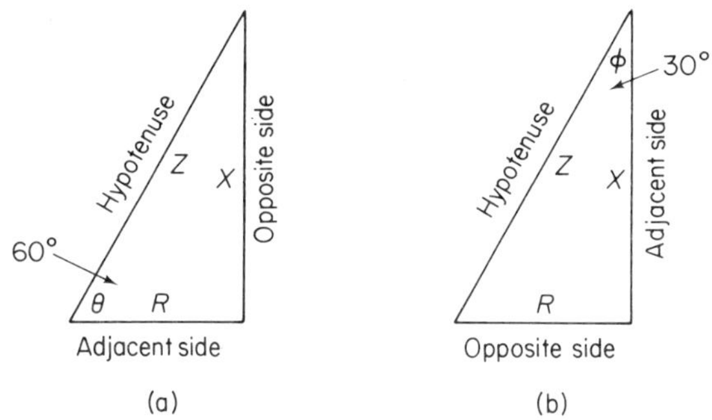

Figure 8-16. Sides in Respect to θ and Φ

In triangle (a), we are dealing with angle θ. *Z* is the hypotenuse; *R* is the adjacent side; and *X* is the opposite side. In triangle (b), we are dealing with angle Φ. *Z* is still the hypotenuse, but now *X* is the adjacent side; and *R* is the opposite side.

trigonometric functions

A trigonometric function is a *ratio between any two sides* of the triangle. Trigonometry includes all types of triangles, but the discussion here will be limited to the triangle that is most often encountered in ac calculations, the right triangle.

Since there are three sides, and any two form a function, there are *six* trigonometric functions (or ratios) with respect to each angle. Using the acute angle θ as an example, the functions are:

$$\text{Sine (sin) } \theta = \frac{\text{opposite side}}{\text{hypotenuse}}$$

$$\text{Cosine (cos) } \theta = \frac{\text{adjacent side}}{\text{hypotenuse}}$$

$$\text{Tangent (tan) } \theta = \frac{\text{opposite side}}{\text{adjacent side}}$$

$$\text{Cotangent (cot) } \theta = \frac{\text{adjacent side}}{\text{opposite side}}$$

$$\text{Secant (sec) } \theta = \frac{\text{hypotenuse}}{\text{adjacent side}}$$

$$\text{Cosecant (csc) } \theta = \frac{\text{hypotenuse}}{\text{opposite side}}$$

Example 1 (Figure 8-17): Calculate the six functions of angle θ.

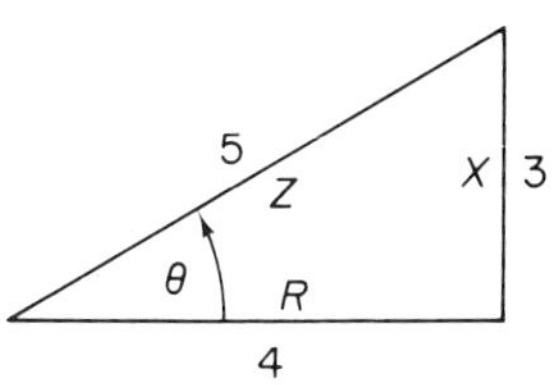

Figure 8-17. Calculate the Functions

$$\sin\theta = \frac{X}{Z} = \frac{3}{5} = 0.6$$

$$\cos\theta = \frac{R}{Z} = \frac{4}{5} = 0.8$$

$$\tan\theta = \frac{X}{R} = \frac{3}{4} = 0.75$$

$$\cot\theta = \frac{R}{X} = \frac{4}{3} = 1.333$$

$$\sec\theta = \frac{Z}{R} = \frac{5}{4} = 1.25$$

$$\csc\theta = \frac{Z}{X} = \frac{5}{3} = 1.667$$

In many cases, we know the values of the function, but we must determine the angle. This is done using an *inverse trigonometric function.* Inverse functions are designated by putting the word *arc* before the name of the function. For example, the sine of 30° is 0.5. Therefore, *arc sin* 0.5 is 30°. In plainer talk, this says that the arc (or angle) which has a sine of 0.5 is 30°.

Although, there are other methods of solving right triangles, very few are as flexible as the use of trigonometric functions. It is strongly recommended that you take time to *memorize* the first three; sine, cosine, and tangent. Once you have accomplished this, you can remember all six because of their reciprocal relationships. The following paired functions are reciprocals of one another: sine and cosecant; cosine and secant; tangent and cotangent.

The functions of the two complementary angles in a right triangle also hold definite relations to one another. Let's consider the other acute angle of the triangle in Figure 8-17. This time we are dealing with angle Φ as shown in Figure 8-18.

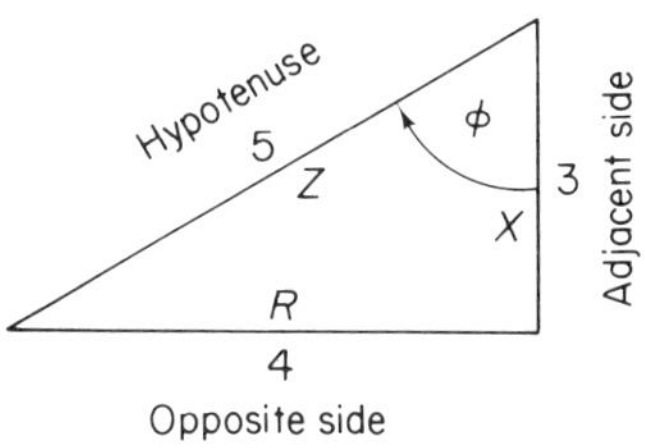

Figure 8-18. Functions of an Angle Φ

Example 2: Calculate the six functions as we did before:

$$\sin \Phi = \frac{R}{Z} = \frac{4}{5} = 0.8$$

$$\cos \Phi = \frac{X}{Z} = \frac{3}{5} = 0.6$$

$$\tan \Phi = \frac{R}{X} = \frac{4}{3} = 1.333$$

$$\cot \Phi = \frac{X}{R} = \frac{3}{4} = 0.75$$

$$\sec \Phi = \frac{Z}{X} = \frac{5}{3} = 1.667$$

$$\csc \Phi = \frac{Z}{R} = \frac{5}{4} = 1.25$$

Careful comparison of the functions of angle θ and angle Φ will reveal the following relations:

1. The *sine* of either angle is the *cosine* of the other.
2. The *tangent* of either angle is the *cotangent* of the other.
3. The *secant* of either angle is the *cosecant* of the other.

functions of any angle

From this point on, we will limit our discussion of functions to the sine, cosine, and tangent, but we need not limit our thinking on the size of the angle. Our problems, for the most part, will involve right triangles, but we *are not* limited to a 90° segment of the circle. Since our angles may be of any size, and either positive or negative, we need a procedure for reducing our calculations to a right triangle. The procedure is very simple. It is a matter of determining where the terminal side of the angle is located and dropping a vertical line from the vector tip to the horizontal axis. But first we divide the circle into four quadrants as shown in Figure 8-19.

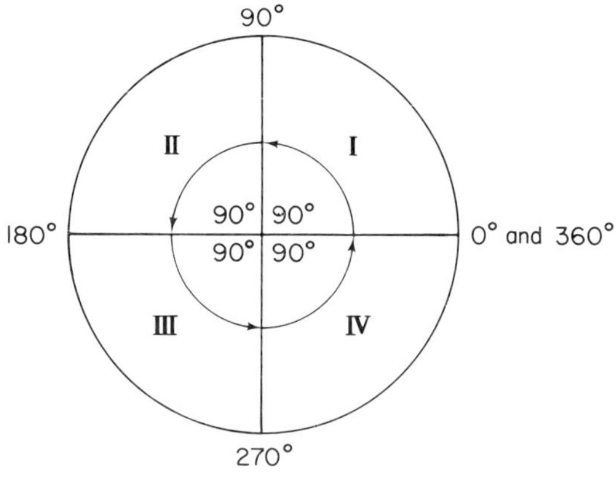

Figure 8-19. Quadrants of a Circle

In order to determine which *quadrant* the terminal side of the vector is in, we subtract the largest multiple of 360° from the total angle. For example, the angle 5422° is in quadrant I. The vector made 15 complete revolutions and 22°, 5422° − 5400° = 22°.

Once you have located the vector within a quadrant, drop a

vertical line from the tip of the vector to the horizontal axis and solve the resulting right triangle. For the previous example, the triangle is shown in Figure 8-20.

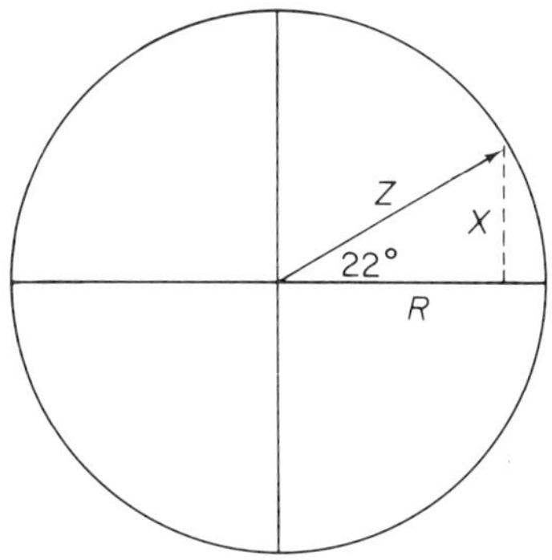

Figure 8-20. 5422° Right Triangle

Triangles for angles that terminate in each quadrant are shown in Figure 8-21.

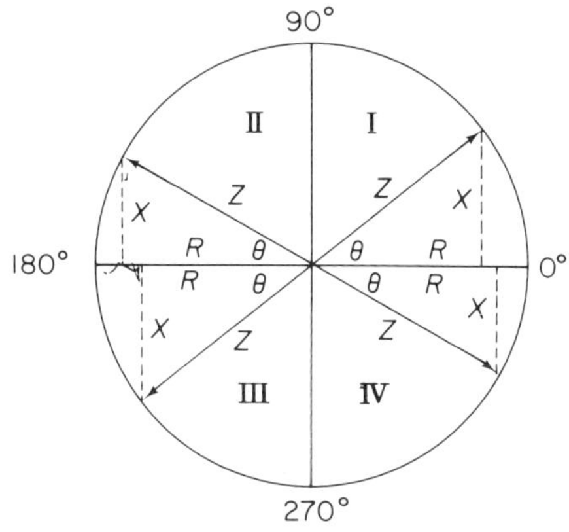

Figure 8-21. Triangles in Each Quadrant

The function of an angle is either positive or negative, and this is determined by the *quadrant* in which the *terminal side* of the angle falls. All functions (sine, cosine, and tangent) are positive in quadrant I. In quadrant II sine is positive, but cosine and tangent are negative. In quadrant III tangent is positive, but the sine and cosine are negative. In quadrant IV the cosine is positive, but the sine and tangent are negative. Figure 8-22 will help you to remember these polarities of functions.

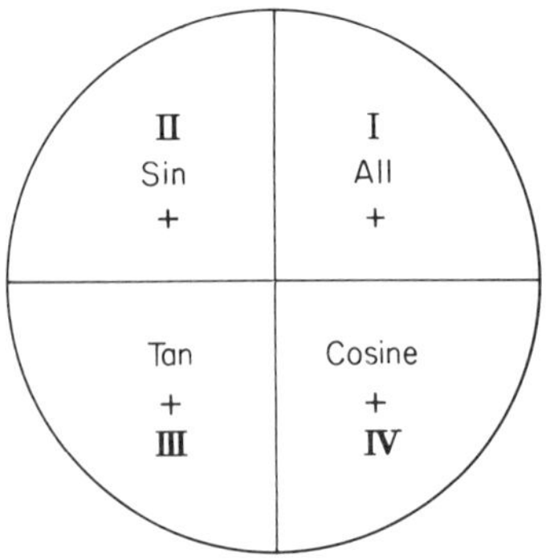

Figure 8-22. Signs of Functions by Quadrant

In mathematical notation, an angle is designated by this mark: $\angle$ which is read *"at an angle of."* Some examples are: $\angle$ 30°, $\angle$ 45°, $\angle$ 12°, $\angle$ —47.3°.

Pythagorean theorem

Many times in circuit problems, the angles are critical. At other times, the amplitudes are critical, and the angles of little importance. The Pythagorean theorem deals with the ratios between the sides without regard to the angles. With the sides designated as *X, R,* and *Z,* as we did previously, the theorem looks like this: $Z^2 = X^2 + R^2$. Resolving this into a more general statement, we find that the square of the hypotenuse is equal to the sum of the squares of the other two sides. The formula generally appears as: $c^2 = a^2 + b^2$.

Example 1 (Figure 8-23): Solve for the hypotenuse.

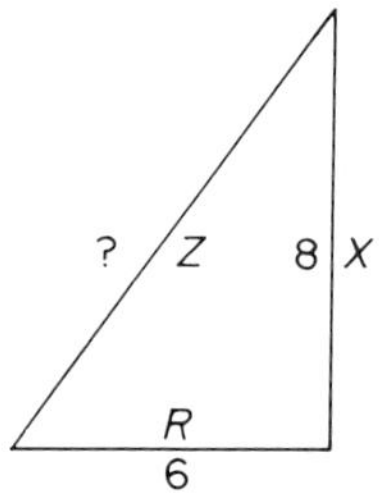

Figure 8-23. Solve for Hypotenuse

Solution:

$$Z^2 = X^2 + R^2 \quad Z = \sqrt{X^2 + R^2} = \sqrt{(8 \times 8) + (6 \times 6)} = \sqrt{100} = 10$$

Example 2 (Figure 8-24): What is the value of X?

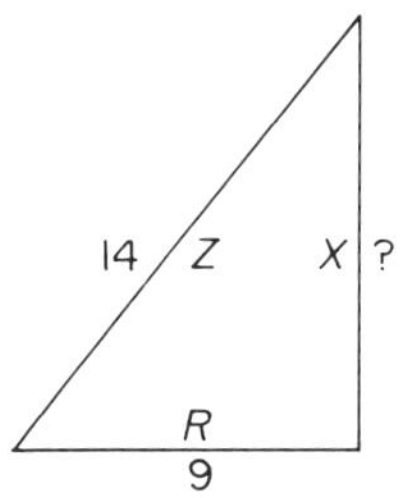

Figure 8-24. Solve for X

$$X^2 = Z^2 - R^2 \quad X = \sqrt{Z^2 - R^2} = \sqrt{(14 \times 14) - (9 \times 9)} = \sqrt{115} = 10.7$$

In these last two examples, we had no angles given, but there are two ways to determine these angles.

1. Draw the triangles to scale and measure the angles with a protractor.
2. Take them from a table of trigonometry functions.

In estimating the size of the complementary angles, we know that their sum must be 90°. We know that the small angle is opposite the small side, and the large angle is opposite the large side. If both sides are the same length, the angles are the same size and each is equal to 45°. But where there is even a slight difference in the lengths of the sides, one angle is less than 45°, and the other is more than 45°.

The tables of trigonometric functions in the appendix show only angles from 0° to 90°, which means that you must remember to affix the proper sign (+ or —) according to the quadrant that the angle is in. The numerical value of the function is in the table, but the sign is not. Sometimes the exact numerical value of a function will not be given in the table, but you should have no trouble locating it to the nearest tenth of a degree.

solving right triangles

Let's use both geometry and trigonometry to solve a few right triangles.

Example 1 (Figure 8-25): Solve for X, Z, and $\angle\ \theta$.

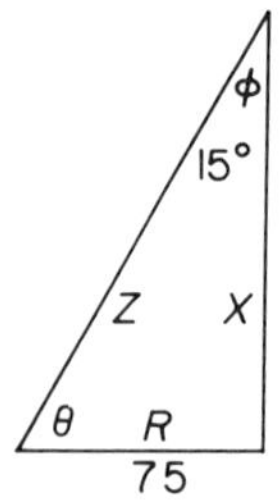

Figure 8-25. Calculate X, Z and $\angle\ \theta$

Angle θ is easy to determine: $\Phi + \theta = 90°$.

$$\theta = 90° - \Phi = 90° - 15° = 75°$$

Now solve for X.

$$X = R \tan\ \theta = (75)(\tan\ 75°) = (75)(3.7321) = 280$$

Now solve for Z.

$$Z = X/\sin\ \theta = 280/0.9659 = 290$$

After finding the value of Z, we could have reverted to the Pythagorean formula:

$$Z^2 = X^2 + R^2 \quad Z = \sqrt{X^2 + R^2} = \sqrt{(75 \times 75) + (280 \times 280)}$$
$$= \sqrt{84{,}025} = 290$$

Example 2 (Figure 8-26): Solve for R, Z, and $\angle\ \Phi$.

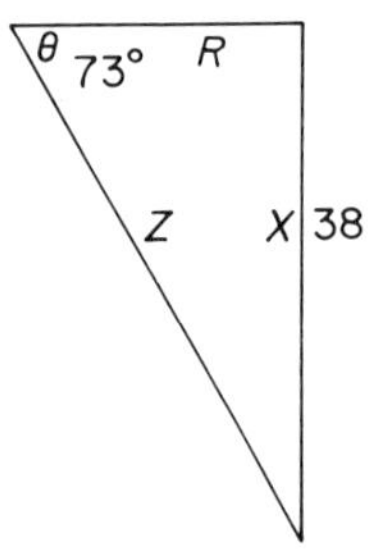

Figure 8-26. Calculate R, Z and $\angle\ \Phi$

$$\Phi + \theta = 90°;\ \Phi = 90° - \theta = 90° - 73° = 17°$$

$$Z = \frac{X}{\sin\ \theta} = \frac{38}{0.9563} = 39.7$$

$$R = Z \cos\ \theta = (39.7)(0.2924) = 11.6$$

Example 3 (Figure 8-27): Solve for both sides.

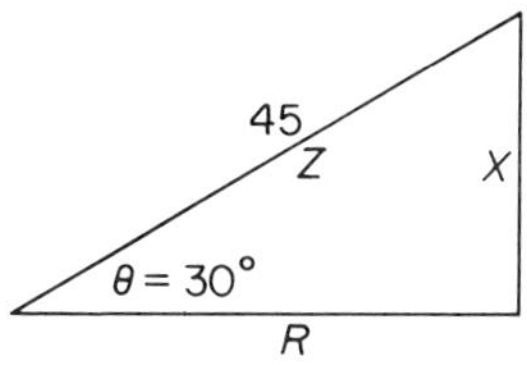

Figure 8-27. Calculate *R* and *X*

$$R = Z \cos \theta = (45)(0.866) = 39$$
$$X = R \tan \theta = (39)(0.5774) = 22.5$$

VECTORIAL FUNCTIONS

In much of our daily lives we encounter quantities that are fully described by a simple number. Such quantities are miles per hour, pages in a book, feet in a yard, and gallons of liquid in a tank. These quantities have *magnitude only.* When the magnitude is expressed, the quantity is fully described. A quantity that can be described by a simple number is a *scalar quantity.*

Since our introduction to ac, we have been dealing with quantities that have both *magnitude and direction.* It is not satisfactory to say that we have 10 amperes of current. We must specify the direction of that 10 amperes of current. In dc, current direction meant flow from negative to positive. An alternating current will have an *angular direction,* which is caused by the components of the circuit. A specified amount of opposition, such as 30 ohms, does not give the complete picture. Opposition to ac has direction, as well as magnitude. Such quantities (i.e., those with magnitude and direction) are called *vector quantities,* and they must be expressed in *complex numbers.*

complex numbers

Complex numbers are no more complex than any other numbers. This is simply a name to distinguish them from numbers that express quantity only. For instance, a vector of 10 units at an angle of 45° is expressed as $10 \angle 45°$. This vector quantity has a horizontal component and a vertical component. The horizontal component is 7.07, and the vertical component is 7.07. The proper notation for this is: 7.07 + j7.07. This is a complex number. 7.07 + j7.07 is actually giving the direction of two vectors that are each 7.07 units in length. The horizontal component is positive, which places that vector at 0°. The vertical component is also positive, which places it at 90°.

Since the vectors are in different directions, one can add them only by taking their resultant. The resultant is 10 at an angle of 45°. This is illustrated in Figure 8-28.

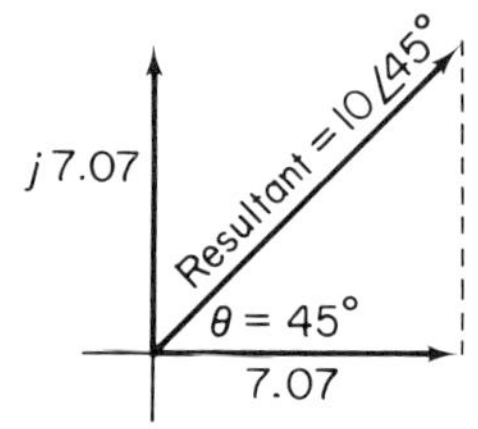

Figure 8-28. Plotting a Complex Number

If you walk six blocks east and, from that point, walk six blocks north, how far and in which direction are you from your starting point?

The distance you traveled expressed in complex numbers is 6 + j6. The resultant of these two vectors places you 8.5 blocks northeast (or 8.5 ∠ 45°). This is illustrated in Figure 8-29.

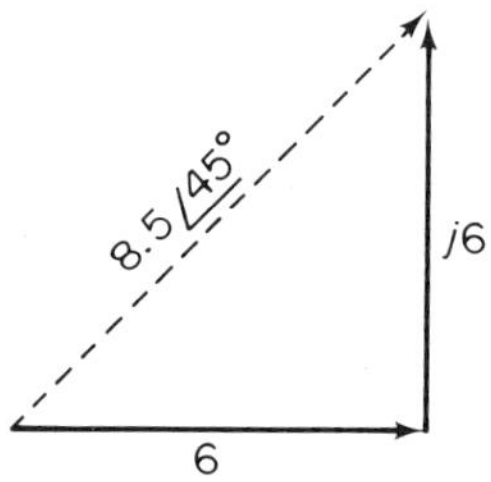

Figure 8-29. Resultant of Two Vectors

In the past few paragraphs, we have shown two methods of denoting vectors and thrown in a notation which mathematicians call an *imaginary quantity.* A unit with a specified direction such as 8.5 ∠ 45° is in the *polar system* of vector notation. The complex number, such as 6 + j6, is in the *rectangular system* of vector notation. The rectangular system specifies *two vectors* each of which has a different direction. The polar system provides a means of expressing the *resultant* of the two vectors.

The *imaginary quantity* is the j factor. It is so named because it is derived from the square root of a negative one. Obviously, there is no such thing as a number that represents the square root of -1. Therefore, it was called an imaginary number and was designated as the i factor. In electricity, the i factor was changed to the j factor in order to avoid confusion with the formulas that use i for current. $j = \sqrt{-1}$, and for that reason it is imaginary. The quantity may be imaginary, but its functions in electrical problems are very real.

Since by definition $j = \sqrt{-1}$, these powers of j may be derived: $j^2 = -1$; $j^3 = -j$; and $j^4 = +1$. Therefore, multiplying a number by a j factor places that number at a specific point in respect to 0°. $j5 = 5$ at 90°; $-j5 = 5$ at $-90°$; $j^2 5 = 5$ at 180°; $j^4 5 = 5$ at 0° or 360°. So, you see, multiplying a vector by a j factor is equivalent to rotating the vector in steps of 90°. This is shown in Figure 8-30.

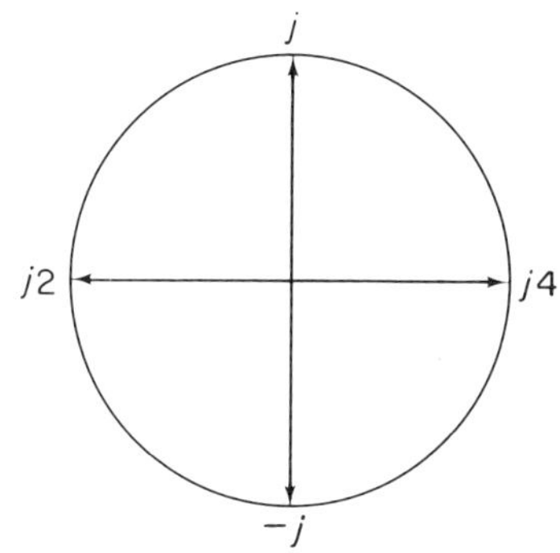

Figure 8-30. Position of Vectors Multiplied by *J* Factors

The majority of our circuit problems will fall in either quadrant I or quadrant IV. Therefore, we are primarily concerned with j and $-j$. Remember that j times a quantity places that quantity at 90°, and $-j$ times a quantity places the quantity at $-90°$.

vector algebra

Quantities may be *added or subtracted in rectangular* form by applying the *ordinary laws* of algebra.

Addition:

$$\text{(A)}\quad \begin{array}{r} 3 + j4 \\ 6 + j2 \\ \hline 9 + j6 \end{array} \qquad \text{(B)}\quad \begin{array}{r} 5 - j9 \\ 4 + j3 \\ \hline 9 - j6 \end{array} \qquad \text{(C)}\quad \begin{array}{r} -6 - j2 \\ 3 + j8 \\ \hline -3 + j6 \end{array}$$

Subtraction:

(A) $\begin{array}{r} 3 + j4 \\ (-)6 + j2 \\ \hline -3 + j2 \end{array}$ (B) $\begin{array}{r} 5 - j9 \\ (-)4 + j3 \\ \hline 1 - j12 \end{array}$ (C) $\begin{array}{r} -6 - j2 \\ (-)3 + j8 \\ \hline -9 - j10 \end{array}$

Multiplication and division may be performed on quantities in *polar form* by using the same rules that we used with the powers of ten: multiply the magnitudes and add the angles to obtain a product, or divide the magnitudes and subtract the angles to obtain the quotient. The rules are the same, but the angles have replaced the powers of ten.

Multiplication:

(A) $(5 \angle 45°)(5 \angle -45°) = 25 \angle 0°$

(B) $(4 \angle 36°)(3 \angle 24°) = 12 \angle 60°$

(C) $(8 \angle -14°)(4 \angle 10°) = 32 \angle -4°$

Division:

(A) $\dfrac{25 \angle 60°}{5 \angle 10°} = 5 \angle 50°$

(B) $\dfrac{18 \angle 45°}{9 \angle -45°} = 2 \angle 90°$

(C) $\dfrac{75 \angle 30°}{15 \angle -60°} = 5 \angle 90°$

(D) $\dfrac{27 \angle 0°}{9 \angle 45°} = 3 \angle -45°$

While all mathematical operations can be performed on figures in either form, it is generally easier to convert to polar form for division and multiplication, and rectangular form for addition and subtraction.

conversion of coordinates

Since most ac circuit problems involve right triangles, many of these problems will contain quantities in both polar and rectangular form. Since rectangular form is easy to add and subtract, and polar form is easy to multiply and divide, we need an easy method of converting from one system to the other. Using X as the vertical component, R as the horizontal component, and Z as the resultant, our conversions are as follows.

Rectangular to polar: $Z = \sqrt{X^2 + R^2}$ and $\theta = \arctan X/R$. This actually amounts to *solving a right triangle* with two sides given.

Example 1 (Figure 8-31): What is the polar form for $8 + j6$?

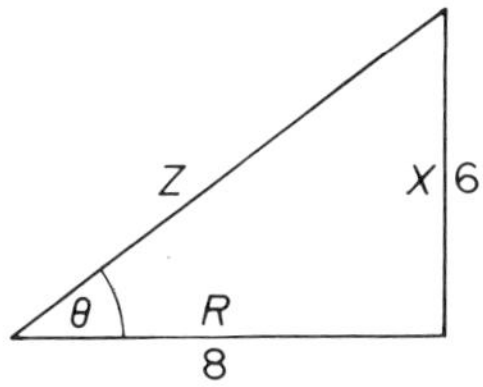

Figure 8-31. Converting to Polar

$$Z = \sqrt{X^2 + R^2} = \sqrt{(6 \times 6) + (8 \times 8)} = \sqrt{100} = 10$$

Solve for θ:

$$\theta = \text{arc tan } X/R = \text{arc tan } 6/8 = \text{arc tan } 0.75 = 36.9°$$

Therefore, $Z = 10 \angle 36.9°$. Polar to rectangular: $R = Z \cos \theta$ and $X = Z \sin \theta$. This is the solution of a right triangle when the hypothenuse, and all angles are given.

Example 2 (Figure 8-32): What are the rectangular coordinates for $10 \angle -53.1°$?

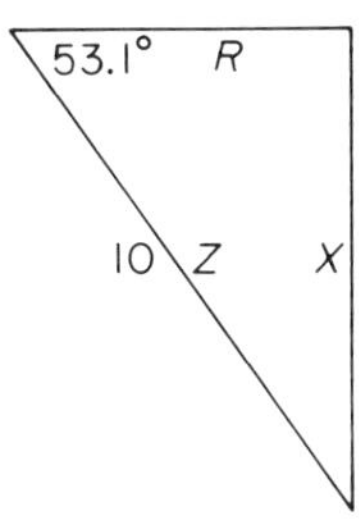

Figure 8-32. Converting to Rectangular Coordinates

Solve for R.

$$R = Z \cos \theta = 10 \cos 53.1° = 10 \times 0.6 = 6$$

Solve for X.

$$X = Z \sin \theta = 10 \sin (-53.1°) = 10\ (-0.7997) = -8$$

Replace the $-$ with a $-j$ and $X = -j8$. Therefore, $10 \angle -53.1° = 6 - j8$.

Example 3 (Figure 8-33): Multiply $(4 + j6)$ by $(6 + j3)$. Construct both triangles as in Figure 8-33.

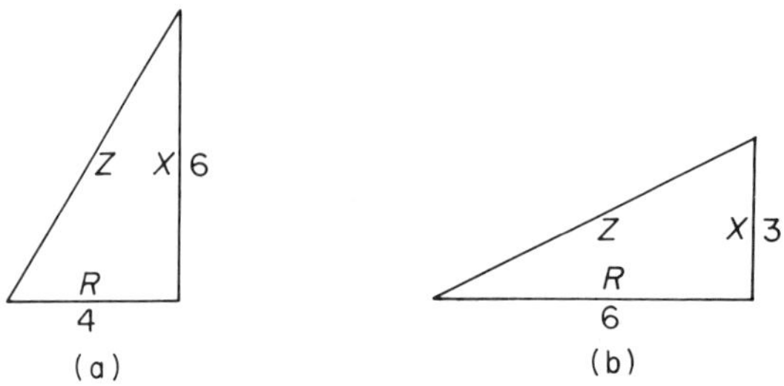

Figure 8-33. Preparing for Conversion to Polar Form

Convert to polar by solving both triangles:

1. $Z = \sqrt{X^2 + R^2} = \sqrt{36 + 16} = 7.2.$
   $\theta = \text{arc tan } X/R = \text{arc tan } 6/4 = \text{arc tan } 1.5 = 56.3°$. Therefore, $Z = 7.2 \angle 56.3°$.
2. $Z = \sqrt{X^2 + R^2} = \sqrt{9 + 36} = 6.7.$
   $\theta = \text{arc tan } 3/6 = 26.6°$. Therefore, $Z = 6.7 \angle 26.6°$.

Multiply the polar coordinates.

$$(7.2 \angle 56.3°)(6.7 \angle 26.6°) = 48.2 \angle 82.9°$$

Example 4 (Figure 8-34): Express the above answer in rectangular form.

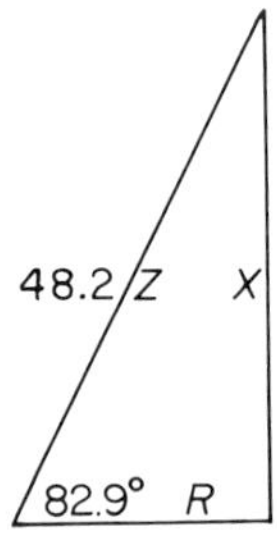

Figure 8-34. Converting Back to Rectangular Coordinates

$$R = Z \cos \theta = (48.2)(0.1236) = 5.95$$
$$X = Z \sin \theta = (48.2)(0.9923) = 47.8$$

Multiply the vertical component by the proper j factor. $X = 5.95 + j47.8$. Therefore, $48.2 \angle 82.9° = 5.95 + j47.8$.

Example 5 (Figure 8-35): Add $20 \angle 36.9°$ to $14 \angle 52°$.

Construct both triangles:

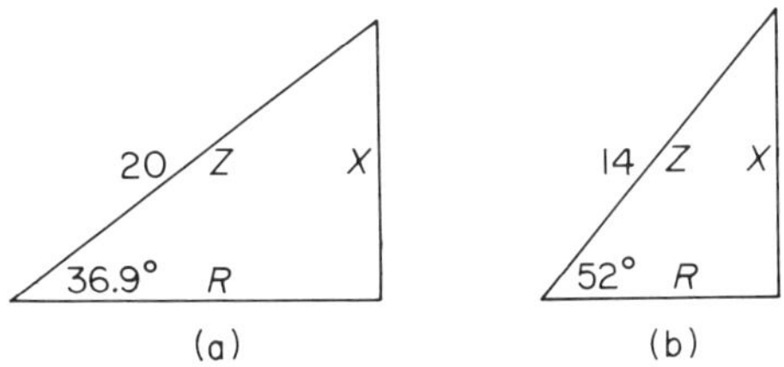

Figure 8-35. Preparing to Convert to Rectangular Coordinates

Solve both triangles.

1. $R = Z \cos \theta = 20(0.7997) = 16$
   $X = Z \sin \theta = 20(0.6) = 12$
   $Z = 16 + j12$
2. $R = Z \cos \theta = 14(0.6157) = 8.62$
   $X = Z \sin \theta = 14(0.788) = 11$
   $Z = 8.62 + j11$

Now add:

$$\begin{array}{r} 16 \quad + j12 \\ 8.62 + j11 \\ \hline 24.62 + j23 \end{array}$$

The mathematics covered here in a very brief fashion will find repeated applications as we go through our analysis of ac circuits.

MEASURING INSTRUMENTS

The ac circuit values are measured with ammeters and voltmeters. These meters are the same as those already studied with the exception of one minor detail: The ac meters have a device called a *rectifier,* which allows current through the movement in only one direction. The rectifier has the same effect on the ac as we observed with the commutator in motors and generators; it changes ac to pulsating dc. In effect, the ac meters are still dc internally, but the rectifier allows ac to be fed into them. The ac ammeter is calibrated to indicate the *effective* value of alternating current in a circuit. It is connected in series as part of the circuit being tested. The ac voltmeter is calibrated to indicate the *effective* value of the alternating voltage. It is still connected parallel to the component across which voltage drop is being measured.

The ohmmeter is used in circuits after the power has been removed. Therefore, it is used on certain components in all types of circuits. Since the circuit power has been removed, the operation is the same as previously described. There are a few special applications with certain components, but they will be taken up at the time that we discuss the application of the component.

introduction to the oscilloscope

The *oscilloscope* is the *principal* measuring device in ac circuits. It consists of a special tube called a cathode ray tube, associated circuits for driving the tube, controls to manipulate the circuits, and provisions for feeding test signals into the circuits. This oscilloscope is the primary piece of test equipment for all except dc circuits. However, it can be used to measure dc voltage.

The oscilloscope produces a graph of voltage wave shapes by using a stream of electrons to trace visible patterns on a viewing screen. These wave shapes can be stopped on the screen for detailed analysis. They can be measured in time, amplitude, and frequency, and they can be compared to other wave shapes.

Some oscilloscopes have the capability of producing two or more visible wave shapes at the same time. This enables the operator to place a good signal on the screen to use as a test standard. Other signals are then classified as good or bad according to how they compare with the standard signal.

These features make the oscilloscope an indispensable, but fairly expensive tool. The general appearance of an oscilloscope is shown in Figure 8-36. This is a *dual beam oscilloscope,* which features up to four complete pictures simultaneously.

tracing the pattern

The cathode ray tube consists of an *electron gun* and a *deflection system* contained in an evacuated glass envelope. The electron gun fires a stream of electrons toward the phosphorescent screen on the inside face of the tube. The deflection system controls and shapes the stream of electrons.

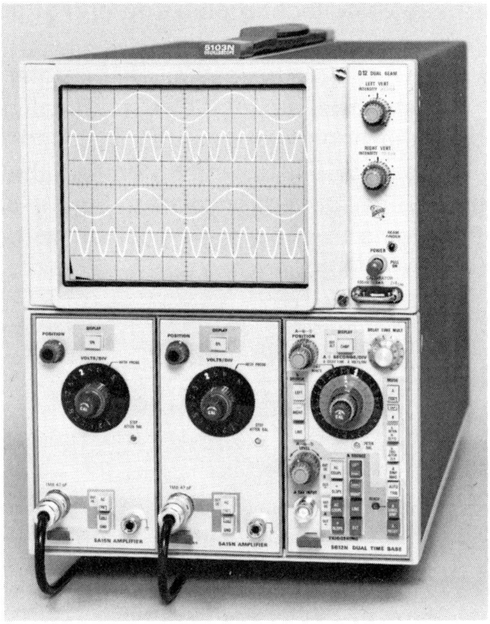

Figure 8-36. A Dual Beam Oscilloscope (Photo courtesy of Tektronix, Inc., Portland, Oregon)

The narrow stream of high velocity electrons is shaped and focused into a beam as it strikes the phosphorescent material which covers the glass on the inside of the screen. The material gives off light at the point where it is touched by the beam. The deflection system moves this beam about on the screen according to the instantaneous values of the voltages being applied. The screen has a high persistency. Several complete wave shapes may be traced by the moving beam before the first portion of the pattern begins to fade. Retracing the same patterns in a periodic fashion keeps the wave shapes bright and clear on the screen for indefinite periods of time.

Cathode ray tubes are utilized in a variety of electronic equipment as well as in the central part of the oscilloscope. Your television picture tube is a cathode ray tube.

the deflection plates

The cathode ray tube (CRT) that is used in the oscilloscope uses electrostatic deflection. Two sets of metallic plates are placed in a position where the electron beam must pass between the plates of each set on its way to the screen. Controllable voltages are placed on these pairs of deflection plates to create an electrostatic field between them.

One set of plates moves the beam left or right (horizontally), while the other set moves it up or down (vertically). The pair of plates that controls the horizontal position is called *horizontal deflection plates.* The pair of plates which controls the vertical position is called the *vertical deflection plates.* Figure 8-37 illustrates a side view of the plates to show their position within the tube.

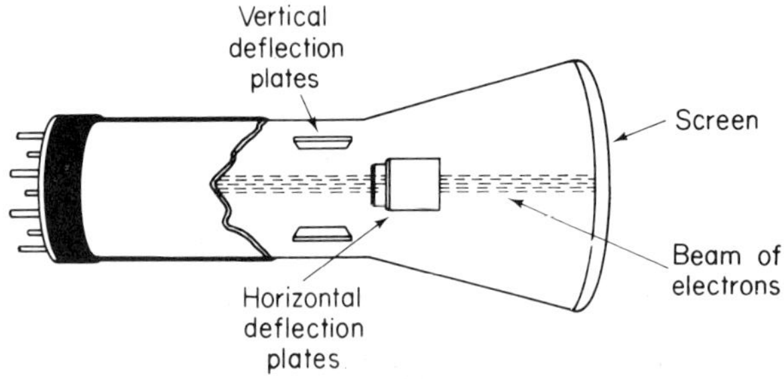

Figure 8-37. Side View of Deflection System

The pairs of plates are physically separated far enough that the electrostatic field between the plates of one pair does not interact with the field between the plates of the other pair. This enables potentials to be placed on the plates to control vertical and horizontal positions independently.

Figure 8-38 represents the deflection plates as viewed from the front of the CRT.

All you need to remember to visualize the functions of the scope is that electrons are negative. They will be attracted to the

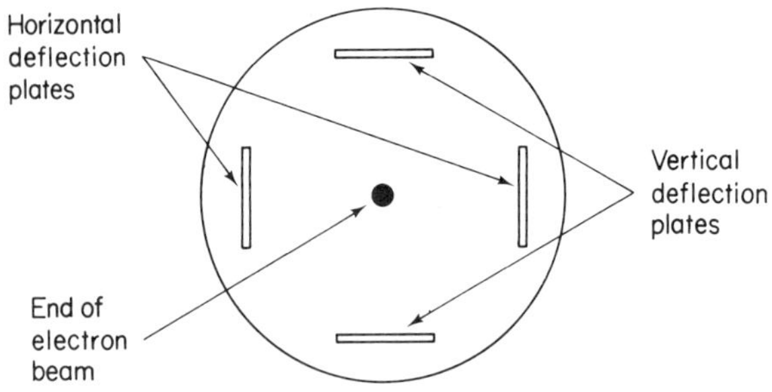

Figure 8-38. Front View of Deflection System

most positive plate and repelled away from the most negative plate. The vertical position of the beam is controlled by the relative potentials on the two vertical plates. The horizontal position of the beam is controlled by the relative potentials on the two horizontal plates. When the two vertical plates have equal potentials, and the two horizontal plates have equal potentials, the beam is centered on the screen as shown in Figure 8-38. The two plates of a pair are connected to the same voltage source so that increasing the potential on one plate decreases the potential on the other.

The individual plates are named according to the *front view* as *left, right, top,* and *bottom*. Varying the potential, so that the left plate is more positive than the right, moves the dot to the viewer's left. Reversing this process, so that the right plate is more positive than the left, moves the dot to the viewer's right. The dot is a small, bright spot showing where the beam is touching the screen. The up and down movements are controlled in the same fashion by varying the potentials on the top and bottom plates.

A certain amount of time is required for the dot to sweep across the screen from left to right. The rate of movement is proportional to the rate of change of the potentials. The horizontal trace then represents time and is called the *time base*. It is, of course, variable.

The vertical position of the dot at any instant in time is proportional to the instantaneous amplitude of the signal applied to the vertical plates. Vertical movement changes with amplitude, while horizontal movement changes with time.

the deflection system

Figure 8-39 is a very basic block diagram of the deflection system.

The *time base generator* provides a sawtooth-shaped voltage wave form for use on the horizontal deflection plates. The sawtooth wave shape looks similar to that in Figure 8-40.

The sawtooth has a linear increase in amplitude from zero to maximum. This is known as *rise time,* and throughout this time, there is a steady increase in potential. This makes amplitude directly proportional to time. The change from maximum amplitude back to zero is almost (but not quite) instantaneous. Notice on the drawing that the drop is almost perpendicular. The time elapsed here is very brief and is called *fall time.* The rise time is frequently called *sweep time* because it is the time used to sweep the dot from left to right across the screen. The fall time is also called *flyback time* because it is the very brief interval of time during which the

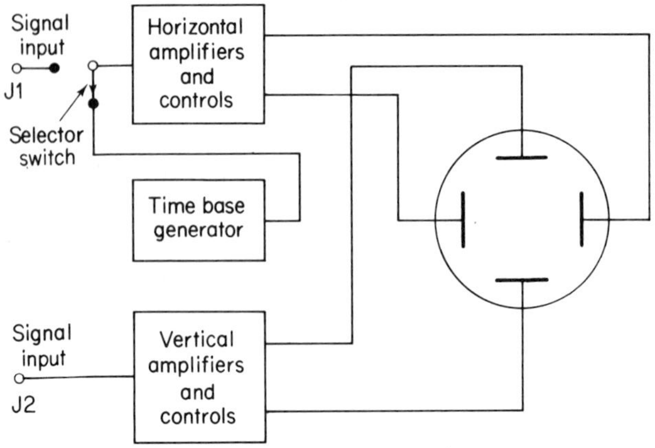

Figure 8-39. Basic Deflection System

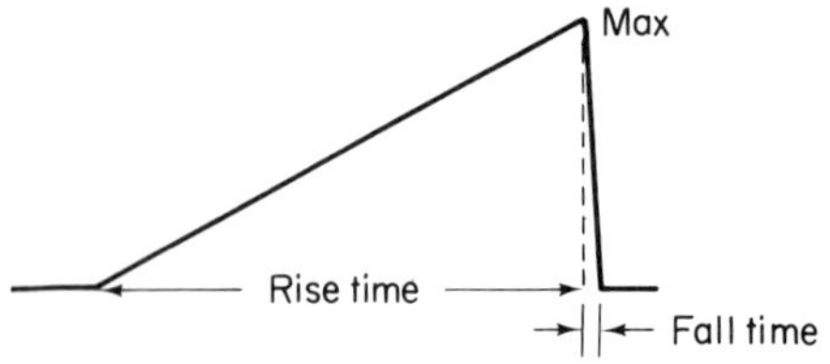

Figure 8-40. One Period of a Sawtooth Voltage

dot moves back from the right side to the left side of the screen. A special feature inside the tube prevents a trace during the flyback time, otherwise there would be interference with the appearance of the wave shapes.

The wave shape in Figure 8-40 is placed on the right deflection plate. The same wave form, after inversion (being turned upside down), is placed on the left plate. The result is a push-pull configuration with the left plate pushing the dot while the right one is pulling it. This action is illustrated in Figure 8-41.

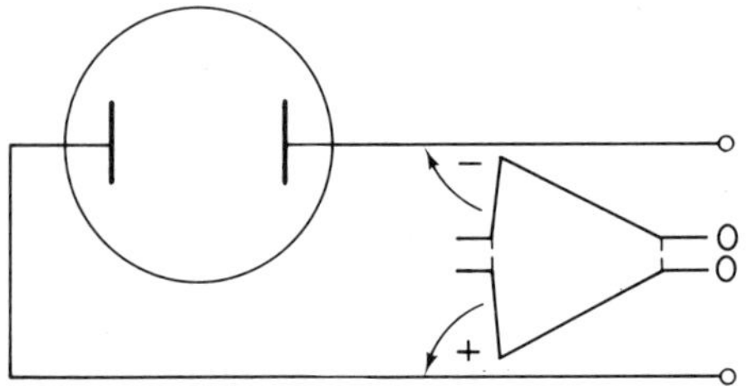

Figure 8-41. Push-Pull Deflection

When a linear sweep is not desired, the selector switch is moved to the signal position. This disconnects the time base generator from the horizontal amplifier and control circuits. In this condition, the sweep voltage is replaced by any type of signal or wave shape that is connected to input jack, J_1.

The vertical amplifiers and control circuits process the input signal for the vertical deflection plates. The signals, or wave shapes, are connected to input jack, J_2. They are processed through the circuits and placed on the vertical plates in a push-pull fashion similar to that of the horizontal sweep voltages.

Another way to express the relationship of voltages on the deflection plates is by the phase relation of the voltages to one another. The voltages on opposite plates are 180° out of phase. Figure 8-42 illustrates sine waves as they are applied to the vertical plates.

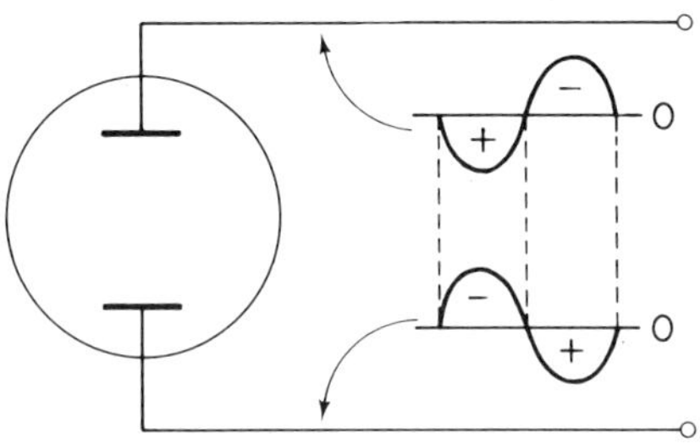

Figure 8-42. Phase Relation on Opposite Plates

When Figures 8-41 and 8-42 are combined, we have the picture shown in Figure 8-43. T_0 indicates that all four of these wave forms start at the same time.

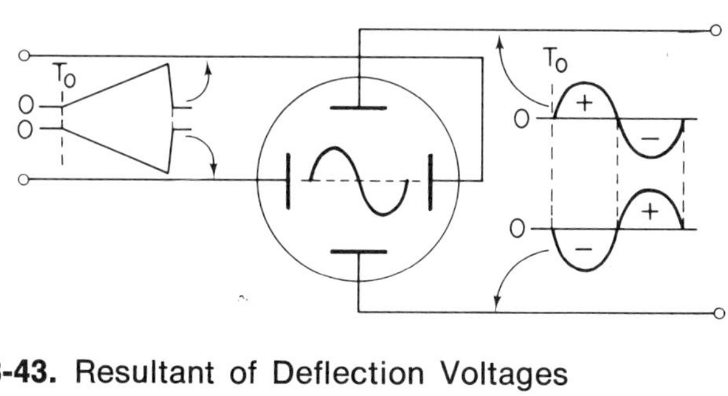

Figure 8-43. Resultant of Deflection Voltages

With the wave shapes as shown, the sawtooth waves move the dot from left to right; quickly back to the left, and then repeat. If there is no potential (or if there are equal potentials) on the vertical plates during this time, the result is a bright line across the center of the screen.

The sine waves on the vertical plates move the dot up and down as their amplitudes change. If the potential on the horizontal plates is zero (or if there are equal potentials) during this time, the result is a bright line running vertically on the center of the screen.

When all plates are activated at the same time, the resultant of all deflections determines the position of the dot at that instant. The sawtooth waves move the dot left to right at the same time that the sine waves move it up and down. The resultant is a trace on the screen identical to the signal, or wave shape, that is connected to the vertical input jack. The mechanics of pattern tracing are illustrated in Figure 8-44.

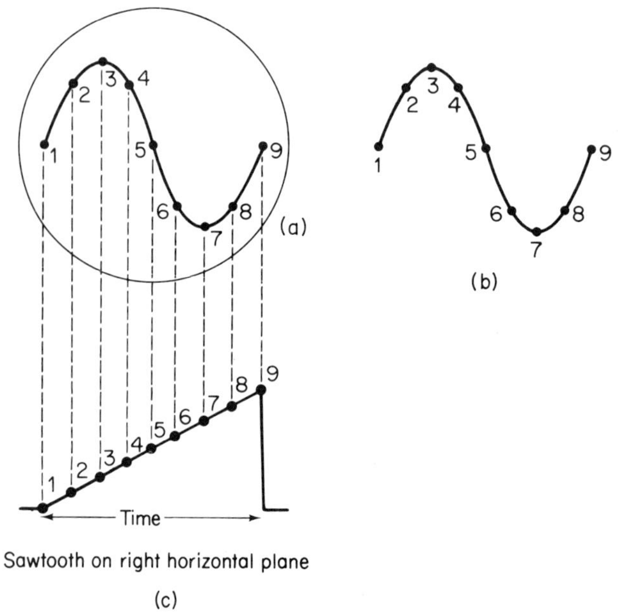

Figure 8-44. Plotting the Pattern

uses of the oscilloscope

The deflection of the dot is proportional to input amplitude. When the oscilloscope is properly calibrated, each division marker on the grid of the screen represents a specific quantity of voltage.

Since dc is a fixed voltage, it will not cause a trace on the screen, but it will deflect the horizontal trace to an off-center position. Direct voltage can be measured by noting the sweep position, applying dc to the vertical input, and counting the marks from the original to the final position of the sweep.

As an ac *voltmeter,* the reading is in any value that the operator desires to use; peak, peak to peak, effective, or average. The sine wave is there to be viewed and analyzed in any fashion that suits the need of the moment. It is generally more accurate to measure peaks and calculate any other desired values from these measurements.

The oscilloscope can be used to measure *frequency.* The horizontal sweep has a known frequency. When the input signal is synchronized (stopped), the number of cycles on the screen is a ratio of vertical signal frequency to horizontal sweep frequency. In Figures 8-43 and 8-44, the ratios are one-to-one, both of the same frequency. If there were three complete cycles on the screens, the ratio would be three-to-one. This would mean that the frequency of the vertical input signal would be three times the sweep frequency.

The oscilloscope can be used to measure *time.* The setting of certain controls determines the time required for one sweep. This enables an accurate *time* measurement between any two points on the screen. Most oscilloscopes have markers that can be placed on the screen to enable a more accurate time count. The space between any two markers represents a specific amount of time. Once either time or frequency has been measured, the other can be determined by $f = 1/t$.

The oscilloscope can be used to measure the *phase-angle* between two wave shapes of the same frequency. With a single trace oscilloscope one does this by placing one wave shape on the horizontal plates and the other on the vertical plates. The shape and position of the resultant pattern give the phase difference.

When two wave shapes are applied to an oscilloscope for the purpose of comparing frequency and phase, some interesting patterns are developed. These patterns are called *lissajous* figures. These figures have as many variations as there are differences between the wave shapes. Figure 8-45 shows a lissajous figure.

Lines *A* and *B,* superimposed on the screen in this fashion, provide a method of determining one frequency when the other is known. Notice that vertical line *A* just touches the edge of each horizontal peak. Line *B* just touches the top of each vertical peak.

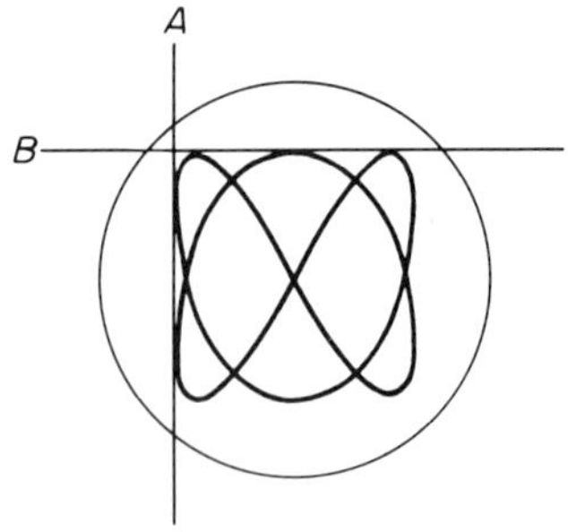

Figure 8-45. Lissajous Figure Showing Frequency Difference

A ratio of the number of times the pattern touches the two lines is a ratio of the frequencies of the two wave shapes.

$$\frac{f_v}{f_h} = \frac{B}{A}$$

Where: f_v is the vertical frequency, f_h is the horizontal frequency, B is the number of times the pattern touches the horizontal line, and A is the number of times the pattern touches the vertical line.

Using values from Figure 8-45, $f_v/f_h = B/A = \frac{3}{2}$ or $2f_v = 3f_h$. If the frequency of either signal is known, calculating the frequency of the other is a simple process. Figure 8-46 shows some sample lissajous figures which indicate the phase relations between signals of the same frequency.

The applied signals are equal both in frequency and in amplitude. All signals are sine waves of voltage. The only difference is the phase angles.

Comparison of phase on the dual beam oscilloscope is a much simpler process. Two traces are obtained on the oscilloscope, and a wave shape is placed on each trace. We then adjust the sweeps until both wave shapes start at the same time. We obtain a static picture and can compare the two signals in any desired fashion—phase, frequency, amplitude, etc. If we desire exact comparisons, we can place markers on the screen and count the number of microseconds of the difference in phase.

This concludes Chapter 8. The following review exercises will reemphasize some of the key points.

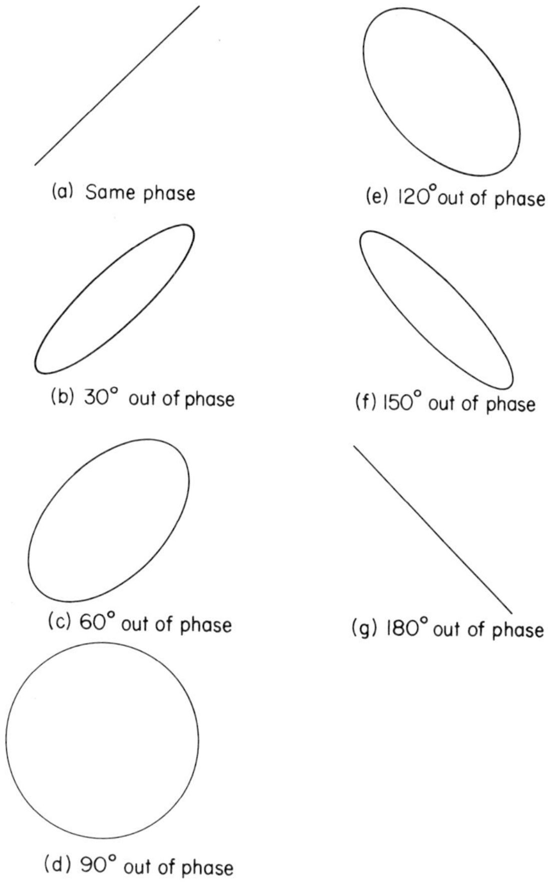

Figure 8-46. Patterns Showing Phase Difference

REVIEW EXERCISES

1. Alternating voltage or current can be described by vectors because their amplitudes, polarities, and frequencies are comparable to the ________, ________, and ________ of a vector.
2. Positive angles are generated when a vector rotates in a ________ direction. Negative angles are created by ________ rotation.

3. The loop in a generator is rotating at 1800 revolutions per minute. What angle will it generate in $4\frac{1}{2}$ seconds?
4. When an angular distance is expressed in degrees, what system of measurement is being used?
5. The circular system of measurement uses the ________ as a unit of measure.
6. In the formula, $C = 2\pi r$, the 2π is derived from the fact that the circle contains ________ ________.
7. One alternation of a sine wave is equivalent to ________ radians.
8. Periods per unit time is ________.
9. One hertz is equivalent to ________ ________ per second.
10. The formula $\omega = 2\pi f$ expresses angular ________ in ________ per second.
11. What is the formula for angular distance in terms of angular velocity and time?
12. What is the wave length of a sine wave that has a frequency of 60 Hz?
13. What is the difference between a sine wave and a cosine wave?
14. What is the phase relation of the sine and cosine waves?
15. What is the maximum number of degrees that two sine waves can be out of phase with each other?
16. The right triangle contains two ________ angles whose sum is ________ degrees.
17. What is the angle of arc sin 0.707?
18. The adjacent side of a triangle is equivalent to the ________ side divided by the ________ of the angle.
19. The sine of one acute angle is equivalent to the ________ of its complementary angle.
20. Sine, cosine, and tangent are all positive in the ________ quadrant.
21. Only the tangent is positive in which quadrant?
22. $Z^2 = X^2 + R^2$ is an expression of the ________ theorem.
23. What is the hypotenuse of a right triangle with one side equal to 4 and the other 6?

24. When the two sides of a triangle are anything other than exactly the same size, the large angle is more than ______ ______, and is opposite the ____________ side.
25. The hypothenuse of a right triangle is 80 at an angle of 30°. What is the length of the: opposite side? adjacent side?
26. A vector quantity contains both ____________ and ______ ______.
27. $10 + j5$ is a ____________ number which expresses a ____________ quantity.
28. What are the horizontal and vertical components of $16 \angle -53°$?
29. j^2 is equivalent to a rotation of ____________ degrees.
30. Express $3 + j4$ in polar form.
31. Addition and subtraction are best performed in ______ ______ form, while multiplication and division should be done in ____________ form.
32. Converting from either system of notation to the other is essentially identical to solving a __________ __________.
33. What is the difference in construction between a dc meter and an ac meter?
34. Ac ammeters and voltmeters are calibrated to indicate ____________ values.
35. The ____________ is the principal measuring device in ac circuits.
36. In addition to measuring amplitudes, the oscilloscope can also be used to measure ____________, ____________, and ____________.
37. The oscilloscope screen is coated with a ____________ material which gives off light when struck by a beam of ____________.
38. The pattern is traced by moving the ____________ beam about on the ____________.
39. The beam of electrons is bent toward the most ____________ plate and away from the most ____________ plate.
40. Vertical movement of the dot represents ____________, while horizontal movement is proportional to ____________.
41. The time base generator provides a ____________ voltage for____________ deflection.

42. The sweep voltage as well as the signal voltage is placed on the plates in a ____________-____________ arrangement.
43. Patterns that result from comparison of frequency or phase of two signals are called ____________ figures.
44. A circular trace is formed when comparing two signals with the ____________ amplitude and frequency and ______ ______ degrees out of phase.
45. All ac values are ____________ unless otherwise specified.
46. Twenty volts ac has a peak value of ____________ volts.
47. What is the peak power in a circuit where 110 volts produce a current of 10 milliamperes?

9

reactance

All circuit opposition can be reduced to one of two categories: *resistive* opposition or *reactive* opposition. The reactive category has two elements: *inductive* opposition and *capacitive* opposition. All of these types of circuit opposition are present to some extent in any electrical circuit. The reactive elements become especially active in an ac circuit. In this chapter, we shall analyze inductive reactance and capacitive reactance separately, and then, in resistive-reactive combinations. The mixing of these components provides some very interesting combinations that suggest a variety of practical applications.

INDUCTIVE CIRCUITS

review of inductors

Earlier we discussed the fact that a magnetic field is generated about any current carrying conductor. We then wound conductors into coils and called these coils inductors. Inductance (*L*) was defined as the ability of a coil to produce a *counter electromotive force* (CEMF). The unit of inductance was specified as the henry (H); the henry was explained as follows:

A coil which has one henry of inductance will produce a CEMF of one volt when the current through the coil is changing at the rate of one ampere per second.

It should be re-emphasized that inductance refers to a *built-in characteristic* that gives the inductor a specific ability. When the inductor is on the shelf, it has the same inductance as it has when used in a circuit. Many people tend to become confused on this point because of the dependency (in the above definition) on the rate of change of current.

Inductors have a small resistance which is measurable on an ohmmeter. This resistance is actually the total resistance of the length of conductor from which the coil was formed. It is generally called *dc resistance* of the coil to distinguish it from the much greater opposition that the inductor offers to ac. The dc resistance is so small that it can be ignored in most calculations. When we speak of a pure inductor, we simply mean that the dc resistance may be ignored.

The opposition that a coil offers to ac is called *inductive reactance.* The induced voltage causes a current which sets up a magnetic field that opposes every change in current. This is a description of inductance (L). When an inductor of some specified quantity of inductance (in henries) is placed in an ac circuit, it will offer an opposition to that ac that is directly proportional to the rate of change times the inductance. This rate of change is the angular velocity. Inductive reactance is measured in ohms, is symbolized by X_L, and is equal to *angular velocity times the inductance.*

$$X_L = \omega L$$

Breaking the angular velocity down into its basic components, we have: $X_L = 2\pi fL$. Regardless of which formula is used, X_L (in ohms) is proportional to the product of the rate of change times the inductance in henries. X_L is a fixed quantity as long as the coil is exposed to a voltage of a fixed frequency. But when frequency changes, X_L changes in the same manner: When frequency increases, inductive reactance increases, and vice versa.

inductors in series

The total circuit inductance (L_t) in a purely inductive series circuit is simply the *summation* of all the individual inductances. This is illustrated in Figure 9-1.

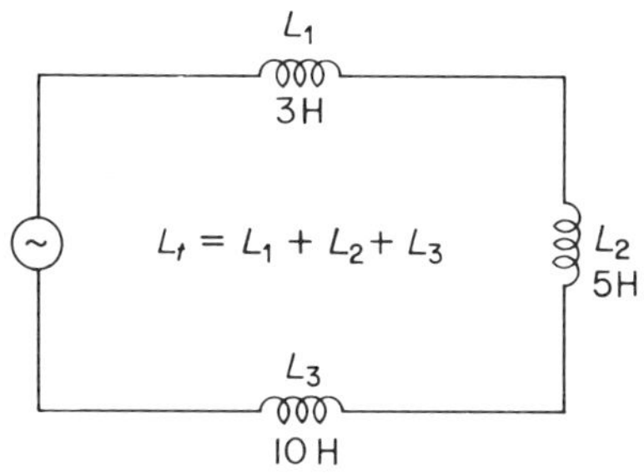

Figure 9-1. Calculating Total Series Inductance

There is no mention of frequency, current, or applied voltage. No matter what the conditions, the total circuit inductance is: 3H + 5H + 10H = 18H.

When we begin consideration of inductive reactances, *frequency* will be an important factor unless the X_L is already specified. The inductive reactance of a coil can be calculated from its inductance and frequency. Frequency may be calculated from inductive reactance and inductance.

Example 1 (Figure 9-2): Calculate the inductive reactance.

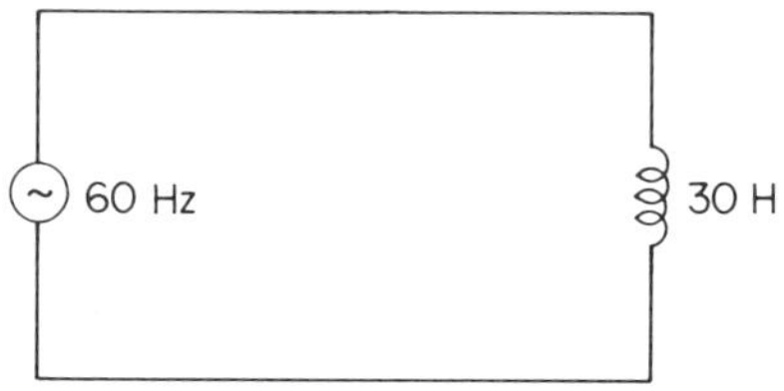

Figure 9-2. Use *F* and *L* to Determine X_L

$$X_L = 2\pi fL = (6.28)(60\text{Hz})(30\text{H}) = 11{,}304\Omega$$

Total inductive reactance in a series circuit may be determined by adding the individual inductive reactances:

$$X_{L_t} = X_{L1} + X_{L2} + X_{L3}$$

Example 2 (Figure 9-3): Calculate the inductive reactance of each inductor and the total circuit reactance.

Reactance of L_1:

$$X_L = 2\pi fL = (6.28)(400\text{Hz})(10 \times 10^{-3}) = 25\Omega$$

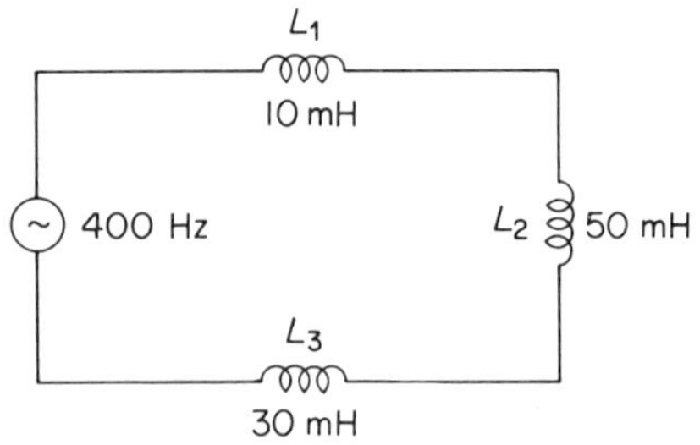

Figure 9-3. Solve for Reactance

Reactance of L_2:

$$X_L = (6.28)(400)(50 \times 10^{-3}) = 125.6\Omega$$

Reactance of L_3:

$$X_L = (6.28)(400)(30 \times 10^{-3}) = 75\Omega$$

Total reactance:

$$X_t = X_1 + X_2 + X_3 = 25\Omega + 125.6\Omega + 75\Omega = 225.6\Omega$$

Once the inductive reactance has been determined, Ohm's law can be applied to solve for voltage or current. Conversely, if voltage and current are known, Ohm's law may be used to solve for inductive reactance.

Example 3 (Figure 9-3): This circuit has a current of 0.5 ampere. What is the applied voltage?

$$E_a = I \times X_L = (0.5)(227) = 113.5\text{V}$$

The above values of E and I are, of course, effective ac values. But notice that X_L is treated exactly as any other opposition to current.

There is a *difference in phase* between the voltage across an inductor and the current through it. Since the inductor opposes every change of current, the current will lag behind the voltage. When a sine wave of voltage is applied to a coil, the coil opposes the increase of current until the voltage has passed its positive peak. As voltage passes peak and starts to decrease, the coil reacts in a fashion which tends to prevent the decrease in current. This accompanies every change in voltage. As a result, the voltage reaches peak just as current begins to flow. The voltage returns to a value of zero when the current reaches peak. The current through a pure inductor is always exactly 90° behind the voltage. We usually express this phase relation by saying that the *current lags the voltage* by 90°. Of course, it would be just as accurate to say that the voltage leads the current by 90°. Here is an acronym that may help to recall the relationship:

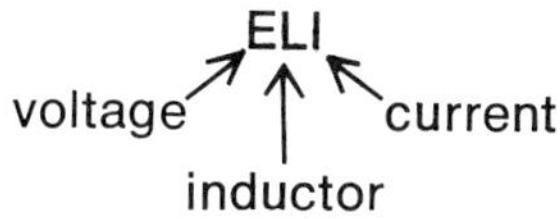

The voltage across an inductor precedes the current through the inductor. All you need to remember is the word, ***ELI,*** and the fixed 90° difference. This phase relation is illustrated in Figure 9-4.

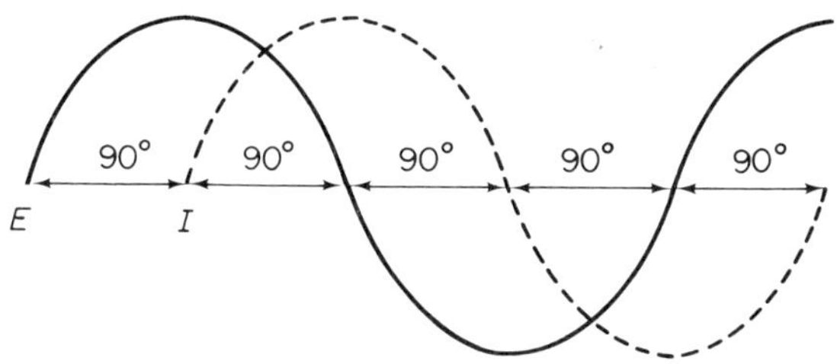

Figure 9-4. Phase Relation of Voltage Across and Current Through an Inductor

The wave shapes are of the same amplitude, but this does *not* imply the existence of any relationship between the quantity of voltage and the current. This should become apparent when you consider the fact that they are measured in different units. The same amplitudes were used to better depict the phase difference between any two points on the wave shapes. Figure 9-5 shows the phase relation in a vector diagram.

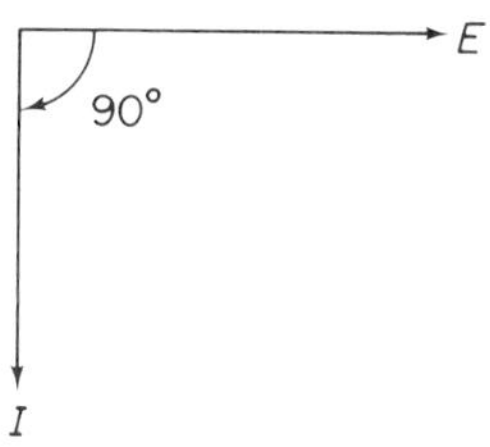

Figure 9-5. Phase Relation, *ELI*

Both vectors are rotating counterclockwise; they are of a common frequency; and the phase relation is fixed.

Example 4 (Figure 9-6): What is the total inductance (L_t) in this circuit?

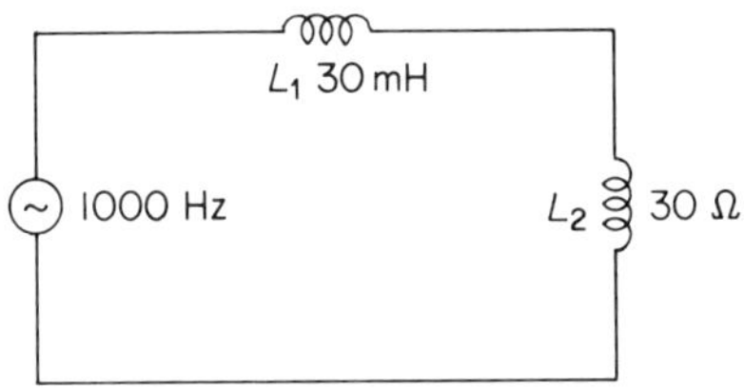

Figure 9-6. Relation of *f*, *L*, and X_L

Calculate the inductance of L_2:

$$L = \frac{X_L}{2\pi f} = \frac{30\Omega}{(6.28)(1000)} = 4.8\text{mH}$$

Add L_1 and L_2:

$$L_t = 30\text{mH} + 4.8\text{mH} = 34.8\text{mH}$$

Example 5 (Figure 9-6): What is the total reactance in this circuit?

Calculate reactance of L_1:

$$X_1 = 2\pi fL = (6.28)(1000)(30 \times 10^{-3}) = 188\Omega$$

Add reactance of L_1 and L_2:

$$X_t = X_1 + X_2 = 188\Omega + 30\Omega = 218\Omega$$

Example 6 (Figure 9-6): What angular velocity would be required to raise the ohmic value of L_2 to 60Ω?

$$\omega = \frac{X_L}{L} = \frac{60\Omega}{4.8\text{mH}} = \frac{60}{4.8 \times 10^{-3}} = \frac{60 \times 10^3}{4.8}$$
$$= 12.5 \text{ kiloradians per second}$$

Example 7: Convert the angular velocity just calculated to frequency.

$$\omega = 2\pi f$$

$$f = \frac{\omega}{2\pi} = \frac{12.5 \times 10^3}{6.28} = 2.2\text{kHz}$$

inductors in parallel

Inductors in parallel are treated in the same manner as resistors in parallel. Total inductance is found:

(1) by the product over the sum formula,

$$L_t = \frac{L_1 L_2}{L_1 + L_2}$$

or (2) by the reciprocal formula,

$$L_t = \frac{1}{\frac{1}{L_1} + \frac{1}{L_2} + \frac{1}{L_3}}$$

Inductive reactance is also handled in the same fashion as it was with resistors.

$$X_t = \frac{X_1 X_2}{X_1 + X_2} \quad \text{or} \quad X_t = \frac{1}{\frac{1}{X_1} + \frac{1}{X_2} + \frac{1}{X_3}}$$

Frequency, voltage, current, inductance, and inductive reactance bear the same relationship to one another as those described for these components in series circuits. These will hold true when dealing with either individual inductors or with the circuit as a whole. The total applied voltage appears across each branch, and the current divides among the branches. This is illustrated in Figure 9-7.

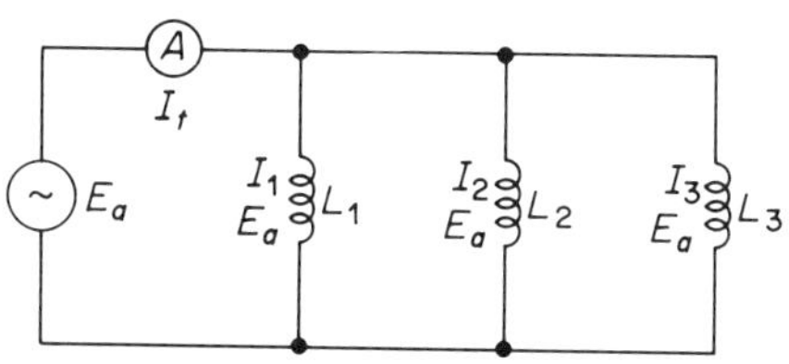

Figure 9-7. Inductors in Parallel

Since these inductors are considered to be pure inductors, the voltage across a branch (E_a) leads the current through that branch by 90°. By the same token, the applied voltage leads the total circuit current by 90°.

Example 1 (Figure 9-8): Calcuate the circuit values of inductance and inductive reactance.

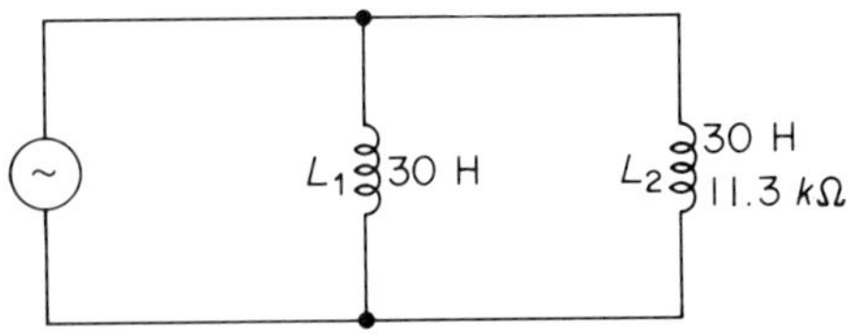

Figure 9-8. Calculate Frequency, Inductance, and Reactance

Use values of L_2 to calculate frequency:

$$f = \frac{X_L}{2\pi L} = \frac{11.3 \times 10^3}{(6.28)(30)} = 60\text{Hz}$$

Solve for L_t:

$$L_t = \frac{30 \times 30}{30 + 30} = \frac{900}{60} = 15\text{H}$$

Since L_1 and L_2 have the same inductance and the same frequency is applied to both, they must have the same inductive reactance: $X_t = (11.3)(11.3)/11.3 + 11.3 = 5.65\text{k}\Omega$.

The total reactance could have also been determined by using frequency with total inductance; that is,

$$X_t = 2\pi f L_t = (6.28)(60)(15) = 5.65\text{k}\Omega$$

Example 2 (Figure 9-8): What would happen to the total circuit inductance if another 15-H branch were added to this circuit?

Total inductance and reactance would both decrease by 50 percent. L_t becomes 7.5H, and X_t becomes 2.82Ω.

Example 3 (Figure 9-9): What is the current in each branch?

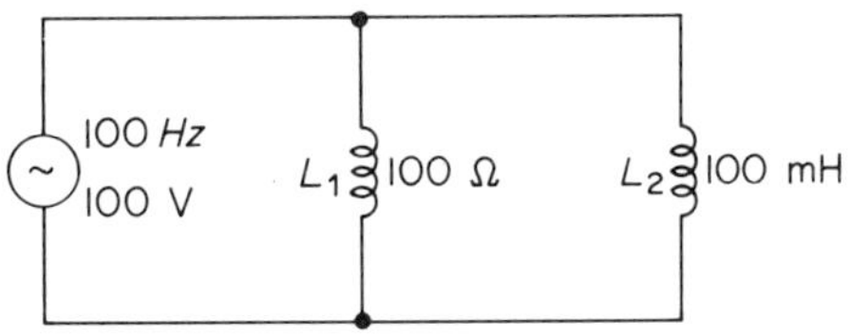

Figure 9-9. Calculate Branch Currents

First determine reactance of L_2:

$$X_L = 2\pi fL = (6.28)(100)(100 \times 10^{-3}) = 62.8\Omega$$

Now, I_1:

$$I_1 = \frac{100\text{V}}{100\Omega} = 1\text{A}$$

Then, I_2:

$$I_2 = \frac{100\text{V}}{62.8\Omega} = 1.6\text{A}$$

Example 4 (Figure 9-9): What is the inductance of L_1?

$$L = \frac{X_L}{2\pi f} = \frac{100}{(6.28)(100)} = 159\text{mH}$$

Perhaps you are wondering why power has not been mentioned in connection with inductive circuits. The reason is quite simple—these circuits use no energy. That's right; a pure inductor would use no energy. When the inductor is *charging* (magnetic field expanding), it is *taking energy* from the circuit and storing it in the magnetic field. When the inductor is *discharging* (magnetic field decaying), it is *returning the stored energy* to the circuit. The pure inductor stores energy during one quarter of a cycle and returns it to the circuit during the next quarter cycle. Therefore, no energy is used.

In ac circuits there are *two types* of power: *apparent power* and *true power. Apparent power* is the product of the voltage and the current; it is expressed in volt-amperes rather than in watts. The *true power* is apparent power times cosine θ, where θ is the phase angle between voltage and current. In all the reactive circuits that we have seen thus far (and in all other purely reactive circuits) the phase angle has been 90°. The cosine of 90° is zero. Therefore, regardless of the apparent power, the true power in such circuits will always be zero.

The *cosine* of the angle between voltage and current is called the *power factor* (PF). In a purely resistive circuit the phase angle is zero which yields a power factor of one. In such a hypothetical case, apparent power is the same as true power. The smaller the angle, the higher the power factor, and coincidentally, the more efficient the circuit.

The only type of component that uses energy is a resistive component. Thus, true power is still given by I^2R, E^2/R, or E_RI_R. This is not a deviation from previous statements because the voltage across a resistor is always in phase with the current through it.

inductors in series-parallel

So far, we have observed that inductance and inductive reactance behave exactly like resistance. There is no reason to suppose that they will react differently from one another in a series-parallel (SP) circuit. In fact, the same rules already discussed along with series circuits and parallel circuits are still applicable. One simply applies parallel rules to the parallel portion of the circuit, and series rules to the series portion. For instance, let's consider the circuit in Figure 9-10. The equivalent parallel inductance (L_p) is:

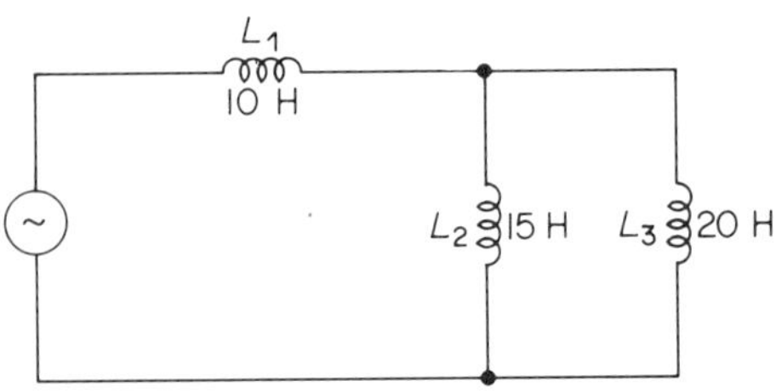

Figure 9-10. Inductors in *SP*

$$L_p = \frac{L_2 L_3}{L_2 + L_3} = \frac{(15)(20)}{15 + 20} = \frac{300}{35} = 8.6\text{H}$$

Total circuit inductance (L_t) is series inductance plus parallel inductance. This is calculated:

$$L_t = L_s + L_p = 10\text{H} + 8.6\text{H} = 18.6\text{H}$$

Inductive reactance is handled in exactly the same way as just demonstrated for inductance. Ohm's law is still applicable to both the individual parts of the circuit and the entire circuit. The relationships between current and voltage, frequency and inductance, frequency and reactance, and inductance and reactance are all still valid.

You might ask about phase relations since some components have total current and others only a portion of total current. The power supply does not provide for the components as individual entities; it views the entire circuit. As a whole, the circuit is inductive; therefore, E_a leads I_t by 90°. The same effect occurs for each component: the voltage across the inductor leads the current through it by 90°.

Example 1 (Figure 9-11): What is the circuit reactance?

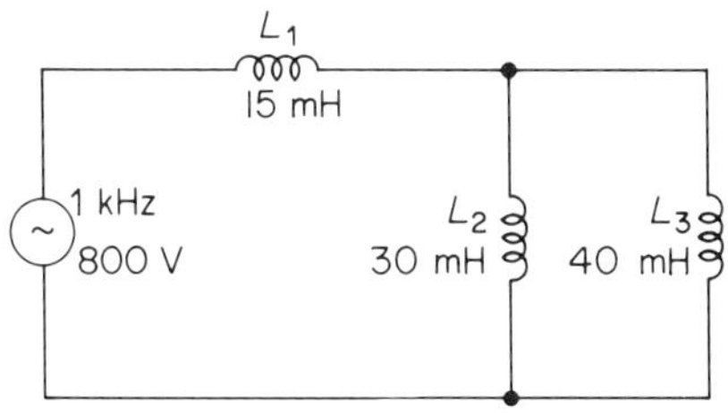

Figure 9-11. Calculate Circuit Reactance

Determine parallel inductance:

$$L_p = \frac{(30\text{mH})(40\text{mH})}{30\text{mH} + 40\text{mH}} = \frac{1200 \times 10^{-6}}{70 \times 10^{-3}} = 17\text{mH}$$

Find total inductance:

$$L_t = L_s + L_p = 15\text{mH} + 17\text{mH} = 32\text{mH}$$

Calculate circuit reactance:

$$X_t = 2\pi f L_t = (6.28)(1 \times 10^3)(32 \times 10^{-3}) = 201\Omega$$

Example 2 (Figure 9-11): What is the total circuit current?

$$I_t = \frac{E_a}{X_t} = \frac{800\text{V}}{201\Omega} = 4\text{A}$$

Example 3 (Figure 9-11): What is the current through each branch? (Review laws for parallel circuits in Chapter 4.)

Current division is inversely proportional to inductance. Three sevenths of I_t passes through L_3, and four sevenths of I_t through L_2. L_3 has 1.7 amperes, and L_2 has 2.3 amperes. These current figures can be checked by Ohm's law, but this would require solving for the reactance of each coil.

Let's consider what would happen if we were to move inductors closer together.

interaction of inductors

When inductors in a circuit, or in adjoining circuits, are connected closely enough for their fields to interact, another factor must be considered. This factor is *mutual inductance* (*M*). This has been covered pretty thoroughly previously, but let's take a quick look at two such circuits.

If the coils are wound in the same direction, the fields aid one another. This is shown in Figure 9-12.

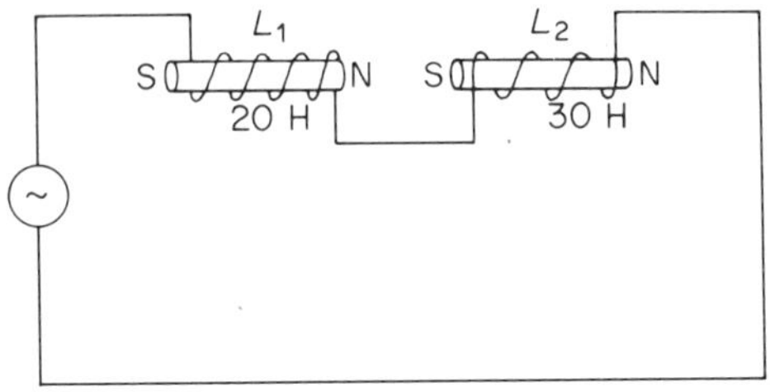

Figure 9-12. Inductors Connected Series Aiding

During one alternation the polarities of the coils are as shown. All polarities reverse during the next alternation. The interaction of the fields results in L_1 inducing current into L_2, and vice versa. The total inductance for this arrangement is: $L_t = L_1 + L_2 + 2M$. Where $2M$ is two times the mutual inductance.

Reversing the windings on one of the inductors results in the arrangement shown in Figure 9-13.

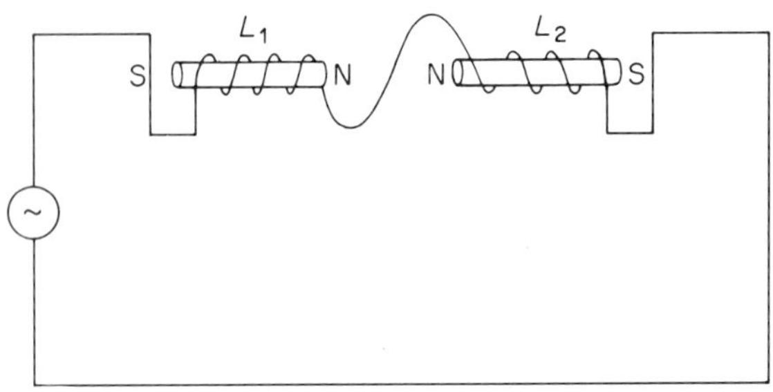

Figure 9-13. Inductors Connected Series Opposing

Here, mutual inductance is still present, but the fields are opposing rather than aiding. Now the total circuit inductance is: $L_t = L_1 + L_2 - 2M$.

If the interacting inductors are in different circuits, the total inductance for each circuit is affected by the amount of the mutual inductance. If the inductors are aiding, the inductance of each circuit is increased by *M.* If they are opposing, the inductance of each circuit is reduced by *M.*

CAPACITIVE CIRCUITS

There is a certain amount of capacitance present when any two current carrying conductors are in the vicinity of one another. This is due to the fact that the two conductors are actually charged, and a dielectric field of force exists between them. This phenomena is not always desirable, but designers have turned it to advantage by creating the capacitor.

structure of capacitors

Simply stated, a *capacitor* consists of two conductors separated by an insulator. The two conductors are called *plates,* and the insulating material between the plates is a *dielectric.* The symbol is 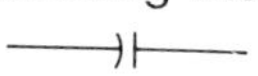. At one time capacitors were improperly called condensers, and some texts still use the term.

Capacitance is the ability of a capacitor to store electrical energy. The energy in this case is stored in the dielectric (electrostatic) field between the two plates. Capacitance (or capacity) is symbolized by *C,* and it is measured in *farads.* The farad is denoted by the symbol *F.* A capacitor has one farad of capacitance when one volt across the capacitor stores a charge of one coulomb.

The farad is an extremely large unit, and for that reason, most capacitors are designated as having a capacitance in millionths or even trillionths of a farad. The microfarad (μF) has a capacitance of 10^{-6} farad; the picofarad (pF) capacitor has a capacitance of 10^{-12} farads. (Remember that pico- and micromicro- are interchangeable terms. A picofarad capacitor may very well be called a micromicrofarad ($\mu\mu$F) capacitor.)

The relationship among charge, voltage, and capacitance is expressed as: $Q = CE$, where Q is the charge in coulombs, C is the capacitance in farads, and E is the applied voltage in volts.

In the case of a manufactured capacitor, the plates are two pieces of metal. The amount of charge that a given voltage can impress on these plates is directly proportional to the *area* of the plates.

The *distance between* the plates also has an effect on the amount of charge. This is an inverse ratio: The capacitance changes inversely with the distance between the the plates; e.g., doubling the distance between the plates reduces the capacitance by 50 percent.

The third factor which affects the capacitance is the *dielectric constant* of the material between the plates. As in electrostatics, air is considered to have a dielectric constant of 1, and other insulators are measured relative to air. Most insulating materials have a dielectric constant much higher than that of air. The capacitance is directly proportional to the dielectric constant of the insulation between the plates.

> Summarizing the previous three statements: Capacitance is directly proportional to the area of the plates and to the dielectric constant, but it is inversely proportional to the distance between the plates, or $C \propto AK/d$.

By adding a constant, we can convert this proportionality into a formula: $C = 0.2248AK/d$, where C is capacitance in farads, A is area of one plate in square inches, K is the dielectric constant, and d is the distance between the plates in inches. (0.2248 is a constant used to convert the results from an electrostatic system of measurement to a more practical system.)

When a voltage source is connected to the plates of a capacitor, an electrostatic field is produced between the plates. This is illustrated in Figure 9-14.

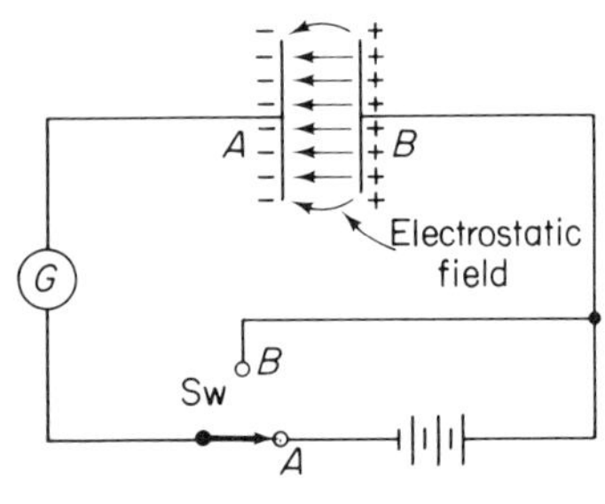

Figure 9-14. Charging a Capacitor

The arrows in the figure indicate the direction of the electrostatic field. When the switch is placed in position *A,* the galvanometer registers current for a short period of time. During this time, plate *A* is collecting a surplus of electrons while plate *B* is losing electrons to the power supply. When the potential between the plates (the charge) is equal to the battery potential, current ceases. A *charge* of a specified quantity of electrical energy has been stored in the electrostatic field. The switch can now be opened and the charge will remain on the capacitor.

Moving the switch to position *B* will remove the battery and place a direct line between the two plates. The meter will register a strong current, which quickly dies out to zero. During this brief interval, the surplus electrons on plate *A* travel around the circuit to plate *B* and equalize the charge. This is called *discharging.* The current ceases when the distribution of electrons on the two plates has equalized. The electrostatic field dissipates as part of the discharging action. When the capacitor reaches full charge (voltage on the capacitor equals the applied voltage), the current ceases. This means that the dc is blocked. Replacing the battery with an ac source allows current to be maintained, as illustrated in Figure 9-15.

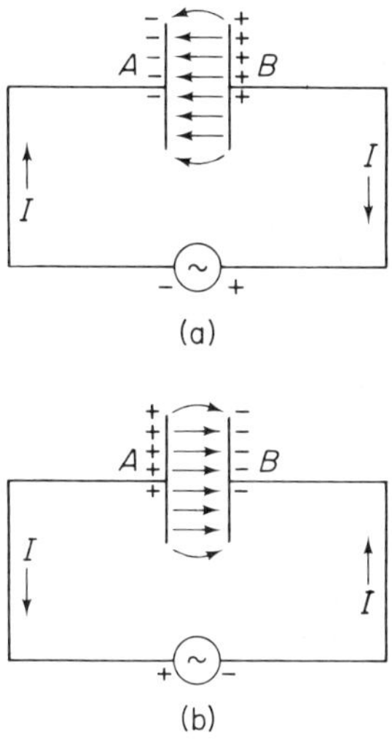

Figure 9-15. Capacitive Action with ac

During the positive alternation, the current from the power source is piling electrons onto plate *A* and removing them from plate *B.* The capacitor is charging with a field direction as indicated. During the negative alternation, current reverses; the capacitor discharges, then recharges in the opposite direction. No electrons pass through the dielectric, but the charging and discharging actions allow a normal ac in the circuit.

As previously stated, any two metal plates separated by a dielectric can constitute a capacitor. Two sheets of aluminum foil separated by oil (or paraffin) impregnated paper make a good capacitor for use in certain types of circuits. The structure of such a capacitor is shown in Figure 9-16.

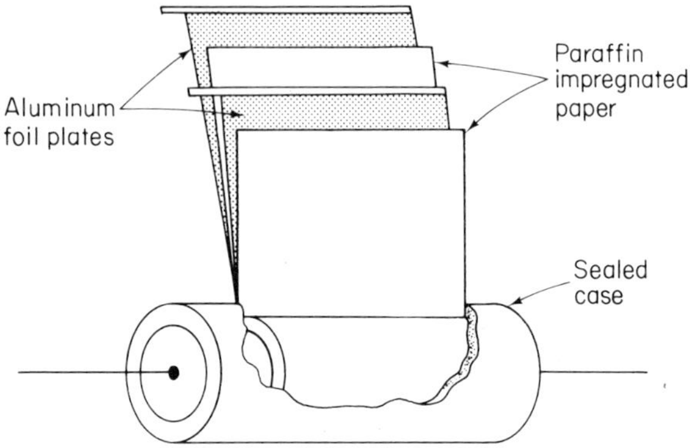

Figure 9-16. Paper Capacitor

Considerable plate area can be made available with this type of capacitor. The foil plates are long strips, which, along with the separating layers of paper, are wound into cylinders and sealed in a moisture-proof container.

The *type* of capacitor refers to the material used for the dielectric. The types most frequently encountered are mica, ceramic, paper, air, and electrolytic.

One method of constructing a mica capacitor is to interleave mica and metal sheets. The alternate metal sheets are then connected to an external lead so that half of the plate area is connected to each lead. This arrangement is shown in Figure 9-17.

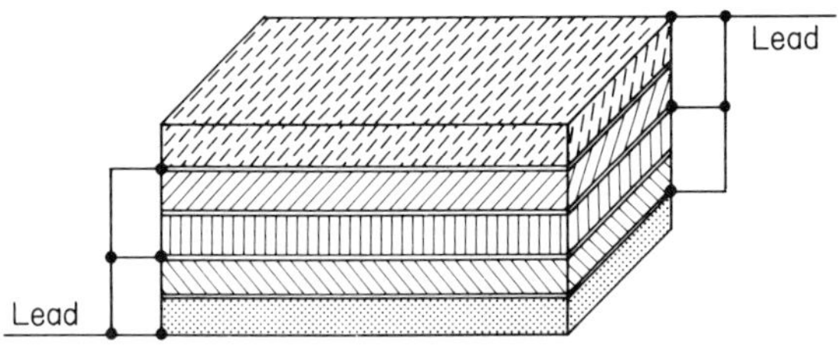

Figure 9-17. Internal Arrangement of Layered Mica Capacitor

Capacitors come in many sizes and shapes. Some serve general purposes, and others are constructed to perform a specific job under carefully controlled conditions. One of the most common capacitors is the *variable air capacitor,* which is commonly used as a frequency selector for radios and television receivers. The symbol for a variable capacitor is an ordinary capacitor symbol with an arrow drawn through it:

Figure 9-18 illustrates the general shape of some types of capacitors.

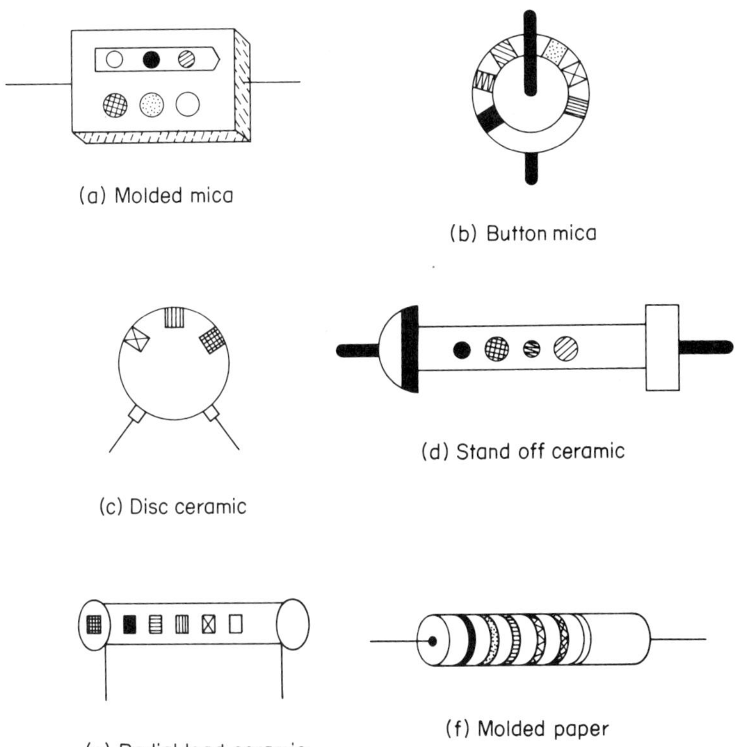

(a) Molded mica

(b) Button mica

(c) Disc ceramic

(d) Stand off ceramic

(e) Radial lead ceramic

(f) Molded paper

Figure 9-18. Some Capacitor Shapes

pacitors. Don't rely too much on the apparent sizes represented here: Some are much smaller than the ones depicted, while others are a great deal larger. Figures 9-19 and 9-20 will provide some perspective of capacitor sizes.

Figure 9-19. Relatively Large Paper Capacitors (Photograph Courtesy of Sprague Electric Company, North Adams, Massachusetts)

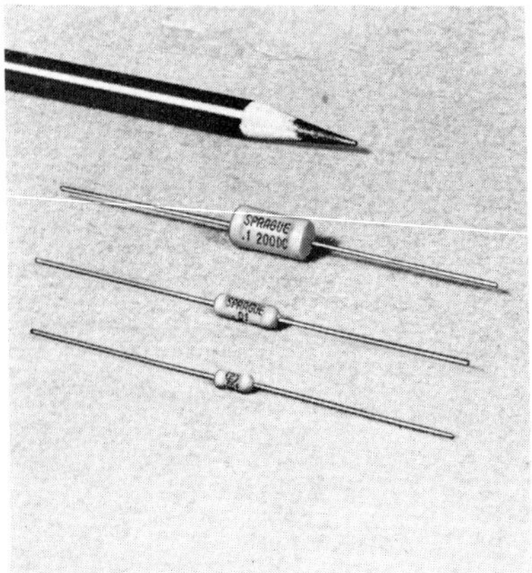

Figure 9-20. Relatively Small Film and Ceramic Capacitors (Photograph Courtesy Sprague Electric Company, North Adams, Massachusetts)

capacitor color codes

Most capacitors are color coded in a fashion similar to the coding of resistors, but these color codes provide more information. The colors are the same as used with resistor color coding, but there is some deviation in the codes from one type of capacitor to another; some types have no color code. In addition to variations in codes, there are two standard systems of code: the military standard system (MIL) and the Electronics Industries Association Standard System (EIA).

Space does not permit full treatment of the capacitor color codes, and there is no real need for the student of electricity to commit them to memory at this time. Color code charts are *available* wherever the capacitors are sold. It is much simpler to refer to these charts than to attempt to memorize them.

Some of the information that is supplied in most color codes are capacitance, working volts dc, tolerance, characteristics, and failure rate. Some of these characteristics are expressed in code that enables location of specific details on charts.

The *working voltage dc* refers to the amount of dc that the capacitor will withstand under its specified operating conditions. One of the critical operating conditions is the temperature range. Working voltage ac is sometimes given, and this is specified in effective value. Capacitors are tested with *peak voltages* in excess of the dc voltage rating. Most capacitors are not harmed by voltage surges above this dc-rated value.

Breakdown results when capacitors are subjected to too much voltage or even when subjected to normal voltage when temperatures are too high. A capacitor is said to be broken down (or shorted) when electrons pass freely between the plates through the dielectric. Once enough voltage has been applied to rupture the dielectric, the electrons can escape through this opening, and the capacitor will not charge.

One type of capacitor is *self-healing* after breakdown, provided the breakdown current is not destructively large. This is the *electrolytic capacitor,* so named because the dielectric is an electrolytic compound. The metal plates are separated by a thin layer of this substance. This capacitor is principally for use with dc even though some are made for use with ac. The polarity must be carefully observed when using a dc electrolytic capacitor since current in the wrong direction will ruin the capacitor. The ac electrolytic capacitor finds its most widespread use in starting motors, such as those on refrigerators.

Since other components we have encountered have proved less than perfect, we might expect the same from a capacitor. All capacitors have some leakage current through (or around) the dielectric. As long as this leakage is kept within the limits of the circuits that we are dealing with, no harm is done. Capacitors may be constructed so that they will have very little leakage, and as a result, they will hold a charge for long periods of time. However, in most cases the improved performance of capacitors so constructed is not sufficient to offset the additional cost.

An ohmmeter is a useful tool for checking capacitors. There are three conditions detectable with the ohmmeter that render the capacitor unsatisfactory. These are (1) short, (2) open, and (3) partial short.

The *short* is the description applied to a capacitor that has had a dielectric breakdown. The ohmmeter needle will show a full scale deflection and remain on a zero reading when it is placed across a shorted capacitor.

The *open* capacitor is probably caused by a broken internal connection. When the ohmmeter is placed on a capacitor with this fault, the needle remains on the infinity mark. Care should be exercised with this test since very small capacitors charge so quickly that the ohmmeter needle may not react. In such cases, open indications may be obtained with capacitors that are actually good.

The *partial short* occurs when the capacitor has an excessive amount of leakage current. This condition will cause a definite resistance indication on the ohmmeter. Again, the type of capacitor must be considered. The electrolytic capacitor allows some current and is supposed to show a definite resistive reading when it is functioning normally.

When making capacitor checks with the ohmmeter, it is a good idea to check with current in both directions. Connect the ohmmeter leads in a selected manner and observe the results; then reverse the leads and check again. Otherwise, a partially charged capacitor may give an erroneous indication.

When making a capacitor check with the ohmmeter, one should use the highest resistance range. The indication of a good capacitor is a momentary deflection of the needle with a subsequent quick return to infinity (no deflection) as the capacitor charges.

capacitors in series

Placing capacitors in series has the effect of increasing the distance between the plates; the total circuit capacitance becomes less than the capacitance of the smallest capacitor. The formula for calculating the capacitance of such an arrangement is derived from Ohm's law and is similar to the parallel resistance formulas:

$$C_t = \frac{C_1 C_2}{C_1 + C_2} \quad \text{or} \quad C_t = \frac{1}{\frac{1}{C_1} + \frac{1}{C_2} + \frac{1}{C_3}}$$

Example 1 (Figure 9-21): What is the total circuit capacitance?

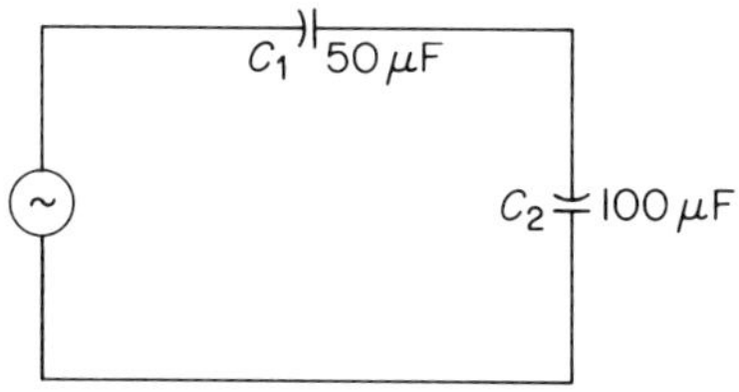

Figure 9-21. Equivalent Capacitance

$$C_t = \frac{C_1C_2}{C_1 + C_2} = \frac{(50 \times 10^{-6})(100 \times 10^{-6})}{(50 \times 10^{-6}) + (100 \times 10^{-6})} = \frac{5 \times 10^{-9}}{1.5 \times 10^{-4}}$$

$$= 3.3 \times 10^{-5} = 33 \times 10^{-6} = 33\mu\text{F}$$

Example 2 (Figure 9-22): Determine total circuit capacitance.

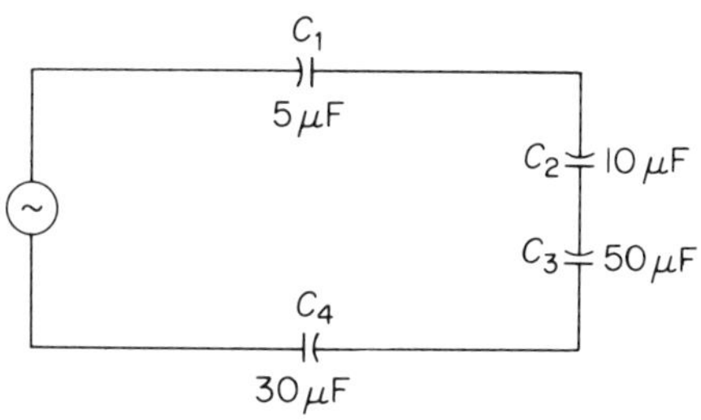

Figure 9-22. Determine Total Capacitance

$$C_t = \frac{1}{\frac{1}{C_1} + \frac{1}{C_2} + \frac{1}{C_3} + \frac{1}{C_4}}$$

$$= \frac{1}{\frac{1}{5 \times 10^{-6}} + \frac{1}{10 \times 10^{-6}} + \frac{1}{50 \times 10^{-6}} + \frac{1}{30 \times 10^{-6}}}$$

$$= \frac{1 \times 10^{-6}}{\frac{30 + 15 + 3 + 5}{150}} = \frac{1 \times 10^{-6}}{\frac{53}{150}} = 2.83 \times 10^{-6} = 2.83\mu\text{F}$$

When subjected to ac, the capacitor exhibits a reactive quality. This is called *capacitive reactance;* it is *opposition to a change in current.* A symbol for capacitive reactance is X_C, and it is measured in ohms. The quantity of X_C in a given capacitor is dependent upon its capacitance and the angular velocity of the applied voltage. This is the reciprocal relationship: $X_C = 1/\omega C$. Since ω is $2\pi f$, the formula is frequently given as $X_C = 1/2\pi fC$.

Example 3 (Figure 9-22): With an angular velocity of 2500 radians per second, what is the reactance of C_1?

$$X_C = \frac{1}{(2500)(50 \times 10^{-6})} = \frac{1 \times 10^3}{125} = 8\Omega$$

This seems to be a relatively small reactance, but remember that reactance is inversely proportional to frequency—2500 radians per second is nearly 400 hertz. Let's try the same capacitor at a frequency of 60 Hz.

$$X_C = \frac{1}{2\pi fC} = \frac{1}{(6.28)(60)(50 \times 10^{-6})} = \frac{1 \times 10^6}{18{,}840} = 53\Omega$$

Once reactance has been determined, it may be handled as any other opposition to current—but there is one difference. The voltage and current are 90° out of phase. In this case, the current leads the voltage by 90°.

During the first instant after voltage is applied across a capacitor, there is no opposition to current. The current jumps almost instantly from zero to peak value. After a short time, electrons begin to accumulate on one plate and become deficient on the other. The charge is beginning to build up, and this charge has a polarity opposite to that of the power source. As time passes, the charge increases, and opposition builds up. As the opposition builds up, current decreases. The current will reach zero at the same time that the voltage reaches peak. The phase relation is illustrated in Figure 9-23.

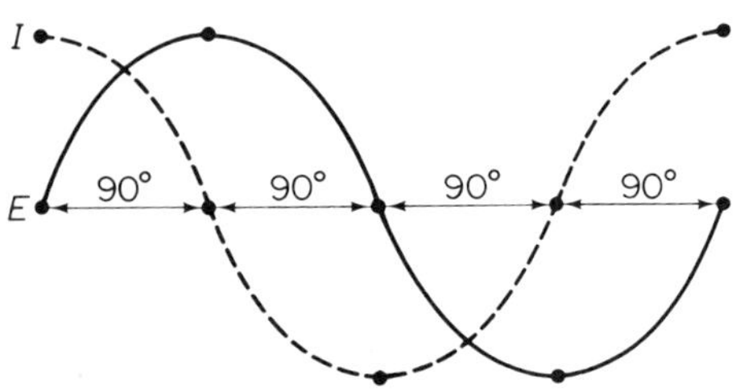

Figure 9-23. Phase Relation of Capacitive Current and Voltage

At any point on these wave shapes, *current is leading voltage* by exactly 90°. Here is another acronym to help you remember this phase relation:

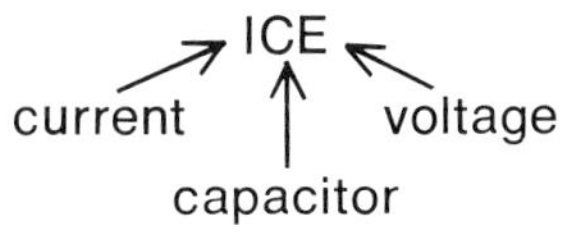

Current in a capacitive circuit precedes voltage. Since the difference is always 90°, the key word ***ICE*** should help you to recall the complete picture. The phase difference is depicted vectorially in Figure 9-24.

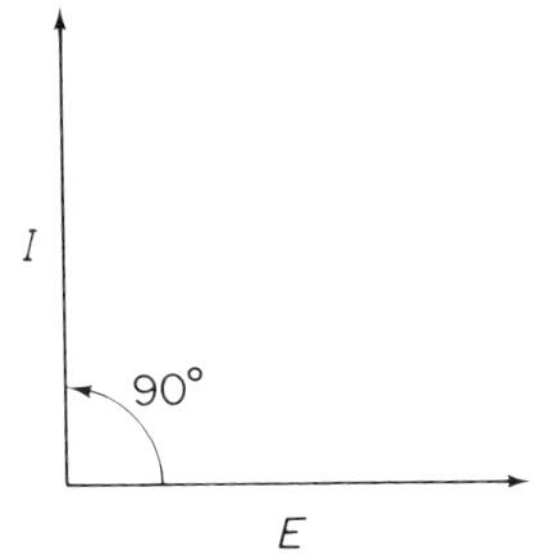

Figure 9-24. Vectors Showing ICE

These vectors are shown with voltage in the start position. Current has already jumped ahead, and the 90° phase angle is fixed. Both vectors rotate counterclockwise at the same rate. In this position, θ for current is 90°. The instantaneous value of current is $I_{max} \times \sin \theta$. The sine of 90° is one; therefore, I is maximum. The voltage angle θ is zero. The sine of 0° is zero; therefore, the voltage is zero.

Example 4 (Figure 9-25): Calculate the total circuit reactance.

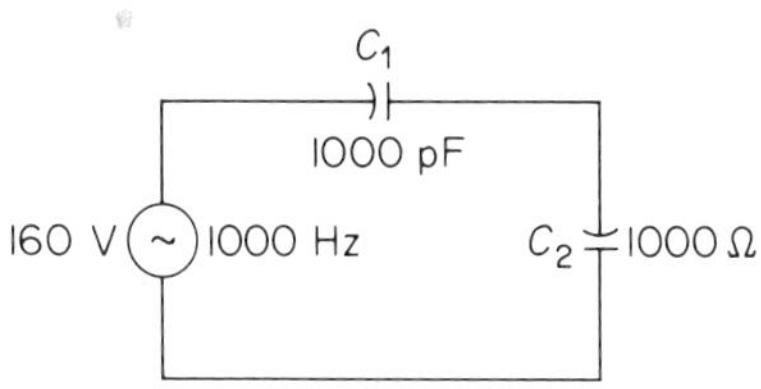

Figure 9-25. Calculate Circuit Reactance

First the reactance of C_1:

$$X_C = \frac{1}{2\pi fC} = \frac{1}{(6.28)(1000)(1000 \times 10^{-12})} = \frac{0.159 \times 10^{6}}{1} = 159\text{k}\Omega$$

Now, circuit reactance:

$$X_t = X_1 + X_2 = 159\text{k}\Omega + 1000 = 160\text{k}\Omega$$

Example 5 (Figure 9-25): What is the quantity of current in the circuit?

$$I_t = \frac{E}{X_t} = \frac{160\text{V}}{160\text{k}\Omega} = 1\text{mA}$$

Example 6 (Figure 9-25): What is the peak voltage across each capacitor?

First, determine the effective voltage on C_1:

$$E_1 = IX_1 = 1\text{mA} \times 159\text{k}\Omega = (1 \times 10^{-3})(159 \times 10^3) = 159\text{V}$$

Find the peak voltage across C_1:

$$E_{\text{peak}} = 1.414E_{\text{eff}} = (1.414)(159) = 225\text{V}$$

Calculate the effective voltage on C_2:

$$E_2 = E_a - E_1 = 160\text{V} - 159\text{V} = 1\text{V}$$

Determine the peak voltage across C_2:

$$E_{\text{peak}} = 1.414E_{\text{eff}} = (1.414)(1) = 1.414\text{V}$$

What is the character of the power in a capacitive circuit? Apparent power we have—it is simply the product of voltage and current, and it is expressed in volt-amperes; true power we *do not* have. The capacitor absorbs energy during charge and returns it to the circuit during discharge. No energy is used by a pure capacitor. Of course, in practice there is some resistance in all components, but when dealing with capacitors, it can be considered negligible.

Example 7 (Figure 9-26): What is the angular velocity of the applied voltage?

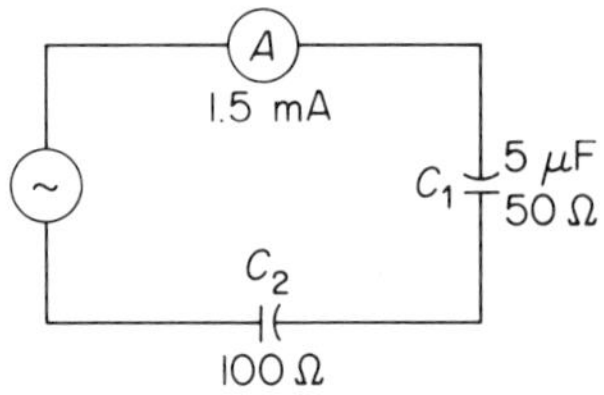

Figure 9-26. Calculate Angular Velocity

$$\omega = \frac{1}{X_C C} = \frac{1}{(5 \times 10^{-6})(50)} = \frac{1 \times 10^6}{250} = 4 \times 10^3$$
$$= 4{,}000 \text{ radians per second}$$

Example 8 (Figure 9-26): What is the capacitance of C_2?

$$C = \frac{1}{\omega X_C} = \frac{1}{(4000)(100)} = 2.5 \times 10^{-6} = 2.5\mu\text{F}$$

This could have been determined by a ratio. Since both capacitors are in the same circuit, C_2 has twice the reactance of C_1; therefore, it must have half the capacitance.

Example 9 (Figure 9-26): Determine the amplitude and frequency of the applied voltage.

The circuit has 1.5mA and 150Ω. This makes calculation of amplitude of voltage fairly easy:

$$E = IX_C = (1.5\text{mA})(150\Omega) = 225\text{mV}$$

With angular velocity known, the determination of frequency is no problem at all. $\omega = 2\pi f$, therefore, $f = \omega/2\pi = 4000/6.28 = 637\text{Hz}$.

capacitors in parallel

Connecting capacitors in *parallel* has the effect of *enlarging* the area of the plates. If one capacitor has a plate area of one square inch, two such capacitors in parallel will have plate area of two square inches. The total plate area of each capacitor is also exposed to the total voltage; that is, the applied voltage appears across each capacitor. This being the case, adding capacitors in parallel increases the capacitance. In fact, total capacitance is given by: $C_t = C_1 + C_2 + C_3$. This is shown in Figure 9-27.

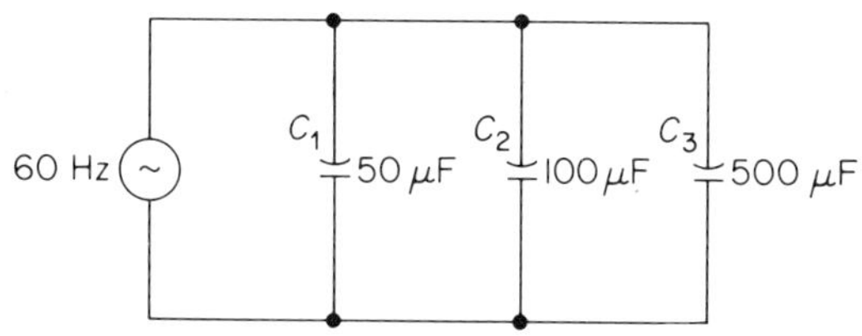

Figure 9-27. Capacitors in Parallel

To solve for total circuit capacitance:

$$C_t = C_1 + C_2 + C_3$$

$$C_t = (50 \times 10^{-6}) + (100 \times 10^{-6}) + (500 \times 10^{-6})$$

$$= 650 \times 10^{-6} = 650\mu F$$

The total reactance of this circuit can be calculated directly from the capacitive reactance formula as follows:

$$X_t = \frac{1}{2\pi fC} = \frac{1}{(6.28)(60)(650 \times 10^{-6})} = 4\Omega$$

The *largest* capacitance offers the *least* opposition to current. Thus, C_3 has the lowest reactance and the largest current. Let's calculate the capacitive reactance of C_3 and find the other reactances by comparison.

$$X_3 = \frac{1}{2\pi fC} = \frac{1}{(6.28)(60)(500 \times 10^{-6})} = \frac{100}{18.84} = 5.3\Omega$$

C_2 has one fifth the capacitance of C_3 and must have five times as much capacitive reactance. Therefore, $X_2 = (5)(5.3\Omega) = 26.5\Omega$.

C_1 has one tenth the capacitance of C_3 and must have ten times the reactance. So, $X_1 = (10)(5.3\Omega) = 53\Omega$.

The total capacitive reactance of the circuit is less than the reactance of any branch. This is the pattern that we found with resistors in parallel, and it holds true for all circuit components in parallel. Total (equivalent) parallel opposition is less than the opposition of any branch. The same formulas used previously can be applied with only minor variations.

$$X_t = \frac{X_1 X_2}{X_1 + X_2} \quad \text{or} \quad X_t = \frac{1}{\frac{1}{X_1} + \frac{1}{X_2} + \frac{1}{X_3}}$$

Current in any branch is found by dividing the applied voltage by the reactance of that branch. Therefore, the parallel circuit is still maintaining the role of current divider.

capacitors in series-parallel

This arrangement could become confusing if care is not exercised. First, we will discuss capacitance only. The equivalent capacitance for the parallel portion of the circuit is given the same treatment as we have just given parallel circuits in the preceding section: $C_p = C_1 + C_2 + C_3$.

The equivalent parallel capacitance is then used with the series capacitance in the product over the sum formula to solve for total circuit capacitance. This is illustrated in Figure 9-28.

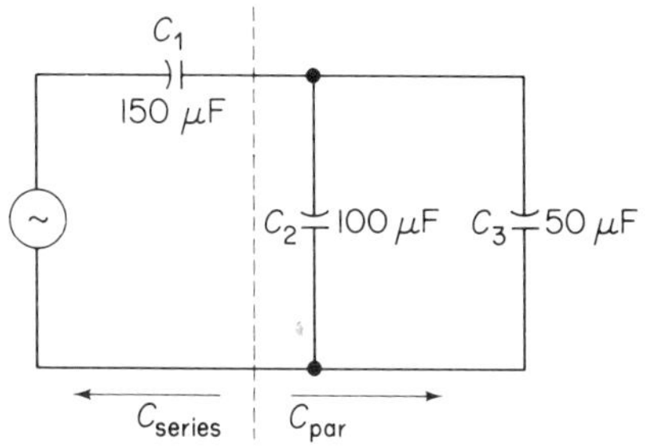

Figure 9-28. Capacitors in SP

$$C_p = C_2 + C_3 = (100 \times 10^{-6}) + (50 \times 10^{-6})$$
$$= 150 \times 10^{-6} = 150\mu F$$

Now, using the series capacitance with the parallel equivalent capacitance, we substitute into the product over the sum formula and solve for total circuit capacitance.

$$C_t = \frac{C_s C_p}{C_s + C_p} = \frac{(150 \times 10^{-6})(150 \times 10^{-6})}{(150 \times 10^{-6}) + (150 \times 10^{-6})} = \frac{225 \times 10^{-10}}{3 \times 10^{-4}}$$
$$= 75 \times 10^{-6} = 75\mu F$$

When more than one capacitor is in the series portion of the circuit, confusion can be avoided by using three separate operations:

1. Either reciprocal or product formulas to solve for series capacitance.
2. Summation formula to solve for parallel capacitance.
3. Product formula with the results of (1) and (2) to solve for total circuit capacitance.

The reactance, frequency, or capacitance can be calculated using any capacitor where two factors are known. Voltage and current bear the same relationship to capacitive reactance that you have already learned.

When dealing with reactance, treat it the same as you would any other opposition to current.

Example 1 (Figure 9-29): Calculate the total circuit capacitance.

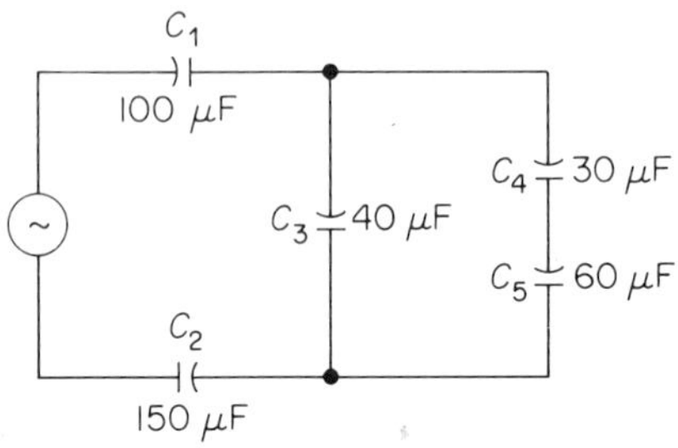

Figure 9-29. Calculate *CT*

First, determine series capacitance:

$$C_s = \frac{C_1C_2}{C_1 + C_2} = \frac{(100 \times 10^{-6})(150 \times 10^{-6})}{(100 \times 10^{-6}) + (150 \times 10^{-6})} = \frac{150 \times 10^{-10}}{25 \times 10^{-5}}$$
$$= 6 \times 10^{-5} = 60 \times 10^{-6} = 60\mu F$$

C_4 is in series with C_5. We use the product over the sum formula again to solve for the capacitance of this branch (C_b).

$$C_b = \frac{C_4C_5}{C_4 + C_5} = \frac{(30 \times 10^{-6})(60 \times 10^{-6})}{(30 \times 10^{-6}) + (60 \times 10^{-6})} = \frac{18 \times 10^{-10}}{9 \times 10^{-5}}$$
$$= 2 \times 10^{-5} = 20 \times 10^{-6} = 20\mu F$$

The summation of this result and C_3 will give us the equivalent parallel capacitance.

$$C_p = C_3 + C_b = (40 \times 10^{-6}) + (20 \times 10^{-6}) = 60 \times 10^{-6} = 60\mu F$$

The total circuit capacitance can be calculated by using the results for series capacitance and the results for parallel capacitance and substituting them into the product over the sum formula:

$$C_t = \frac{C_1C_2}{C_1 + C_2} \quad \text{or} \quad \frac{C_sC_p}{C_s + C_p}$$

$$C_t = \frac{(60 \times 10^{-6})(60 \times 10^{-6})}{(60 \times 10^{-6}) + (60 \times 10^{-6})} = \frac{36 \times 10^{-10}}{12 \times 10^{-5}}$$
$$= 3 \times 10^{-5} = 30 \times 10^{-6} = 30\mu F$$

Example 2 (Figure 9-29): With a frequency of 400Hz applied to this circuit, what will be the capacitive reactance of the circuit?

$$X_C = \frac{1}{2\pi fC} = \frac{1}{(6.28)(400)(30 \times 10^{-6})} = \frac{1 \times 10^{3}}{(6.28)(12)} = 13.3\Omega$$

This concludes Chapter 9. The following section is provided for review.

REVIEW EXERCISES

1. The two categories of circuit oppositions are __________ and __________.
2. What are the two types of reactance?
3. Inductance is the ability of a __________ to produce a __________.
4. Opposition which a coil offers to a changing current is called __________ __________. It is measured in ohms and symbolized by __________.
5. Inductive reactance is directly proportional to __________ and __________.
6. What is the inductive reactance of a 30-mH coil with a frequency of 30kHz?
7. The current through an inductor __________ the voltage across it by __________ degrees.
8. Write two formulas for calculating equivalent inductance of inductors in parallel.
9. In a parallel inductive circuit, the current __________ among the branches.
10. What is the phase relation between applied voltage and total circuit current in a parallel inductive circuit?
11. No power is used by an inductor because it __________ energy during one-quarter cycle and __________ it during the next quarter cycle.
12. An inductor stores energy in a __________ __________.
13. The only component which uses power is the __________.
14. When the fields of two series coils are aiding, total inductance is increased by a factor of twice the __________ inductance.

15. "Two conductors separated by an insulator" is a simplified description of a ____________.

16. A capacitor stores energy in a ____________ ____________.

17. What are the three factors which affect the capacitance of a capacitor?

18. Paraffin impregnated paper is a good ____________ for some capacitors.

19. Working voltage dc indicates the dc voltage which the capacitor is expected to withstand under ____________ working conditions.

20. An ohmmeter can be used to detect three types of capacitor failures. They are ____________, ____________, and partial short.

21. When capacitors are placed in series the total capacitance is ____________ than the capacitance of the smallest.

22. The reciprocal of angular velocity times capacitance is ____________ ____________.

23. The X_C of a given capacitor will increase when frequency ____________.

24. The current through a capacitor ____________ E_a by ____________ degrees.

25. Capacitors ____________ dc and effectively ____________ ac.

26. What angular velocity will cause a 133-pF capacitor to exhibit 500kΩ of X_C?

27. Adding capacitors in parallel will ____________ the total capacitance.

10

elementary impedance

When either or both of the reactive components are placed in a circuit along with resistance, some interesting, intricate, and at times surprising, things happen. We'll start our mixture with resistors and inductors in series circuits. This will create a type of opposition to current that we haven't seen yet.

RESISTIVE-INDUCTIVE (*RL*) CIRCUITS

The opposition that is created when resistance and reactance are combined is called *impedance.* Impedance is symbolized by the letter *Z,* and is measured in ohms. Impedance implies a *direction as well as a magnitude,* and it is generally expressed as $Z\Omega \angle \theta$.

series circuits

When resistors and inductors are placed in the same series circuit, they share a common current. This is perhaps obvious since there is only one path for current. The series *RL* circuit is a voltage divider, as are all series circuits, but the applied voltage is now *divided vectorially.* The phase angle between current and voltage is no longer 90°, but some angle of compromise between 0° and 90°. A totally resistive circuit has a phase angle of 0°. A totally inductive circuit has the voltage leading the current by 90°. A series circuit that is half resistance and half inductive reactance has the voltage leading the current by 45°. This phase angle is more than 45° when reactance is predominant; it is less than 45° when resistance is predominant. In a series circuit, the component with the largest ohmic value will have the greater portion of the applied voltage and will, therefore, be the predominant circuit component. The impedance, as well as most other factors, in these circuits is calculated by solving right triangles. We will use Figure 10-1 to illustrate these points.

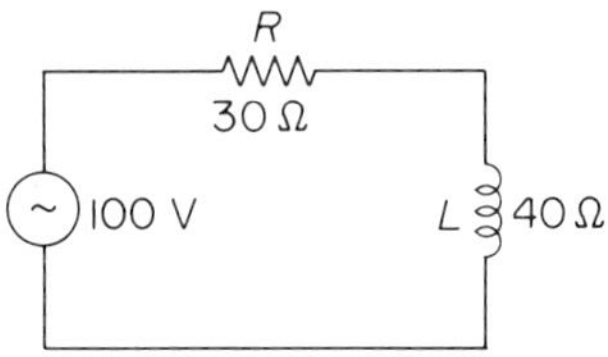

Figure 10-1. Resistor and Inductor in Series

Here we have a 30-Ω resistor in series with a 40-Ω inductive reactance. Vectorial resistance is plotted at zero degrees (the horizontal start position), and inductive reactance is placed at 90°. Since each quantity is a vector quantity, all additions must be performed vectorially. Vectorial addition of 30Ω of resistance and 40Ω of inductive reactance produces a sum of 50Ω at an angle of 53.1°. How did we determine that?—by placing the vectors for R and X_L into a triangle and solving for the hypotenuse. This is illustrated in Figure 10-2.

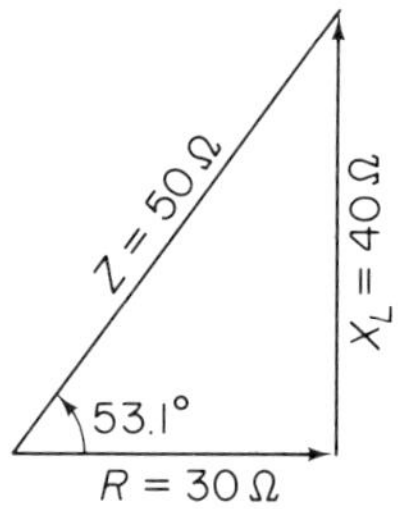

Figure 10-2. Vectorial Addition of Series Components

$$\text{Tan } \theta = \frac{\text{opposite side } (X_L)}{\text{adjacent side } (R)} = \frac{40}{30} = 1.33$$

$$\theta = \text{arc tan } 1.33 = 53.1°$$

$$\text{Sin } \theta = \frac{\text{opposite side } (X_L)}{\text{hypotenuse } (Z)}$$

$$Z = \frac{X_L}{\sin 53.1°} = \frac{40}{0.8} = 50 = 50\Omega \angle 53.1°$$

The phase angle between voltage and current in an *RL* circuit is always of the same magnitude as the angle between *Z* and *R*. But the angles of *Z* and *I* are in different quadrants, so that the signs of the angles are always different. In this case, *Z* is at an angle of 53.1°, therefore, the circuit current (I_t) is at an angle of −53.1°.

The total circuit current is determined by: $I = E/Z$. In this circuit, $I = 100\text{V} \angle 0°/50\Omega \angle 53.1° = 2\text{A} \angle -53.1°$.

Voltages are also vectorial quantities. These can best be handled by plotting the applied voltage at zero degrees. This places the voltage across the resistor in phase with the current through the resistor, which constitutes a true relationship. Since I_t is common to both components, E_R and I_t are plotted together, and E_L is placed 90° ahead of I_t as shown in Figure 10-3.

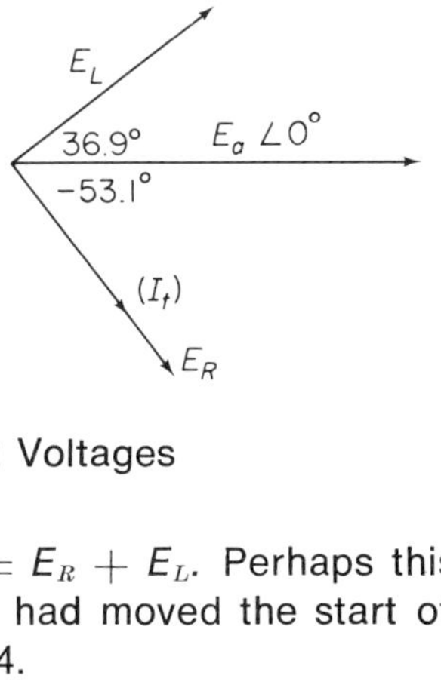

Figure 10-3. Plotting Voltages

Vectorially, $E_a = E_R + E_L$. Perhaps this fact would have been more apparent if we had moved the start of E_L to the tip of E_R as shown in Figure 10-4.

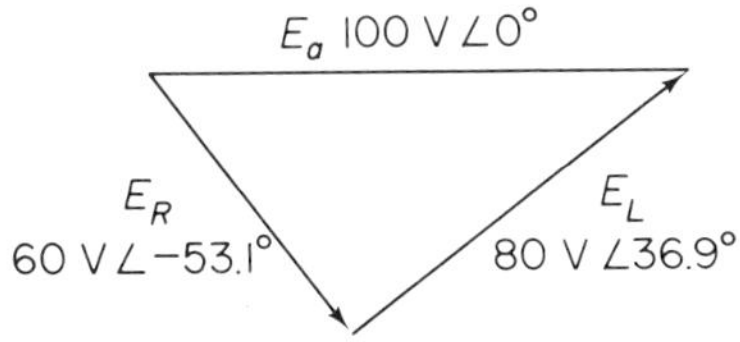

Figure 10-4. Voltage Triangle

E_R is determined by:

$$E_R = I_t R = (2\text{A} \angle -53.1°)(30\Omega \angle 0°) = 60\text{V} \angle -53.1°$$

E_L is determined by:

$$E_L = I_t X_L = (2\text{A} \angle -53.1°)(40\Omega \angle 90°) = 80\text{V} \angle 36.9°$$

What about power in an ac circuit? The voltage times the current formula is still valid, but we must now multiply the result by a power factor. The power factor is always the cosine of the angle between total current (I_t) and the applied voltage (E_a). In Figure 10-1, the phase angle has been calculated as −53.1°. The cosine of this angle is 0.6, so the power factor (PF) for that circuit is 0.6. In reality, the power in this circuit is:

$$P = E_a \times I_t \times \cos\,\theta = 100\text{V} \times 2\text{A} \times 0.6 = 120\text{W}$$

$E_a \times I_t \times$ PF is known as true power and is expressed in watts. Throughout the remainder of this text, any reference to power will mean *true power* unless otherwise specified. When discussing the quantity of apparent power, the volt-amperes or the abbreviation *va* will be used. *P,* power, and true power will be expressed in watts, and the terms will be used interchangeably. In Figure 10-1, we have 60 volts across a resistor, and there are two amperes of current through that resistor. 60V × 2A = 120W (which is the same results as obtained by $E_a \times I_t \times$ PF).

The other power formulas that have been used previously are still good as long as we apply them only to the resistive component: $P = I^2R$; $P = (E_R)^2/R$; and $P = E_R I_R$.

Example 1 (Figure 10-5): How much voltage is applied to this circuit?

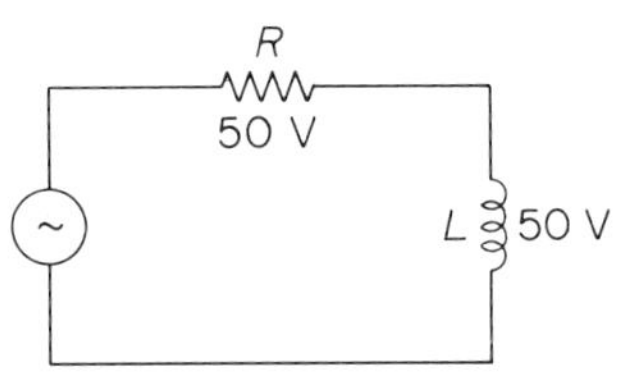

Figure 10-5. Solve for *Ea*

Draw the voltage vectors as in Figure 10-6:

Both angles must be 45° because E_R and E_L are equal. Move E_L to the tip of E_R, as in Figure 10-7.

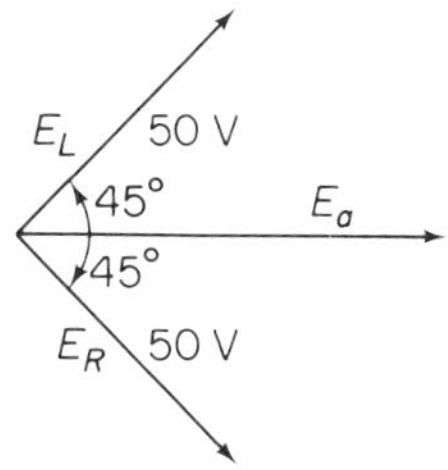

Figure 10-6. Voltages in Circuit of Figure 10-5

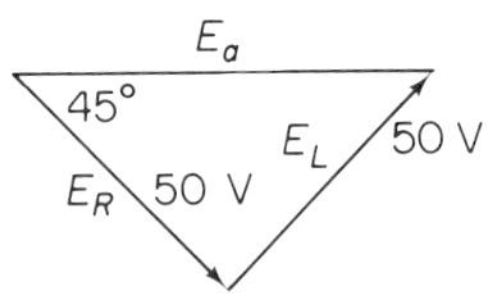

Figure 10-7. Forming a Triangle from the Voltage Vectors

Check the angle:

$$\text{Tan } \theta = \frac{\text{opposite side } (E_L)}{\text{adjacent side } (E_R)} = \frac{50}{50} = 1$$

$$\theta = \text{arc tan } 1 = 45°$$

Solve for E_a:

$$\text{Sin } \theta = \frac{\text{opposite side } (E_L)}{\text{hypotenuse } (E_a)}$$

$$E_a = \frac{E_L}{\sin 45°} = \frac{50}{0.707} = 70.7\text{V}$$

We need more information in order to calculate power, but what is the power factor for the circuit in Figure 10-5? Same as always—cosine θ. In this case, θ is $-45°$, and the cosine of 45° is 0.7 (or to be more precise, it is 0.7071). Angle θ here means the phase angle between E_a and I_t. How do we know that it is $-45°$?—from deduction. We know that the E_a vector is in position 0°. We know that E_R lags E_a by 45°. Since the voltage across a resistor is always in phase with the current through it, I_t must also be located at $-45°$.

The phase relation of the three voltages can be observed on an oscilloscope. The vectors in Figure 10-6 have already shown this. E_L leads E_a by 45°, and it leads E_R and I_t by 90°. This will probably have more meaning if we observe the three wave shapes. These are shown in Figure 10-8.

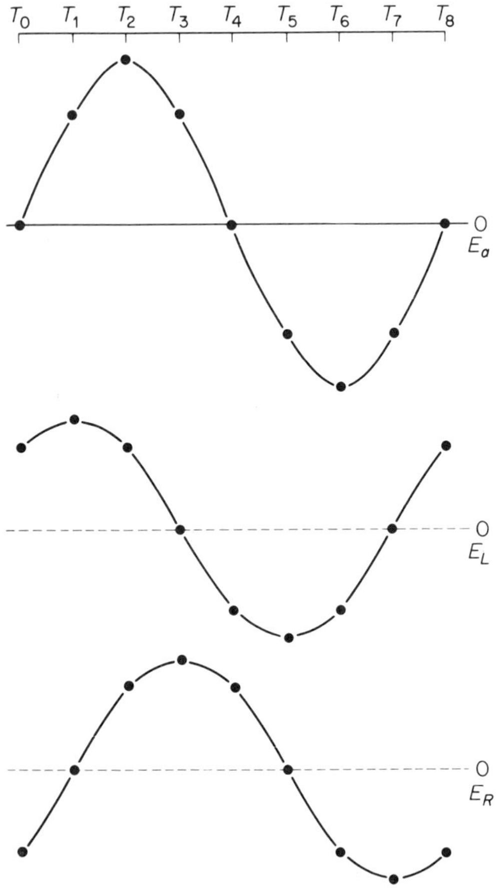

Figure 10-8. Wave Shapes from Figure 10-5

Notice that as time progresses from left to right, E_L reaches a peak in the positive direction at the same time that E_R is zero and E_a is 45°. This same phase relationship can be observed at any point on the wave shapes. Since the current in a series (*RL*) circuit is always in phase with E_R, the E_R wave shape may also be used to represent the phase of circuit current in respect to the voltages. A glance at the wave shapes should show that I_t lags E_L by 90° and lags E_a by 45°.

Another interesting bit of information can be verified from these wave forms. They have been drawn to scale in respect to one another. Notice that at any given instant, $E_a = E_L + E_R$. (This is an algebraic sum.) For instance, at T_4, E_L is negative by the same amount that E_R is positive. This gives a resultant of zero. And at this instant, E_a is zero. E_a is crossing the zero reference line going in a negative direction. Don't forget that we are dealing with the specific values in a given circuit. This circuit is represented by the schematic in Figure 10-5. The principles will hold true for any *RL* circuit, but the values will change according to the circuit's configuration.

Figure 10-9 shows a series circuit with multiple components.

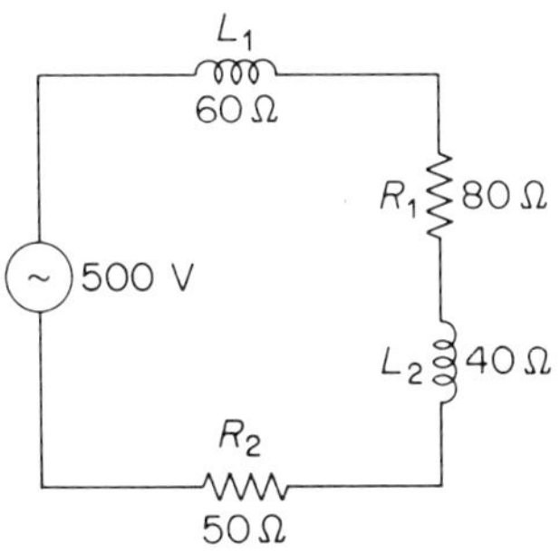

Figure 10-9. Multiple Components

This type of circuit is not complex. We are still dealing with only two types of components, and all the components are sharing a common current.

The procedure in solving a circuit of this type involves adding the ohmic values of like components before placing them into the triangle. In this circuit, we have 60Ω plus 40Ω for 100Ω of inductive reactance. We, also, have 80Ω plus 50Ω for 130Ω of resistance. The total impedance then is $130 + j100\Omega$. Converting this to polar form yields the triangle in Figure 10-10.

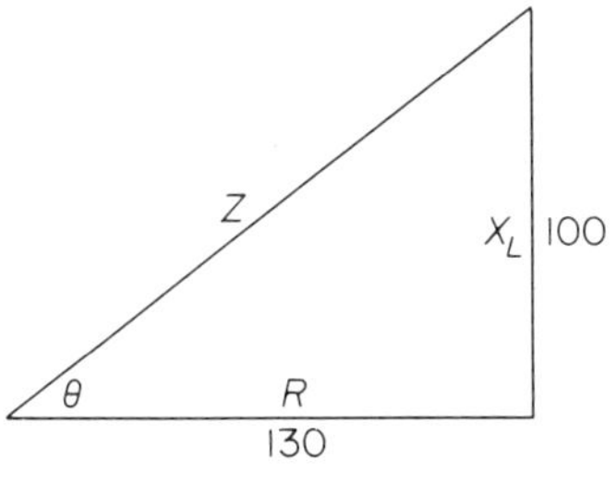

Figure 10-10. Impedance Triangle for Figure 10-9

Solve for θ:

$$\text{Tan } \theta = \frac{\text{opposite side } (X_L)}{\text{adjacent side } (R)}$$

$$\theta = \text{arc tan } \frac{100}{130} = \text{arc tan } 0.77 = 37.6°$$

Solve for impedance:

$$\text{Sin } \theta = \frac{\text{opposite side } (X_L)}{\text{hypotenuse } (Z)}$$

$$Z = \frac{X_L}{\sin 37.6°} = \frac{100}{0.6} = 167\Omega \angle 37.6°$$

The total circuit current is readily determined by:

$$I_t = \frac{E_a}{Z} = \frac{500\text{V} \angle 0°}{167\Omega \angle 37.6°} = 3\text{A} \angle -37.6°$$

The apparent power is $E \times I$ or 1500 volt-amperes. The true power is calculated by:

$$P = I^2R = (3 \times 3)(130) = 1170\text{W}$$

Figure 10-11 shows a circuit with impedance as the known factor. Let's calculate the ohmic value of resistance and the reactance. The problem requires a slightly different approach, but it is essentially the same as most other problems we have encountered with these impedance circuits.

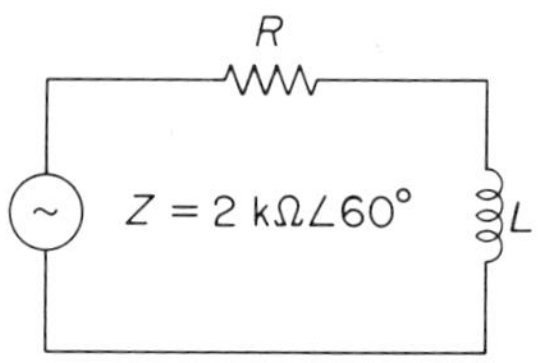

Figure 10-11. Solve for R and X_L

First, we construct the triangle as in Figure 10-12.

Solve for reactance:

$$\text{Sin } \theta = \frac{\text{opposite side } (X_L)}{\text{hypotenuse } (Z)}$$

$$X_L = Z \sin 60° = (2 \times 10^3)(0.866) = 1.7\text{k}\Omega$$

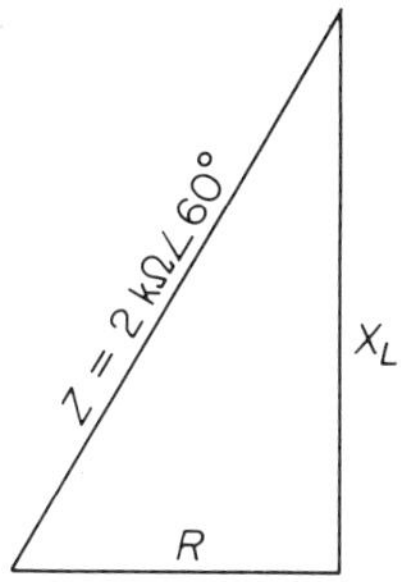

Figure 10-12. Impedance Triangle for Figure 10-11

Solve for resistance:

$$\text{Cos } \theta = \frac{\text{adjacent side } (R)}{\text{hypotenuse } (Z)}$$

$$R = Z \cos 60° = (2 \times 10^3)(0.5) = 1\text{k}\Omega$$

Let's leave the series circuit and examine the results of placing the inductor in parallel with the resistor.

parallel circuits

Figure 10-13 illustrates a parallel *RL* circuit.

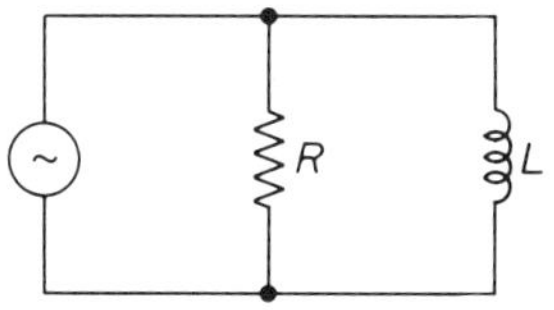

Figure 10-13. Parallel RL Circuit

In this type of circuit, the same voltage (E_a) is impressed across each branch. The circuit impedance is determined by:

$$Z = \frac{E_a \angle 0°}{I_t \angle \theta} \quad \text{or} \quad Z = \frac{(R \angle 0°)(X_L \angle 90°)}{R + jX_L}$$

When more than two branches are involved, it is a good idea to *simplify* the problem. This can be done by reducing the circuit to a two-branch equivalent circuit. The impedance will have some positive angle between 0° and 90°, regardless of the number and relative size of the branches.

The parallel *RL* circuit is a current divider as are all parallel circuits. A branch current is inversely proportional to the opposition of that branch. Current through the resistive branch is calculated by: $I_R = E_a \angle 0°/R \angle 0°$. Current through a pure resistive branch will have an angle of 0°, which places it in phase with the applied voltage.

The current through the inductive branch is determined by: $I_L = E_a \angle 0°/X_L \angle 90°$. The current through a pure inductive branch will be 90° lagging the applied voltage, which is plotted at 0°.

The impedance will have a positive angle (θ), and the total circuit current will have a negative angle ($-\theta$) of the same amplitude. The vectors in Figure 10-14 show the phase relationship of the circuit currents to the applied voltage.

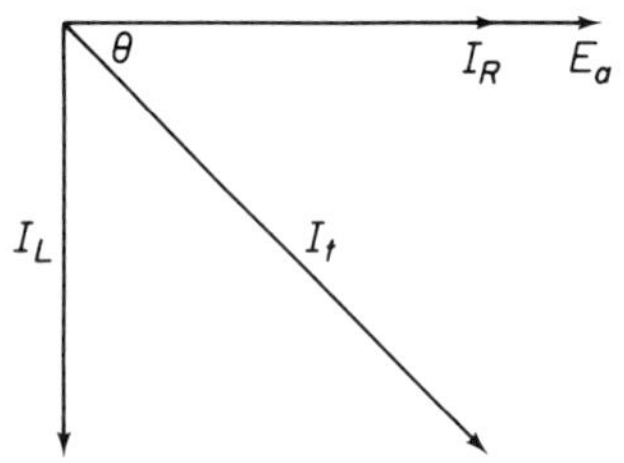

Figure 10-14. Vectors for a Parallel RL Circuit

Let's solve the circuit in Figure 10-15 for all the unknown values.

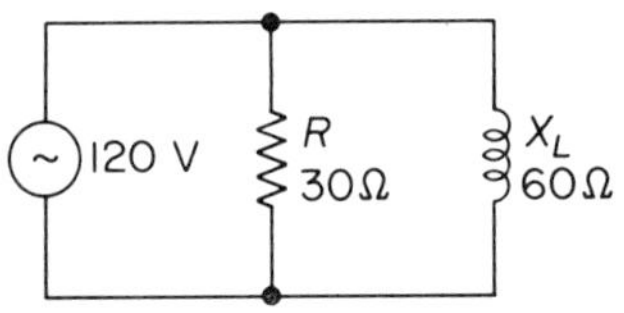

Figure 10-15. Solve for Unknown Value

Solve for impedance:

$$Z = \frac{(R \angle 0°)(X_L \angle 90°)}{R + jX_L} = \frac{(30\Omega \angle 0°)(60\Omega \angle 90°)}{30\Omega + j60\Omega}$$

Add the quantities in the denominator and obtain polar form by solving the triangle in Figure 10-16.

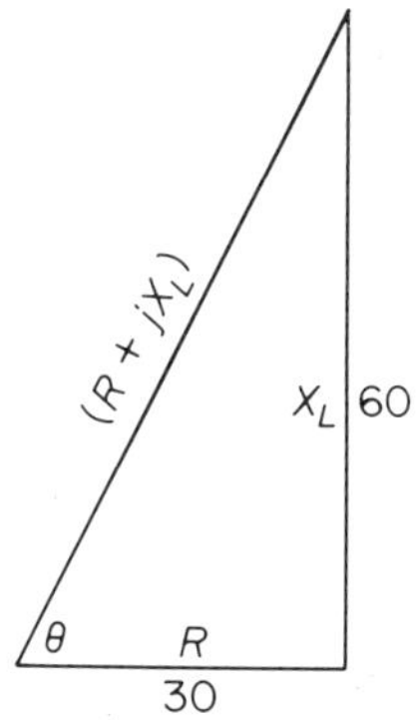

Figure 10-16. Converting to Polar Norm

$$\text{Tan } \theta = \frac{\text{opposite side } (X_L)}{\text{adjacent side } (R)} = \frac{60}{30} = 2$$

$$\theta = \text{arc tan } 2 = 63.4°$$

$$\text{Sin } \theta = \frac{\text{opposite side } (X_L)}{\text{hypotenuse } (R + jX_L)}$$

$$(R + jX_L) = \frac{\text{opposite side } (X_L)}{\sin \theta} = \frac{60}{0.9} = 67\Omega \angle 63.4°$$

Now, back to impedance:

$$Z = \frac{(30\Omega \angle 0°)(60\Omega \angle 90°)}{67\Omega \angle 63.4°} = \frac{1800 \angle 26.6°}{67} = 27\Omega \angle 26.6°$$

Solve for total circuit current:

$$I_t = \frac{E_a \angle 0°}{Z \angle \theta} = \frac{120\text{V} \angle 0°}{27\Omega \angle 26.6°} = 4.5\text{A} \angle -26.6°$$

Let's cross check the results of these calculations by taking a different approach. This time we will start with current:

$$I_R = \frac{E_a \angle 0°}{R \angle 0°} = \frac{120\text{V}}{30\Omega} = 4\text{A} \angle 0°$$

$$I_L = \frac{E_a \angle 0°}{X_L \angle 90°} = \frac{120\text{V} \angle 0°}{60\Omega \angle 90°} = 2\text{A} \angle -90°$$

$$I_t = I_R + I_L = 4\text{A} \angle 0° + 2\text{A} \angle -90°$$

Addition is awkward in polar form, but this resultant quantity can be expressed in rectangular coordinates and placed into a triangle. Solving the triangle will then reveal the quantity and direction of total current. This triangle is shown in Figure 10-17.

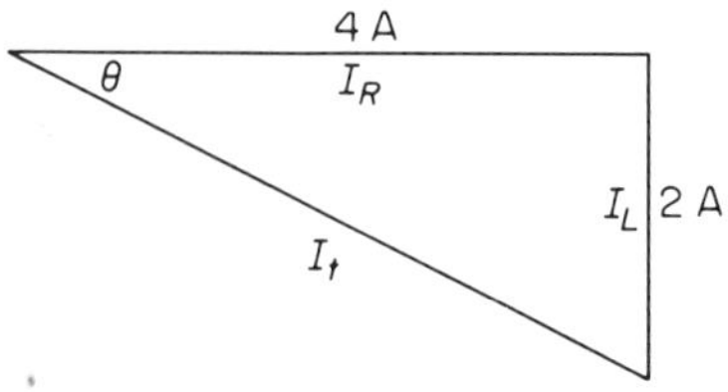

Figure 10-17. Obtaining I_t

$$\text{Tan } \theta = \frac{\text{opposite side } (I_L)}{\text{adjacent side } (I_R)} = -0.5$$

$$\theta = \text{arc tan } -0.5 = -26.6°$$

$$\text{Sin } \theta = \frac{\text{opposite side } (I_L)}{\text{hypotenuse } (I_t)}$$

$$I_t = \frac{I_L}{\sin 26.6°} = \frac{2}{0.45} = 4.5\text{A} \angle -26.6°$$

Solve for impedance:

$$Z = \frac{E_a \angle 0°}{I_t \angle \theta} = \frac{120\text{V} \angle 0°}{4.5\text{A} \angle -26.6°} = 27\Omega \angle 26.6°$$

Since the currents and impedance have checked satisfactorily, let's proceed with power calculations. Cross check the power calculations by using different formulas: $P = E_a I_R = 120\text{V} \times 4\text{A} = 480\text{W}$.

And another: $P = (E_a)^2/R = 14400/30 = 480\text{W}$.

And yet another: $P = (I_R)^2(R) = (4 \times 4)(30) = 480\text{W}$.

Figure 10-18 illustrates a circuit of somewhat more complexity.

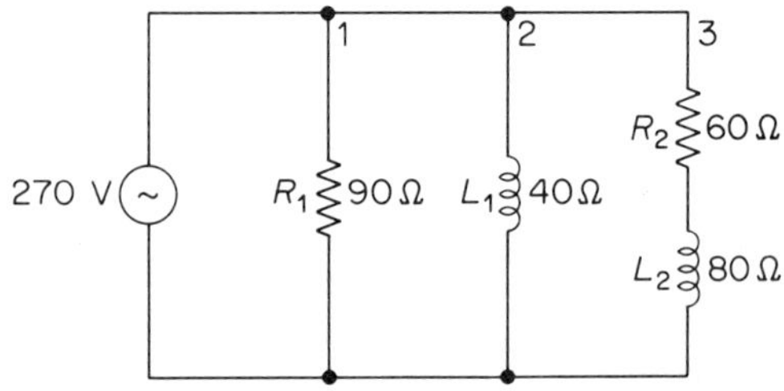

Figure 10-18. Parallel Circuit with a Complex Branch

Actually we have a series circuit as one of the parallel branches. Solving this type of circuit is not complicated. There are more steps to the solution, but each step is relatively simple.

The impedance of the combination branch 3 is: $R_2 + jX_L = 60\Omega + j80\Omega$. The results of this indicated addition can be determined by solving the triangle in Figure 10-19. This will also place the result

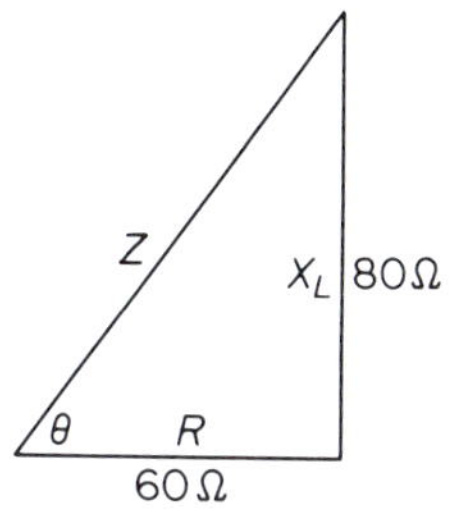

Figure 10-19. Impedance Triangle for Figure 10-18

into polar form—the way in which it will be needed for the next calculation.

$$\text{Tan } \theta = \frac{\text{opposite side } (X_L)}{\text{adjacent side } (R)} = \frac{80}{60} = 1.33$$

$$\theta = \text{arc tan } 1.33 = 53.1°$$

$$\text{Sin } \theta = \frac{\text{opposite side } (X_L)}{\text{hypotenuse } (Z)}$$

$$Z = \frac{80}{\sin 53.1°} = \frac{80}{0.8} = 100\Omega \angle 53.1°$$

This result can now be used with the opposition of another branch to reduce the circuit to a two-branch equivalent circuit. Let's do the inductive branch 2.

$$Z = \frac{(40\Omega \angle 90°)(100\Omega \angle 53.1°)}{(0 + j40) + (60 + j80)} = \frac{4000 \angle 143.1°}{60 + j120}$$

That's as far as we can go without constructing another triangle. The denominator indicates that two vector quantities are to be added. This addition can be performed by solving the triangle in Figure 10-20.

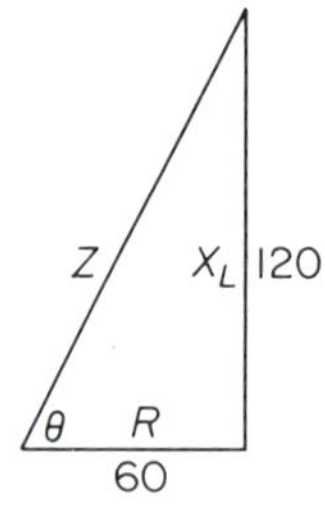

Figure 10-20. Adding 60 + *j*120

$$\text{Tan } \theta = \frac{X_L}{R} = \frac{120}{60} = 2$$

$$\theta = \text{arc tan } 2 = 63.4°$$

$$\text{Sin } \theta = \frac{\text{opposite side } (X_L)}{\text{hypotenuse } (Z)}$$

$$Z = \frac{120}{\sin 63.4°} = \frac{120}{0.9} = 133\Omega \angle 63.4°$$

This is *not* the circuit impedance. It is simply the vectorial summation of the two branch impedances. This result now goes back into our equation: $Z = 4000 \angle 143.1°/133 \angle 63.4° = 30\Omega \angle 79.7°$.

This is the *combined* impedance of branches 2 and 3. We can now use this result with the opposition of branch 1 to solve for total circuit impedance. Branch 1 contains a 90-Ω resistor. Using this 90Ω along with the 30Ω just calculated in the product formula will look like this:

$$Z_t = \frac{(90\Omega \angle 0°)(30\Omega \angle 79.7°)}{(90\Omega \angle 0°) + (30\Omega \angle 79.7°)}$$

Obviously, the next step is to clear the denominator. So, let's leave the main problem at this point and see what we can do about it. The resistive branch is no problem. The rectangular coordinates are obviously 90 + *j*0. But we must solve the triangle in Figure 10-21 in order to obtain the coordinates from the 30Ω at 79.7°.

$$\text{Sin } \theta = \frac{X_L}{Z}$$

$$X_L = Z \sin 79.7° = (30)(0.98) = 29\Omega$$

$$R = Z \cos \theta = (30)(0.18) = 5\Omega$$

The combined rectangular coordinates in the denominator are: (90 + *j*0) + (5 + *j*29). Adding these numbers, we get: 95 + *j*29.

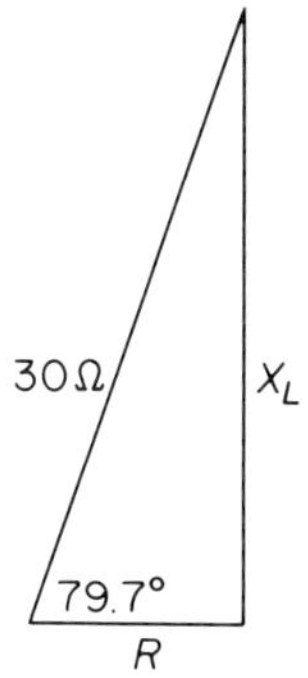

Figure 10-21. Converting to Rectangular Coordinates

We'll use the triangle in Figure 10-22 to convert this back to polar form.

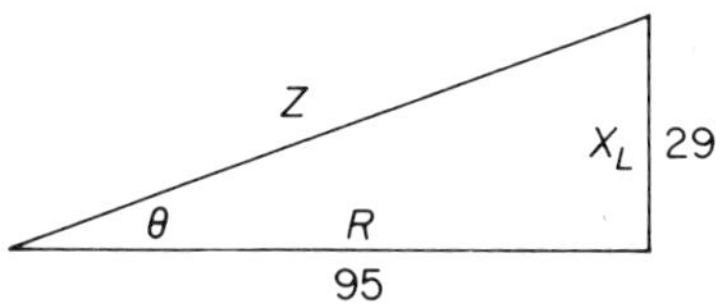

Figure 10-22. Adding 95 + *j*29

$$\text{Tan } \theta = \frac{X_L}{R} = \frac{29}{95} = 0.3$$

$$\theta = \text{arc tan } 0.3 = 16.7°$$

$$\text{Sin } \theta = \frac{X_L}{Z}$$

$$Z = \frac{29}{\sin 16.7°} = \frac{29}{0.29} = 100\Omega \angle 16.7°$$

This is the sum indicated in our previous denominator. Let's go back and pick it up.

$$Z_t = \frac{(90\Omega \angle 0°)(30\Omega \angle 79.7°)}{100\Omega \angle 16.7°}$$

$$= \frac{2700 \angle 63°}{100} = 27\Omega \angle 63°$$

Now that we know the total circuit impedance, the total current can be determined by:

$$I_t = \frac{E_a \angle 0°}{Z \angle \theta} = \frac{270\text{V} \angle 0°}{27\Omega \angle 63°} = 10\text{A} \angle -63°$$

Of course, we can work in the opposite direction just as well. When current and voltage are known, we can solve for opposition. Let's use the circuit in Figure 10-23 to illustrate this procedure.

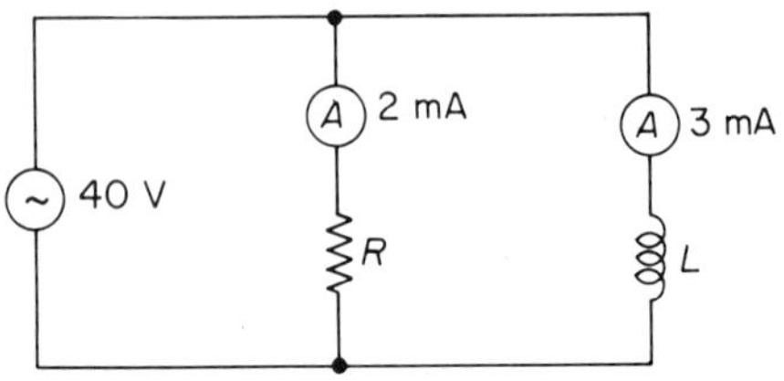

Figure 10-23. Solve for R, X_L, and Z

$$R = \frac{E_a}{I_R} = \frac{40\text{V}}{2\text{mA}} = 20\text{k}\Omega$$

$$X_L = \frac{E_a}{I_L} = \frac{40\text{V}}{3\text{mA}} = 13\text{k}\Omega$$

The total circuit current is 2mA — j3mA. We need to place the current into a triangle as in Figure 10-24 in order to determine its quantity and direction.

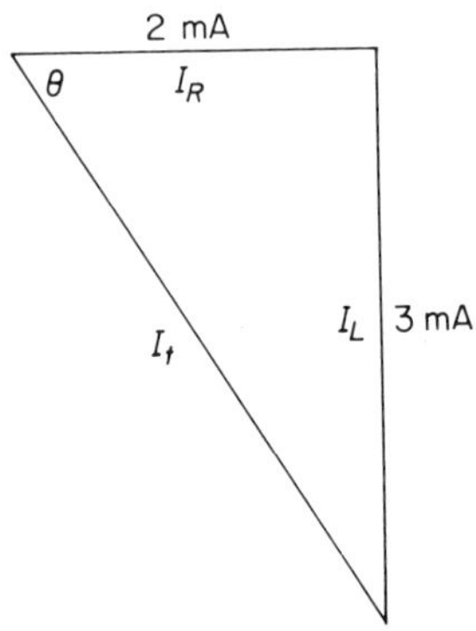

Figure 10-24. Adding Branch Currents

$$\text{Tan } \theta = \frac{\text{opposite side } (I_L)}{\text{adjacent side } (I_R)} = \frac{3\text{mA}}{2\text{mA}} = 1.5$$

$$\theta = \text{arc tan } 1.5 = -56.3°$$

$$\text{Sin } \theta = \frac{\text{opposite side } (I_L)}{\text{hypotenuse } (I_t)}$$

$$It = \frac{I_L}{\sin 56.3°} = \frac{3}{0.832} = 3.6\text{mA} \angle -56.3°$$

The total circuit impedance is calculated by:

$$Z_t = \frac{E_a \angle 0°}{I_t \angle \theta} = \frac{40\text{V} \angle 0°}{3.6\text{mA} \angle -56.3°} = 11.1\text{k}\Omega \angle 56.3°$$

Since the applied voltage appears across *R,* power in this circuit can be determined in this manner:

$$P = E_a I_R = 40\text{V} \times 2\text{mA} = 80\text{mW}$$

Let's leave the parallel circuit for a time and examine the *RL* components in series-parallel arrangements.

series-parallel circuits

Figure 10-25, illustrates a series-parallel *RL* circuit.

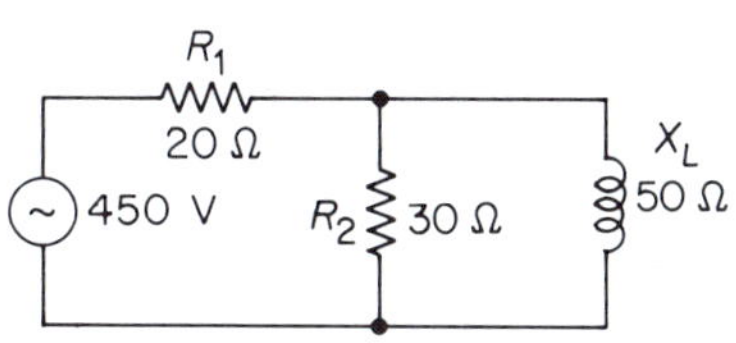

Figure 10-25. Series Parallel *RL* Circuit

We need *no new rules* for this circuit—but we do need the practice in analyzing the actions. The proper procedure for this particular circuit is first to determine the impedance of the parallel section. The familiar parallel circuit formulas will do quite nicely for this purpose. When the equivalent parallel impedance has been obtained, it is added to the series impedance by a vectorial process to produce the total impedance. In Figure 10-25, the equivalent parallel impedance is 26Ω, and the total circuit impedance is 45Ω. These facts were determined by the following formula:

$$Z_p = \frac{(30\Omega \angle 0°)(50\Omega \angle 90°)}{30\Omega + j50\Omega}$$

Clear the denominator by solving the triangle in Figure 10-26.

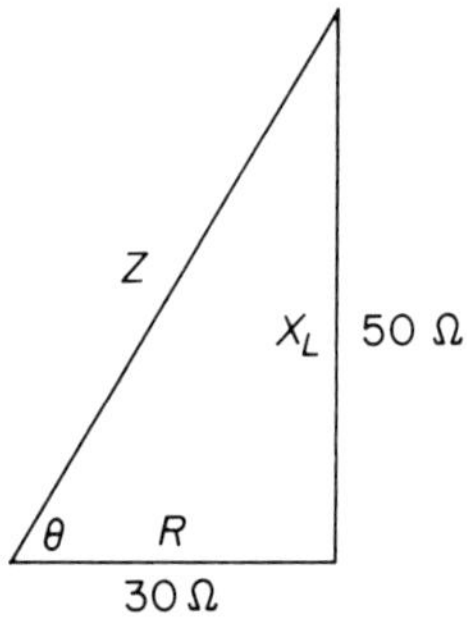

Figure 10-26. Adding the Branch Oppositions

$$\text{Tan } \theta = \frac{X_L}{R} = \frac{50}{30} = 1.67$$

$$\theta = \text{arc tan } 1.67 = 59.1°$$

$$\text{Sin } \theta = \frac{X_L}{Z}$$

$$Z = \frac{50}{\sin 59.1°} = \frac{50}{0.86} = 58\Omega \angle 59.1°$$

This is the denominator for our parallel impedance formula. Let's pick it up.

$$Z_p = \frac{(30\Omega \angle 0°)(50\Omega \angle 90°)}{58\Omega \angle 59.1°} = \frac{1500 \angle 30.9°}{58} = 26\Omega \angle 30.9°$$

We now have the parallel equivalent impedance. Adding this (vectorially) to the series opposition will provide the total circuit impedance. The first step in this addition is converting Z_p to the rectangular form. This is done by solving the triangle in Figure 10-27.

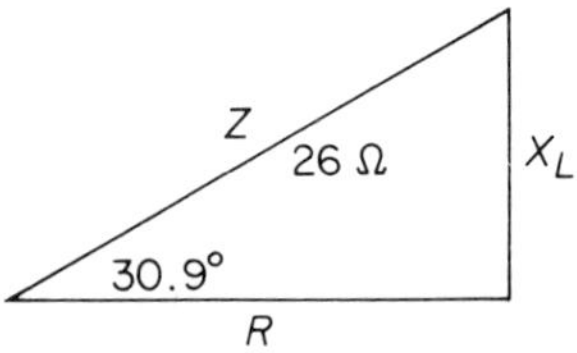

Figure 10-27. Converting Z to Rectangular Form

$$\text{Sin } \theta = \frac{X_L}{Z}$$

$$X_L = 26\Omega \times \sin 30.9° = 26 \times 0.5 = 13\Omega$$

$$R = Z \cos \theta = 26\Omega \cos 30.9° = 26\Omega \times 0.86 = 22\Omega$$

Therefore, $26\Omega \angle 30.9° = 22\Omega + j13\Omega$.

We need to add this parallel impedance to the series resistance. The rectangular form for the resistance is $20\Omega + j0\Omega$. Adding this to $22\Omega + j13\Omega$, we have $42\Omega + j13\Omega$.

The foregoing calculations tell us that the effective impedance of the circuit in Figure 10-25 is the same as that of a series circuit that contains a 42-Ω resistor and a 13-Ω inductive reactance. To convert this number to the polar form, we need to solve one more triangle. This is shown in Figure 10-28.

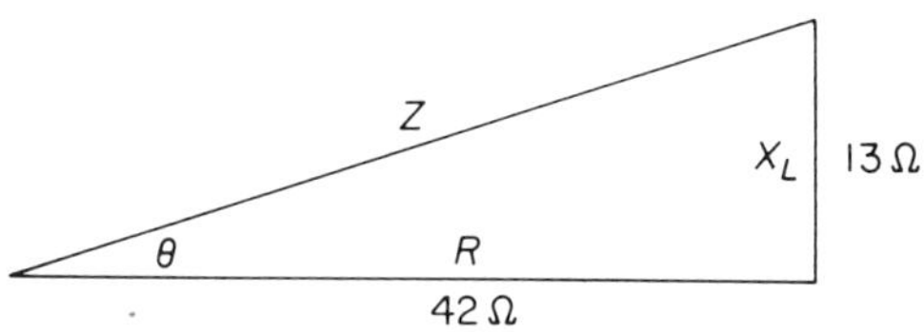

Figure 10-28. Converting to Polar Form

$$\text{Tan } \theta = \frac{X_L}{R} = \frac{13}{42} = 0.3$$

$$\theta = \text{arc tan } 0.3 = 16.7°$$

$$\text{Sin } \theta = \frac{X_L}{Z}$$

$$Z = \frac{13}{\sin 16.7°} = \frac{13}{0.29} = 45\Omega \angle 16.7°$$

We are still dealing with the circuit of Figure 10-25. We have determined that it has an impedance of 45Ω at an angle of 16.7°. The applied voltage was given as 450 volts. Let's calculate the total circuit current:

$$I_t = \frac{E_a \angle 0°}{Z \angle \theta} = \frac{450\text{V} \angle 0°}{45\Omega \angle 16.7°} = 10\text{A} \angle -16.7°$$

There is one more arrangement that we need to consider. How do we cope with several components in the series section of the circuit? We will practice on the circuit in Figure 10-29.

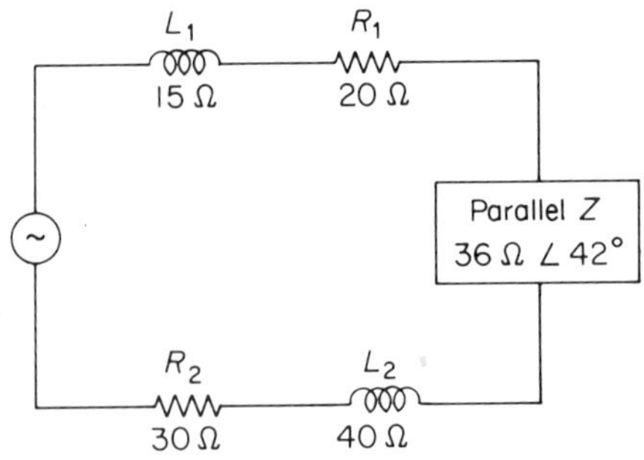

Figure 10-29. Complex Series Section of an *SP* Circuit

The parallel portion of this circuit is represented by a block, which gives us the equivalent parallel impedance. We'll proceed from that point. The first step is to convert the parallel impedance into rectangular form in order to facilitate our adding it to the series components. This is done by placing the value into a triangle as in Figure 10-30 and solving for the two sides of the triangle.

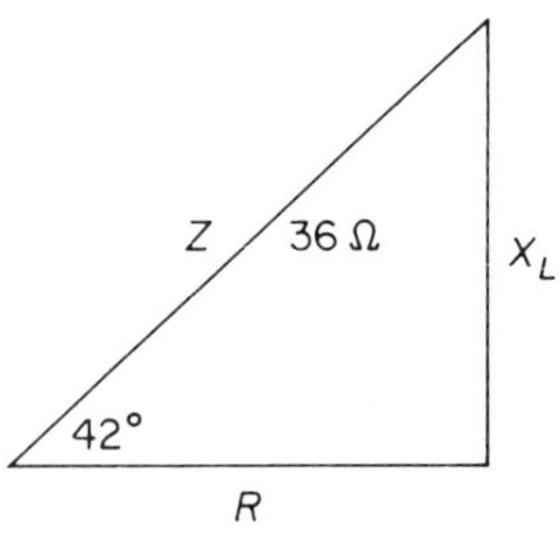

Figure 10-30. Converting to Rectangular

$$\text{Sin } \theta = \frac{X_L}{Z}$$

$$X_L = (36\Omega)(\sin\ 42°) = (36)(0.67) = 24\Omega$$

$$\text{Cos } \theta = \frac{R}{Z}$$

$$R = (36\Omega)(\cos\ 42°) = (36)(0.74) = 27\Omega$$

The parallel portion of the circuit is equivalent to a 27-Ω resistor in series with a 24-Ω inductive reactance. Figure 10-31 is a reproduction of Figure 10-29 with these equivalent components replacing the block. Add resistances: 20Ω + 27Ω + 30Ω = 77Ω. Add inductive reactances: 15Ω + 24Ω + 40Ω = 79Ω. Combine the re-

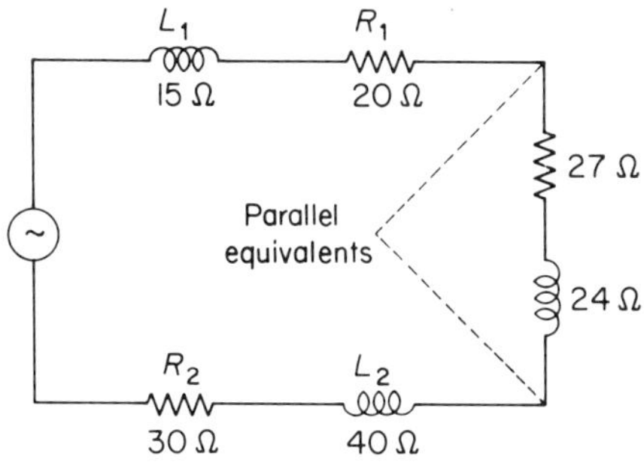

Figure 10-31. Equivalent Circuit of Figure 10-29

sults into a complex number: $77\Omega + j79\Omega$. This is the total circuit impedance in rectangular form. If we want to know the quantity and direction of this impedance, we will need to place it into a triangle and solve for the hypotenuse as in Figure 10-32.

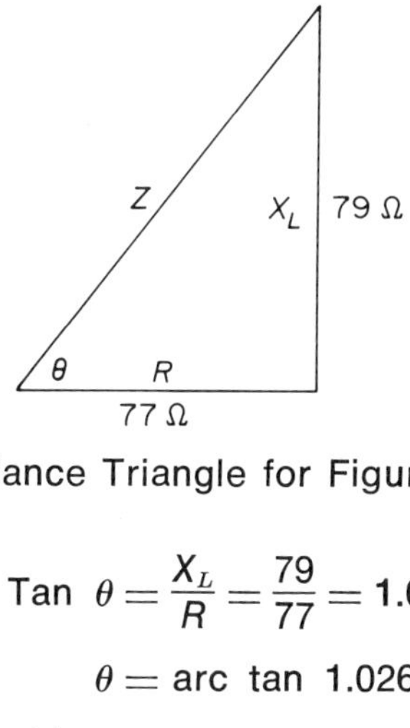

Figure 10-32. Impedance Triangle for Figures 10-29 and 10-31

$$\text{Tan } \theta = \frac{X_L}{R} = \frac{79}{77} = 1.026$$

$$\theta = \text{arc tan } 1.026 = 45.8°$$

The tangent, in this case, was carried out to three decimal places in order to avoid an apparent contradiction. We have stressed the fact that the two sides must be equal in order to have both complementary angles exactly 45°. These sides are almost equal which means that both angles are very close to 45°. If we had rounded off the tangent to a one-digit number, θ would have been accurate enough for most work, but it would have appeared to be 45°. In this instance, we needed to emphasize the small difference. Now, solve for Z:

$$\text{Sin } \theta = \frac{X_L}{Z}$$

$$Z = \frac{79}{\sin \angle 45.8°} = \frac{79}{0.72} = 109.7\Omega \angle 45.8°$$

Let's set aside inductive circuits for a while and consider the results of mixing resistors and capacitors.

RESISTIVE-CAPACITIVE (*RC*) CIRCUITS

Circuit opposition will be known as impedance. This is the same unit that we have been dealing with, and we will treat it in the same manner. However, since *current leads* the applied voltage in a capacitive circuit, the impedance vector will be in the fourth quadrant. By the same token, the total current vector will be in quadrant 1.

We know that current through a capacitor (effectively speaking) leads the voltage across the capacitor by 90°. The current through a resistor is always in phase with the voltage across the resistor. Therefore, when resistors and capacitors are mixed in a circuit, the *total circuit current will lead the applied voltage* by some angle between 0° and 90°.

series circuits

Let's start with the circuit in Figure 10-33.

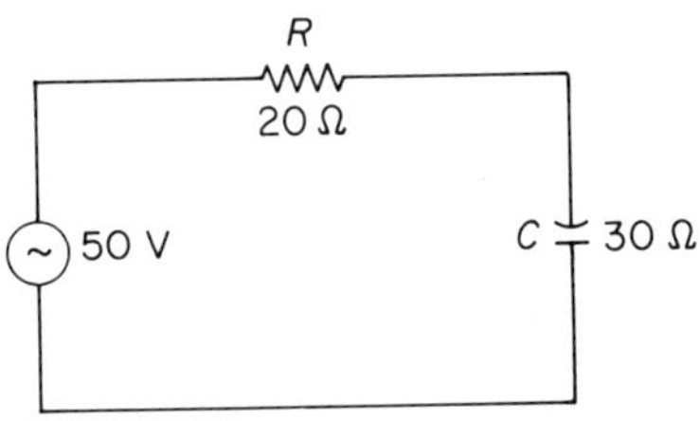

Figure 10-33. Resistor and Capacitor in Series

Don't forget that we are still dealing with vector quantities. The resistance is plotted at 0°, as always, and the capacitive reactance is plotted at −90°. The impedance triangle for the circuit in Figure 10-33 is shown in Figure 10-34.

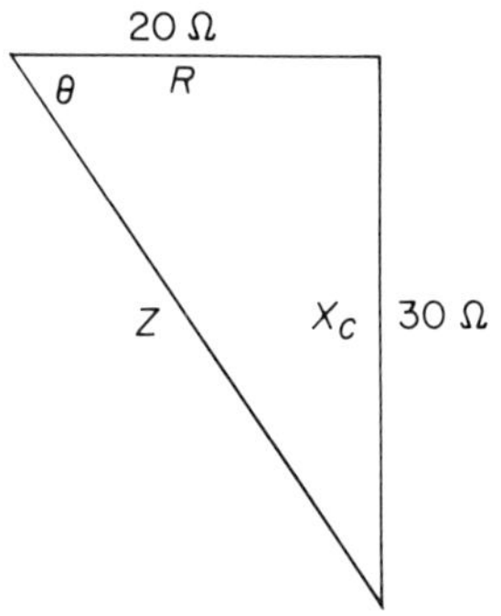

Figure 10-34. Adding R and X_C in Series

$$\text{Tan } \theta = \frac{-X_C}{R} = \frac{-30}{20} = -1.5$$

$$\theta = \text{arc tan } -1.5 = -56.3°$$

(Remember that in the fourth quadrant only the cosine is positive.)

$$\text{Sin } \theta = \frac{X_C}{Z}$$

$$Z = \frac{30}{\sin -56.3°} = \frac{30}{0.832} = 36\Omega \angle -56.3°$$

The current in this circuit can be determined by:

$$I_t = \frac{E_a \angle 0°}{Z \angle -\theta} = \frac{50\text{V} \angle 0°}{36\Omega \angle -56.3°} = 1.4\text{A} \angle 56.3°$$

The plot of the voltages for the circuit in Figure 10-33 is shown in Figure 10-35.

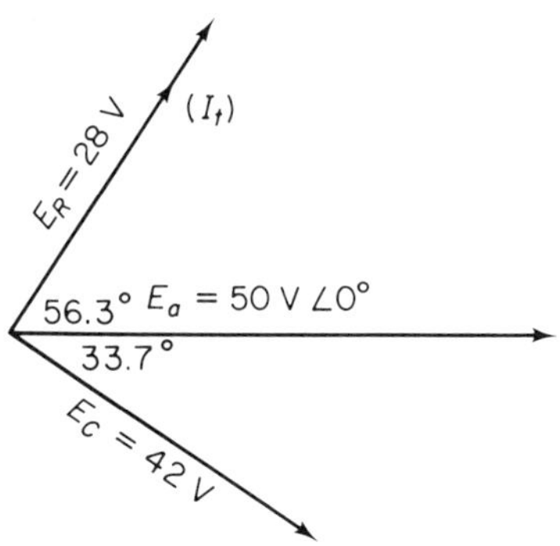

Figure 10-35. Voltage Vectors

The total circuit current (I_t) is shown here to illustrate the fact that it is in phase with E_R. Moving the vectors to form a triangle will produce the diagram in Figure 10-36.

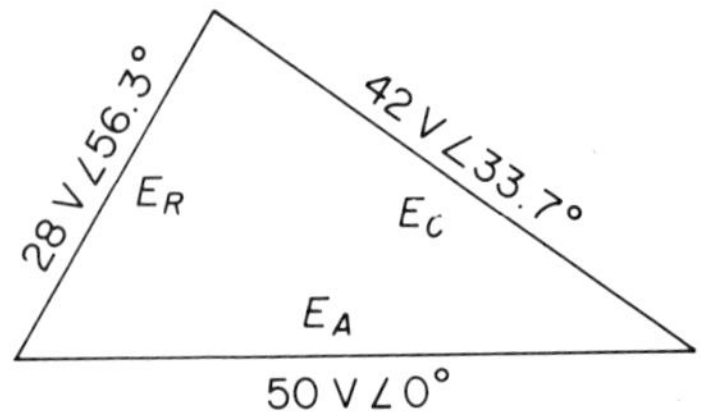

Figure 10-36. Voltage Triangle

Power in *RC* circuits is exactly the same as it is in *RL* circuits. Only the resistive component uses energy. The power factor is determined by the cosine of the phase angle between the total circuit current (I_t) and the applied voltage (E_a). In this case, it is: PF = cos 56.3° = 0.555.

In the *RC* circuit, the angle between total current and the applied voltage will always be positive; it may vary from near zero to near 90°. When the predominant component is resistive, the angle will be small; when the capacitive reactance is predominant, the angle will be large; and when the two components are exactly the *same size,* the angle will be 45°. This means that the circuit current leads the applied voltage by 45°. Figure 10-37 illustrates the voltage wave shapes from a circuit where X_C and R are exactly the same size.

Compare the wave shapes in Figure 10-37 with those in Figure 10-8. In the inductive circuit, E_L was leading E_a by 45°, and E_R (and I_t) was lagging E_a by 45°. But now the circuit is capacitive. E_R (and I_t) now lead E_a by 45° while E_C lags E_a by 45°. Remember that in both of these drawings we are describing the conditions in a circuit where reactance and resistance have the same ohmic value. These are the *only* conditions that create a phase angle between E_a and I_t of exactly 45°.

Figure 10-38 shows a series *RC* circuit with multiple components.

In this type of situation, the procedure is to add like components, and then combine the results. First we add resistances: $R_t = 20\Omega + 30\Omega + 15\Omega = 65\Omega$. Then reactances: $X_C = 10\Omega + 40\Omega$

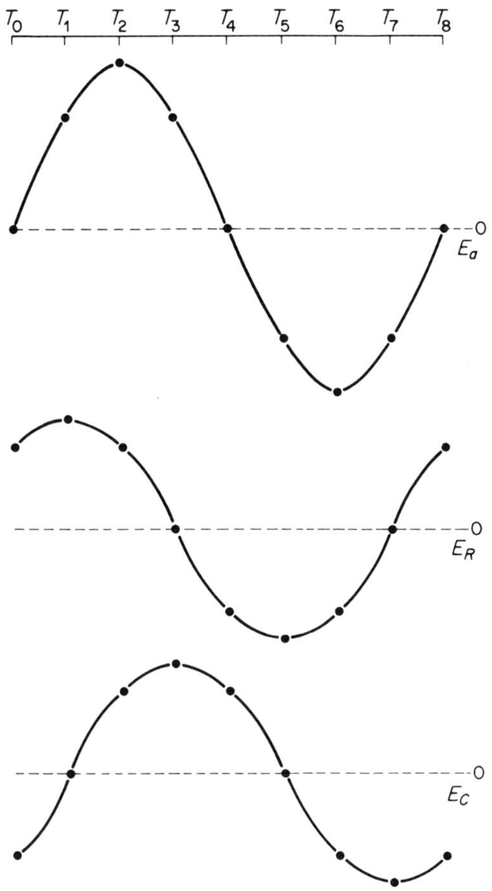

Figure 10-37. Wave Shapes from an RC Circuit

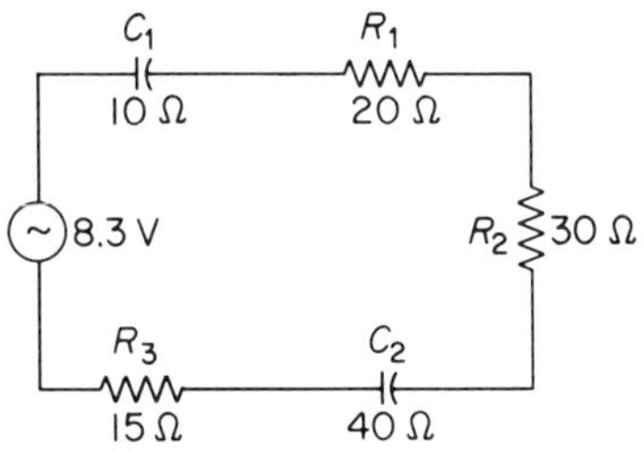

Figure 10-38. Multiple Components

= 50Ω. Expressing the results in a complex number gives us: 65Ω − j50Ω. We place this in a triangle to solve for magnitude and direction of impedance as shown in Figure 10-39.

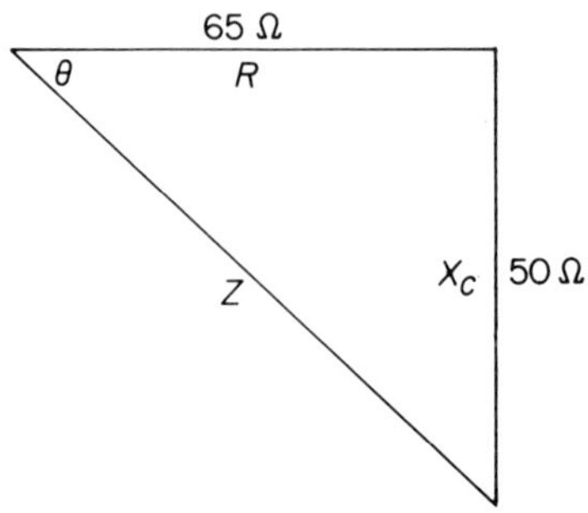

Figure 10-39. Impedance Triangle for Figure 10-38

$$\text{Tan } \theta = -\frac{X_C}{R} = \frac{-50}{65} = -0.77$$

$$\theta = \text{arc tan } -0.77 = -37.6°$$

$$\text{Sin } \theta = \frac{X_C}{Z}$$

$$Z = \frac{50}{\sin 37.6°} = \frac{50}{0.6} = 83\Omega \angle -37.6°$$

The total circuit current has a positive angle of the same magnitude as the negative angle of impedance. It is calculated as follows:

$$I_t = \frac{E_a \angle 0°}{Z \angle -\theta} = \frac{8.3\text{V} \angle 0°}{83\Omega \angle -37.6°} = 0.1\text{A} \angle 37.6°$$

Series circuits have been covered pretty thoroughly. It is time to move on into parallel circuits.

parallel circuits

Figure 10-40 illustrates a parallel *RC* circuit.

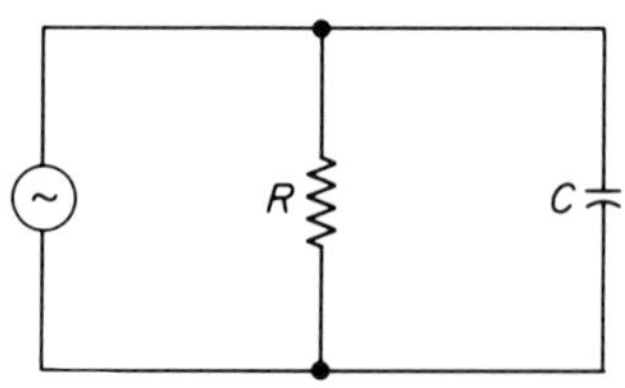

Figure 10-40. Parallel RC Circuit

These parallel circuits will be exactly like parallel *RL* circuits with the exception of the angles for current and impedance. Impedance is:

$$Z = \frac{E_a \angle 0°}{I_t \angle \theta} = \frac{(R \angle 0°)(X_C \angle -90°)}{R - jX_C}$$

Current in the resistive branch is: $I_R = E_a \angle 0°/R \angle 0°$. Current in the capacitive branch is: $I_C = E_a \angle 0°/X_C \angle -90°$. Total circuit current is: $I_t = E_a \angle 0°/Z \angle -\theta$.

Example 1 (Figure 10-41): The caption for Figure 10-41 lists all the

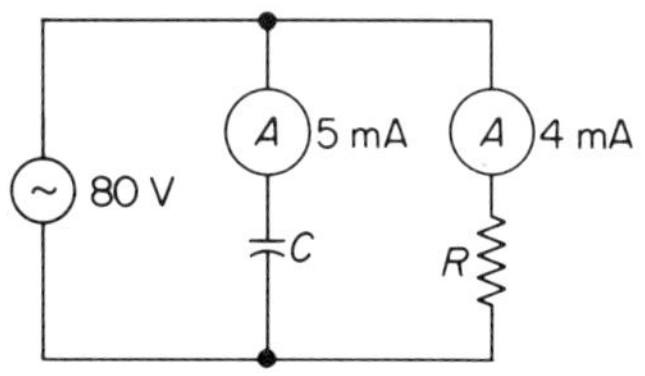

Figure 10-41. Solve for X_C, R, I_t and P

unknown quantities in the circuit. Solving for those will provide needed practice in using the formulas previously listed.

Solving for reactance: $X_C = 80\text{V}/5\text{mA} = 16\text{k}\Omega$. Solving for resistance: $R = 80\text{V}/4\text{mA} = 20\text{k}\Omega$.

Since the branch currents are already given, we can place them into a triangle and solve for total circuit current. This has been done in Figure 10-42.

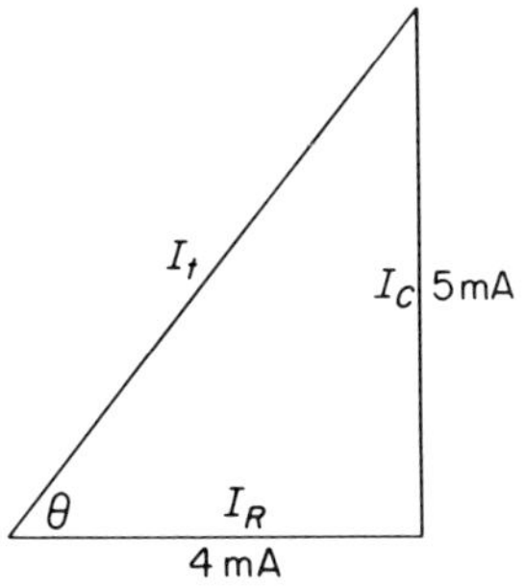

Figure 10-42. Current Triangle

Solve for magnitude and direction of total circuit current:

$$\text{Tan } \theta = \frac{I_C}{I_R} = \frac{5}{4} = 1.25$$

$$\theta = \text{arc tan } 1.25 = 51.3°$$

$$\text{Sin } \theta = \frac{I_C}{I_t}$$

$$I_t = \frac{5}{\sin 51.3°} = \frac{5}{0.78} = 6.4\text{mA} \angle 51.3°$$

Using I_t and E_a, we solve for impedance:

$$Z = \frac{E_a \angle 0°}{I_t \angle \theta} = \frac{80\text{V} \angle 0°}{6.4\text{mA} \angle 51.3°} = 12.5\text{k}\Omega \angle -51.3°$$

Since all the circuit resistance is in one branch, the power is easy. It is: $P = E_a I_R = 80\text{V} \times 4\text{mA} = 320\text{mW}$.

Let's move on into series-parallel circuits.

series-parallel circuits

A series-parallel *RC* circuit is illustrated in Figure 10-43.

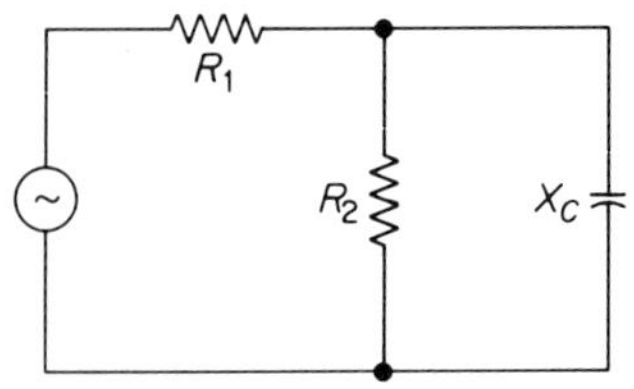

Figure 10-43. Series-Parallel RC Circuit

The equivalent parallel impedance is determined by:

$$Z_p = \frac{(R \angle 0°)(X_C \angle -90°)}{R - jX_C}$$

Series opposition is vectorially added to parallel impedance to obtain the total circuit impedance. The procedures for addition, subtraction, multiplication, and division will not change.

When complex branches are encountered, branch impedances are determined first. When complex series sections are encountered, like components are added and the result expressed as a complex number. Then add this number to the rectangular expression of the equivalent parallel impedance. The result is the circuit impedance in rectangular form. This is then placed into a triangle to obtain the magnitude and direction of the total impedance.

The majority of the errors in solutions of series-parallel circuits are committed as a result of trying to solve the entire circuit without going through all the small procedural steps. In many cases, the whole problem can be greatly simplified by redrawing equivalent circuits as the overall circuit problem is reduced.

All these points will be demonstrated during the solution of the circuit illustrated in Figure 10-44.

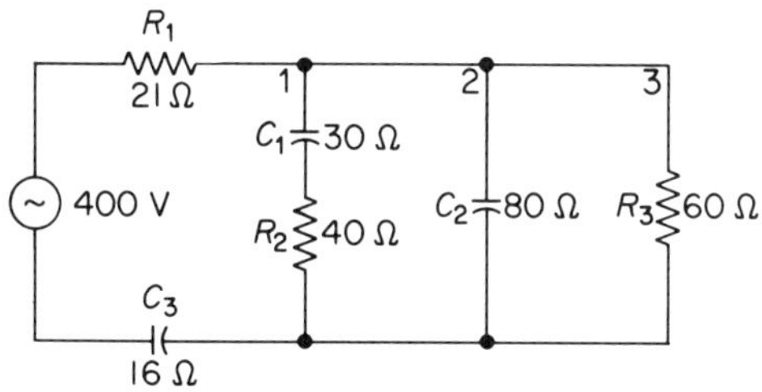

Figure 10-44. Series Parallel RC Circuit (Complex Arrangement)

The impedance of branch 1 is 40Ω — *j*30Ω. We convert that to polar form by solving the triangle in Figure 10-45.

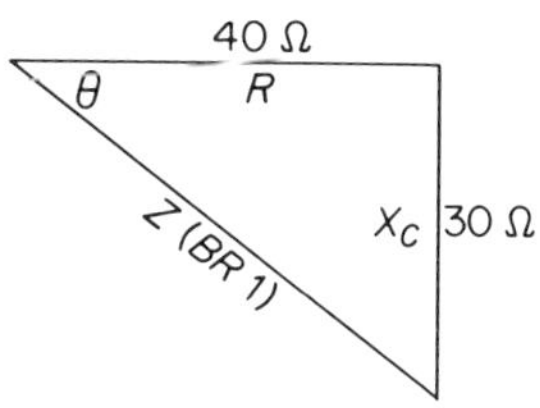

Figure 10-45. Impedance Triangle for Branch 1

$$\text{Tan } \theta = -\frac{X_C}{R} = \frac{-30}{40} = -0.75$$

$$\theta = \text{arc tan } -0.75 = -36.9°$$

$$\text{Sin } \theta = \frac{X_C}{Z}$$

$$Z = \frac{X_C}{\sin 36.9°} = \frac{30}{0.6} = 50\Omega = 50\Omega \angle -36.9°$$

(Branch 1)

We will use the product over the sum formula and calculate the equivalent impedance of branches 2 and 3 (Z_e).

$$Z_e = \frac{(60\Omega \angle 0°)(80\Omega \angle -90°)}{60\Omega - j80\Omega}$$

We need a triangle in order to convert the denominator to polar form. This is provided in Figure 10-46.

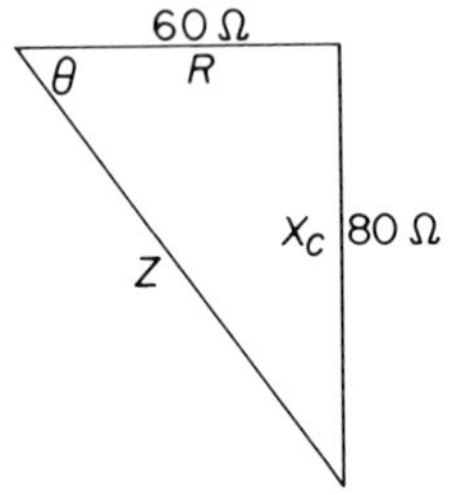

Figure 10-46. Converting to Polar Coordinates

$$\text{Tan } \theta = -\frac{X_C}{R} = -\frac{80}{60} = -1.33$$

$$\theta = \text{arc tan } -1.33 = -53.1°$$

$$\text{Sin } \theta = -\frac{X_C}{Z}$$

$$Z = \frac{80}{0.8} = 100\Omega = 100\Omega \angle -53.1°$$

(Summation of branches 2 and 3)

Substituting this value into the denominator of our equivalent impedance equation produces:

$$Z_e = \frac{(60\Omega \angle 0°)(80\Omega \angle -90°)}{100\Omega \angle -53.1°} = 48\Omega \angle -36.9°$$

(branches 2 and 3)

We can now combine this equivalent impedance with the impedance of branch 1 to calculate the impedance of the parallel section (Z_p):

$$Z_p = \frac{(50\Omega \angle -36.9°)(48\Omega \angle -36.9°)}{(50\Omega \angle -36.9°) + (48\Omega \angle -36.9°)}$$

The denominator must be reduced to rectangular form for addition, then back to polar form for division. The left portion is no problem; we had it in rectangular form in Figure 10-45. It is 50Ω ∠ −36.9° = 40Ω − j30Ω. Solving the triangle in Figure 10-47 will convert the remaining portion of the denominator.

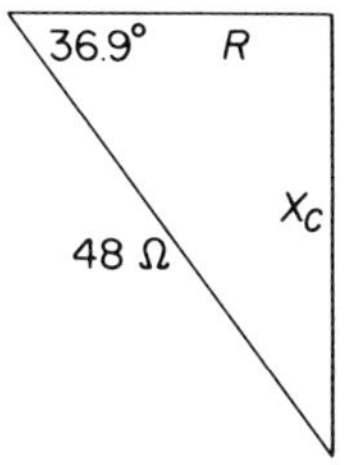

Figure 10-47. Converting to Rectangular Coordinates

$$\text{Sin } \theta = \frac{X_C}{Z}$$

$$X_C = 48 \sin 36.9° = (48)(0.6) = 29\Omega$$

$$\text{Cos } \theta = \frac{R}{Z}$$

$$R = 48 \cos 36.9° = (48)(0.8) = 38\Omega$$

Therefore, 48Ω ∠ −36.9° = 38Ω − j29Ω. (branches 2 and 3)

When we add these two complex numbers, we have performed the addition indicated in the denominator of the Z_p equation: (38Ω − j29Ω) + (40Ω − j30Ω) = 78Ω − j59Ω. Solving the triangle in Figure 10-48 will convert this to polar form.

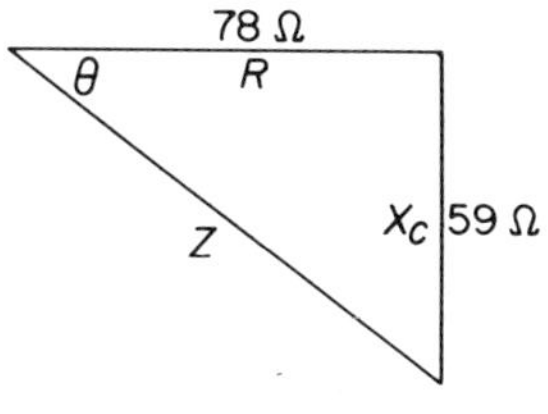

Figure 10-48. Converting to Polar Coordinates

$$\text{Tan } \theta = \frac{-X_C}{R} = \frac{-59}{78} = -0.76$$

$$\theta = \text{arc tan } -0.76 = -37.2°$$

$$\text{Sin } \theta = \frac{X_C}{Z}$$

$$Z = \frac{59}{\sin 37.2°} = \frac{59}{0.6} = 98\Omega = 98\Omega \angle -37.2°$$

Picking up the Z_p equation and substituting, we have:

$$Z_p = \frac{(50\Omega \angle -36.9°)(48\Omega \angle -36.9°)}{98\Omega \angle -37.2°}$$

$$= \frac{2400 \angle -36.6°}{98} = 24\Omega \angle -36.6°$$

Figure 10-49 breaks this down into rectangular form to facilitate adding to the series oppositions.

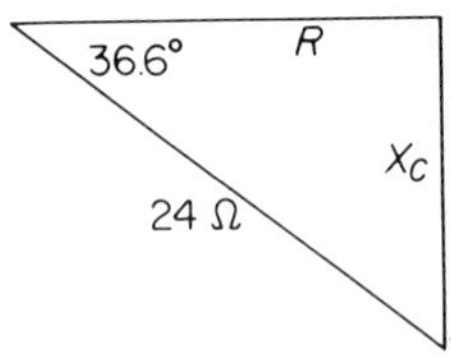

Figure 10-49. Converting Z_p to Rectangular Form

$$\text{Sin } \theta = \frac{X_C}{Z}$$

$$X_C = 24 \sin 36.6° = (24)(0.6) = 14\Omega$$

$$\text{Cos } \theta = \frac{R}{Z}$$

$$R = 24 \cos 36.6° = (24)(0.8) = 19\Omega$$

Therefore: $24\Omega \angle -36.6° = 19\Omega - j14\Omega$.

These figures tell us that the entire parallel section of the circuit in Figure 10-44 is equivalent to a 19-Ω resistor in series with a 14-Ω capacitor. This is illustrated in Figure 10-50.

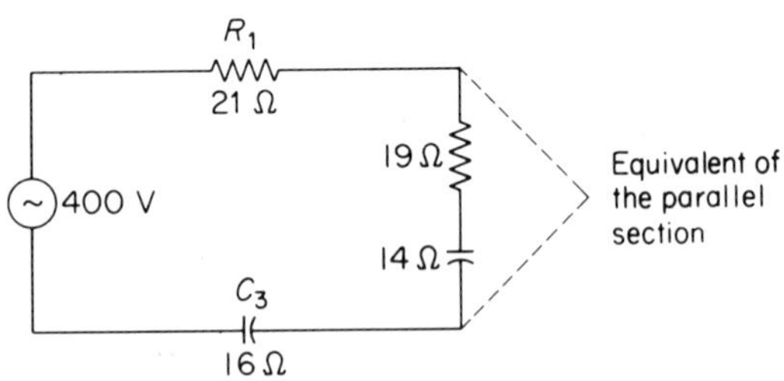

Figure 10-50. Equivalent of Figure 10-44

Combine resistances. $R = 19\Omega + 21\Omega = 40\Omega$. Combine reactances. $X = 14\Omega + 16\Omega = 30\Omega$. Therefore, $Z_t = 40\Omega - j30\Omega$. Using the triangle in Figure 10-51, we convert to polar form.

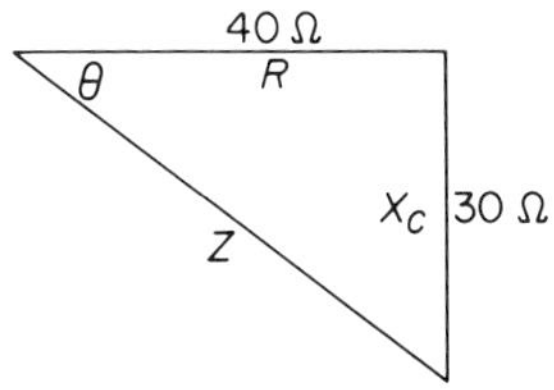

Figure 10-51. Converting to Polar Coordinates

$$\text{Tan } \theta = -\frac{X_C}{R} = \frac{-30}{40} = -0.75$$

$$\theta = \text{arc tan } -0.75 = -36.9°$$

$$\text{Sin } \theta = \frac{X_C}{Z_t}$$

$$Z_t = \frac{30}{\sin 36.9°} = \frac{30}{0.6} = 50\Omega = 50\Omega \angle -36.9°$$

Total circuit current is:

$$I_t = \frac{400\text{V} \angle 0°}{50\Omega \angle -36.9°} = 8\text{A} \angle 36.9°$$

The circuit power is: $P = (400)(8)(0.8) = 2560\text{W}$.

This concludes Chapter 10. The following review exercises will take you back through the material and reinforce your learning.

REVIEW EXERCISES

1. Impedance is a combination of resistance and __________. It implies a __________ as well as a magnitude.
2. In a series *RL* circuit, the __________ is divided vectorially among the components.
3. The voltage vectors for a series *RL* circuit places E_a at __________ degrees, E_L in quadrant one, and E_R in quadrant __________.
4. Volt-amperes is an expression of __________. When multiplied by the power factor this provides __________ in watts.

5. The power factor is the ________ of the phase angle between total current and ________.
6. In an *RL* circuit, when $R = E_L$, E_L leads the applied voltage by ________ degrees, and E_R ________ the applied voltage by 45°.
7. The total circuit current in a parallel *RL* circuit is 3A — *j*4A. Express this in polar form.
8. In all *RC* circuits, the current ________ the applied voltage.

(Items 9 through 11 refer to Figure P10-9.)

9. Draw the current triangle for the circuit in Figure P10-9.

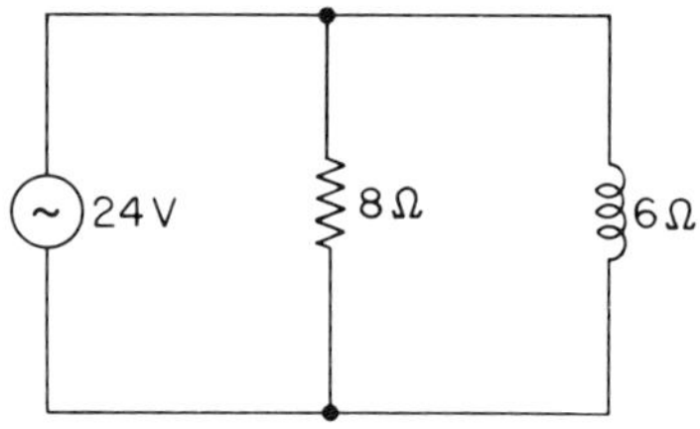

Figure P10-9.

10. What is the amplitude and phase of I_t?
11. What is the magnitude and direction of Z?
12. What is the impedance of a series circuit composed of a 20-Ω resistor and a 15-Ω capacitive reactance?

(Items 13 through 17 refer to Figure P10-13.)

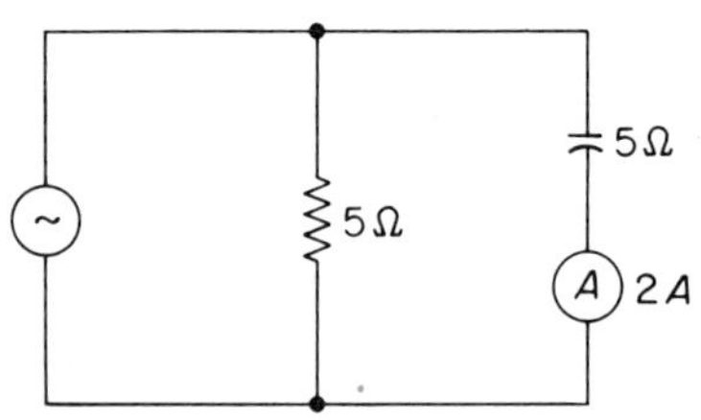

Figure P10-13.

13. What is the circuit impedance?
14. What is the applied voltage?

15. Draw the current triangle showing amplitude and angle of the total circuit current.
16. What is the power in this circuit?
17. What is the power factor?
18. Draw the circuit represented by the impedance triangle in Figure P10-18.

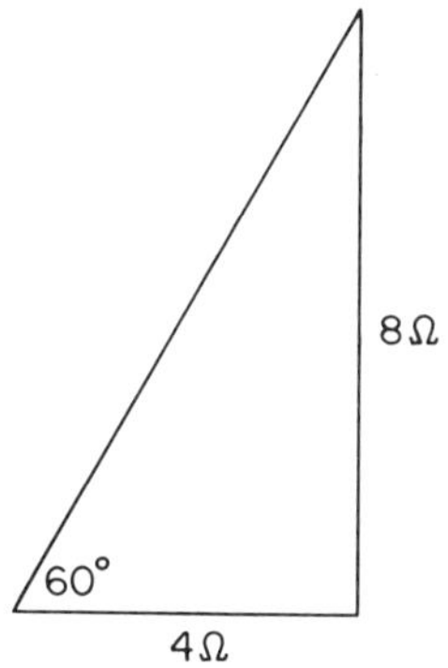

Figure P10-18.

19. Solve the circuit of Figure P10-18 for total impedance.
20. With 37 volts applied to the circuit of Figure P10-18, what will be the total: a. current; b. power?
21. Draw the circuit represented by the current triangle in Figure P10-21.

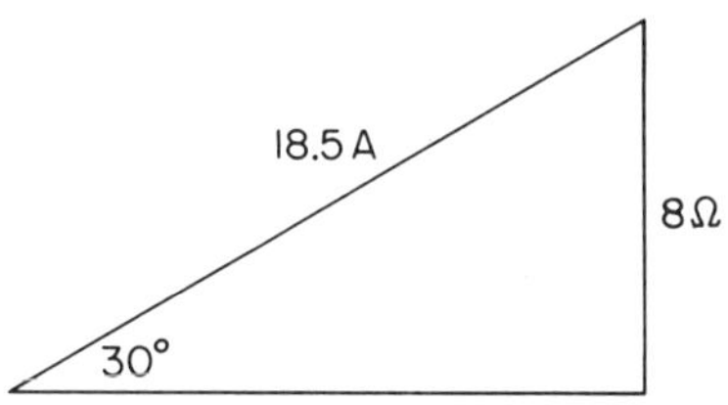

Figure P10-21.

22. With 64 volts applied to the circuit of Figure P10-21, what will be the ohmic value of the: a. resistor; b. capacitive reactance?
23. Solve the circuit of Figure P10-21 for: a. power factor; b. apparent power; c. true power; and d. total impedance.

11

advanced impedance

In this chapter we shall study *RL* and *RC transient networks* and *resistive-inductive-capacitive* (*RLC*) *combinations* in the three types of circuits. These are basic circuits used to illustrate principles of electricity, but they are more than just examples. Circuits of this type are *widely utilized* in electrical equipment.

RL TRANSIENTS

When resistors and coils are connected in a certain manner, they form a network that can be used for *timing* and *wave shaping.* The application of a dc voltage to a pure resistor causes current through the resistor to change instantly from zero to maximum. This is not the case with an inductor. Since the inductor opposes every change in current, there is a definite time factor (a delay) involved between application of voltage and the attainment of maximum current. The inductance and resistance together form a predictable time control on both the *growth and decay* of the current in the circuit.

current growth and decay

In this section we are speaking of current growth in the circuit and we are concerned with both time and amplitude of that growth.

(Figure 11-1 shows an *RL* network powered by a battery that can be switched in and out.) The *first instant* after the switch is placed in position *A* (as shown), there is zero current in the circuit and the entire battery voltage appears across the inductor. The current starts rapidly and gradually builds up to a maximum value. As the

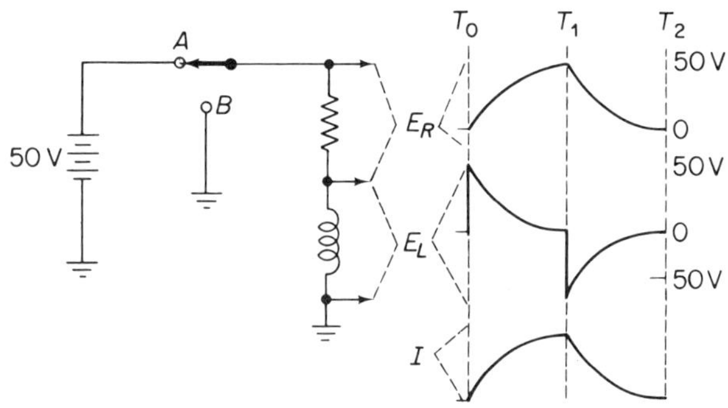

Figure 11-1. Inductor Charging

current grows, the voltage across the inductor decreases while the voltage across the resistor increases. When current reaches its maximum value, the entire battery voltage is across the resistor, and there is zero voltage across the inductor. (In actual practice this condition is never quite attained—a negligible amount of voltage is always detectable on the inductor.) With the inductor fully charged, a stable condition has been reached. This condition will continue as long as the switch remains in position *A*.

Moving the switch to position *B* removes the battery from the circuit and places a ground at each end of the network. It would appear that the current would immediately drop to zero, but this is not the case. The charged inductor becomes a power supply. It is charged to the value of the battery voltage and will gradually lose this charge as the magnetic field decays.

The contracting lines of the decaying field induce a current that opposes the change. The voltage measured across the coil at the first instant is of the same magnitude as the battery voltage, but it is of opposite polarity. As the field contracts back into the inductor, the inductor voltage grows weaker, and the circuit current decays. When the lines of force have died out, inductor voltage is zero, and circuit current is zero.

A mathematical analysis of the current and voltages would show:

1. The voltage across the inductor jumped from zero to E_a at the first instant of the growth cycle. It then dropped at an exponential rate to almost zero. It remained there until the switch changed to position *B.* At that instant, the decay cycle started, and the voltage of the coil jumped from zero to $-E_a$. The coil voltage then decreases exponentially back to zero.
2. The voltage across the resistor was zero at the first instant of the growth cycle, and it increased at an exponential rate to E_a. It held this level until the switch moved to position *B* and started the decay cycle. E_R then decreased exponentially back to zero.
3. The circuit current was zero at the first instant of the growth cycle. Thereafter, it followed the same growth and decay as E_R. Since $E_R = IR$, it could not be expected to act otherwise.

The wave shapes shown in Figure 11-1 could have been plotted *mathematically.* The voltage wave shapes may also be observed on the *oscilloscope.* By either way, we could prove that the wave shapes are approximately as shown. This is more than an interesting fact. It forms the basis for a great many timing and wave shaping circuits.

Time has been accounted for in the figure to indicate that a definite period of time is required for both the growth and decay cycles. In fact, the growth cycle requires exactly the same increment of time as the decay cycle. The time intervals from T_0 to T_1 and from T_1 to T_2 are the same. How is this time period determined? The actual time (in seconds) is dependent upon the resistance and the inductance.

RL *time constant*

An *RL time constant* is defined as the increment of time required for the current to increase to 63% of its maximum value or to decay 63% from any given value (both intervals being the same). For instance, if I_{max} were 10A, during the first time constant (*TC*) of the growth cycle, the current would rise from zero to 6.3A. Of course, this means that E_R rises from zero to 63% of E_a during the same *TC.* How about E_L? It had jumped from zero to E_a at the first instant, but then decreased. After one time constant, it had dropped by 63% (from E_a to 37% of E_a).

The changes continue in the same direction and at an exponential rate. During the second time constant, the current increases by

63% of the difference between 6.3A and 10A. Thus, at the end of the second *TC*, the current is 8.63A. Each *TC* measures 63% of the difference between the level of current when that *TC* starts and the maximum level.

These figures are fairly rough, but a careful analysis will show that *five time constants* are required for the *growth cycle*, and *five time constants* are required for the *decay cycle*. Actually, in five time constants the circuit current grows (or decays) by 99.3% of its maximum value. The same percentages apply to E_R and E_L.

You may well say, "Fine, each cycle requires five time constants, but how long is a time constant?" A time constant (in seconds) is equal to the inductance divided by the resistance. How can we obtain time by dividing resistance into inductance? Inductance is determined by a small change of current in a small increment of time. In this circuit the resistance is a definite factor in the rate of change. Therefore, $TC = L/R$, where *TC* is one time constant in seconds, *L* is inductance in henries, and *R* is resistance in ohms.

Example 1 (Figure 11-2): How long is one time constant for this circuit?

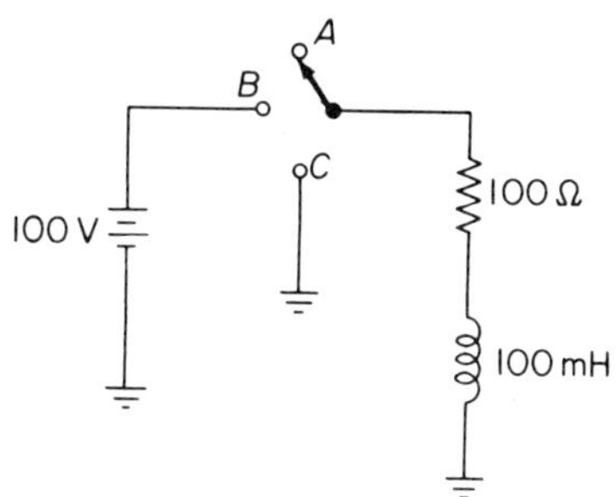

Figure 11-2. Calculate RL Time Constant

$$TC = \frac{L}{R} = \frac{100\text{mH}}{100\Omega} = 1\text{ms} = 0.001\text{s}$$

Example 2 (Figure 11-2): After the switch is moved to position *B*, how many milliseconds are required to raise I_t to its maximum value?

It requires 5 *TC*. One *TC* is 1ms. Therefore, I_t reaches maximum value in 5ms or in 0.005s.

In Figure 11-2, if we place the switch in position *B* for 5 *TC*, then move it to position *C*, we start a decay cycle with a fully charged coil. After one time constant on the decay cycle has expired, these changes have taken place:

1. E_R has dropped from 100V to 37V.
2. E_L has decreased from—100V to —37V.
3. I_t has decayed from 1A to 0.37A.

Regardless of the actual duration of the time constant, the five time constants per cycle is a fixed factor. And the 63% change per time constant will hold true in any situation. These facts have enabled the construction of the time constant chart that is shown in Figure 11-3.

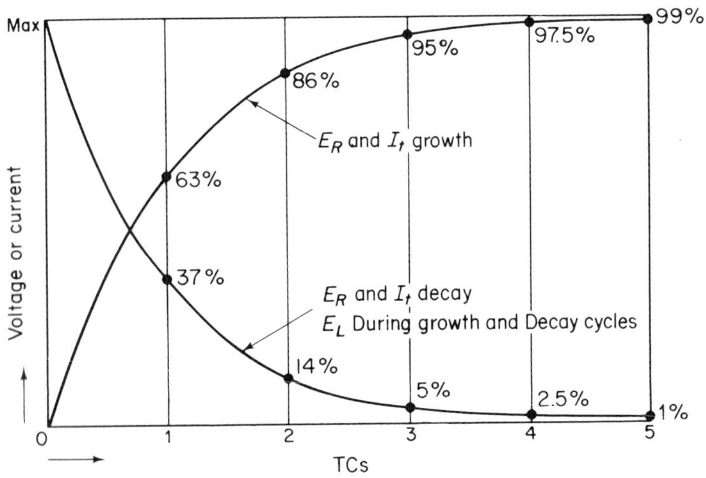

Figure 11-3. RL Time Constant Chart

The percentages reflected on this chart are only rough estimates, but they are accurate enough for most calculations. Barring unusual situations, a chart reading and a simple calculation will produce the desired results. In high precision work where exact values of voltage and current must be known at exacting intervals of time, this method would be unsatisfactory. For such cases, we have fairly exact formulas.

The current in the circuit at any instant during the growth cycle may be determined by: $i = I_{max} \times 1 - \epsilon(-t/L/R)$, where i is the instantaneous value of current in amperes, I_{max} is the maximum circuit current in amperes, 1 is a constant, ϵ (epsilon) is the base of natural logarithms, 2.718, and the exponent $(-t/L/R)$ is the elapsed time in seconds divided by the time of one TC in seconds.

A variation of the same formula may be used to determine the instantaneous circuit current during the decay cycle: $i = I_{max} \times \epsilon(-t/L/R)$. After calculating the instantaneous value of current, the result may be used with Ohm's law to determine the exact value of E_R.

The formula method of calculation gets involved, but it does produce extremely accurate results. However, the difference between the chart and formula solution is *not* significant in most cases.

Now that we are acquainted with *RL* transient circuits, we might ask how they are used. As previously stated, they are used in both wave shaping and timing. First, let's consider their application in timing.

timing

Suppose that we have a square wave with each alternation of a specified duration, and we want to synchronize a circuit with the *leading edge* of each positive alternation. It is a small problem to design a circuit that does nothing until it is triggered into action by a sharp positive pulse. The timing of this circuit is then controlled by the rate of the input triggers. This is illustrated in Figure 11-4.

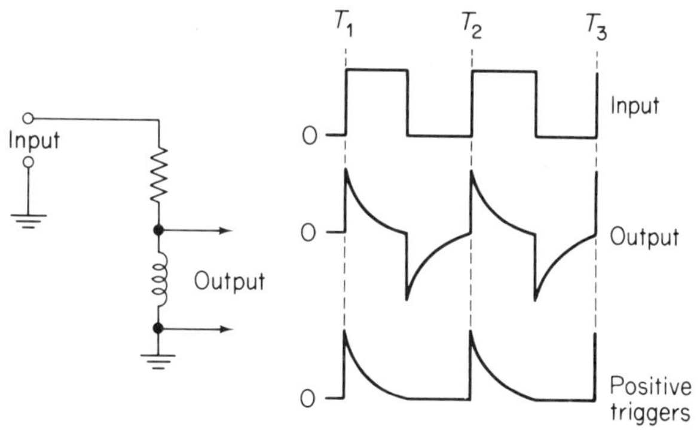

Figure 11-4. RL Transient Used for Timing

The problem here is to generate the positive input trigger pulses for the circuit previously mentioned. We have the square wave and need to alter it in such a manner that the leading edge of each positive alternation becomes a *sharp positive pulse.* In Figure 11-4 the leading edges of the square wave have been designated as T_1, T_2, T_3, etc., because they denote the times that we are interested in. Using the square wave as an input to an *RL* circuit, we take the output *across the inductor.* Each time that the input voltage level changes in a positive direction, the inductor voltage jumps to its maximum positive value, then dies out to zero. Each time that the input voltage level changes in a negative direction, the inductor voltage jumps to its maximum negative value, then dies out to zero. Thus, at the start of each positive alternation of the input square wave, the inductor produces a sharp positive pulse. Assuming that one period of the input is equivalent to 50ms, the positive pulses from the inductor will be 50ms apart. As a byproduct, we also generate a negative pulse, which coincides with the trailing edge of each positive alternation of the input.

From this point, it is a simple matter to design the circuit which is to be synchronized in such a manner that it responds only to the positive trigger pulses. Better still, another circuit may be used to strip away the negative pulses from our transient circuit output chain. This will leave a chain of positive pulses at 50-ms intervals. These pulses can be *triggers* to start or stop our imaginary circuit. In fact, chains of just this sort of trigger pulse are used to start and stop all manner of circuit actions.

Of course, the action described here has involved some wave shaping in the sense that we did alter the shape of the input wave form. Therefore, you could say that this was a wave shaping action. However, in this case it was incidental to our purpose, which was to synchronize a circuit to start operation at specific times. Wave shaping can be a much more intricate process. In fact, almost any shaped signal can be molded into just about any desired shape by our merely passing it through the proper wave shaping circuits. Let's explore this procedure just a bit further.

wave shaping

The wave shapes in Figure 11-5 were shown earlier, but we did not give them a name.

This is in no way to imply that all *TC* circuits produce the same wave shapes that you see here. On the contrary, the relation between the time constant and the time duration of the input signal determines how the outputs will be altered with respect to the shape of the input.

Three types of time constants are referred to: long, short, and medium. These categories are rather loose and somewhat arbitrary. Any *RL* transient circuit has a *long TC* for some frequencies, a *short*

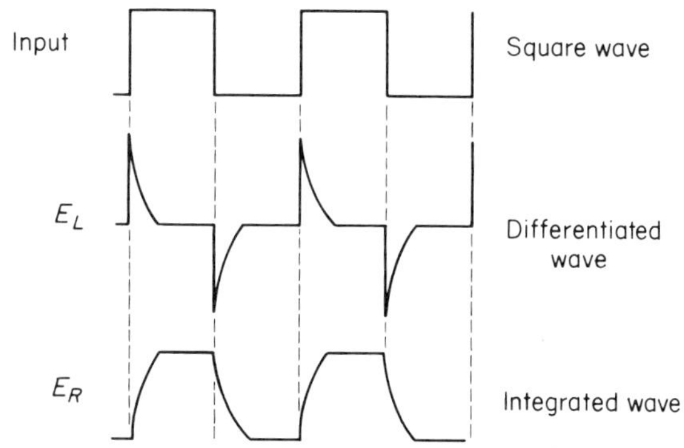

Figure 11-5. Inputs and Outputs of an RL Transient Circuit

TC for other frequencies, and a *medium TC* for still other frequencies. The controlling factor is the relationship between the time duration of the input and the time constant of the circuit. The actual time comparison for this purpose is made between one alternation of the input and one *TC* of the circuit.

A circuit is said to have a *short time constant* when one alternation of the input is at least ten times as long as one time constant. The *time constant is long* when a *TC* is at least ten times as long as one alternation of the input. A *medium time constant* lies anywhere between these two extremes. In other words, when a *TC* fails to qualify as either long or short, it is placed in the medium category.

When *differentiation* is desired, a short time constant circuit is used, and the output is taken across the inductor. For *integration* use a long time constant, and take the output across the resistor. By applying discrete amounts of integration or differentiation, we can mold any wave form into a shape that will fit the circuit needs.

There are many other types of circuits that can be used in shaping waves into precise designs. Among these are *RC* transient circuits.

RC TRANSIENTS

When a capacitor is in series with a resistor, the actions are of a *transient* nature, but they are considerably different from the actions of *RL* transient. In fact, there are enough differences to justify separate treatment of *RC* transients.

growth and decay cycles

Figure 11-6 illustrates an *RC* transient circuit with a battery that can be switched in and out.

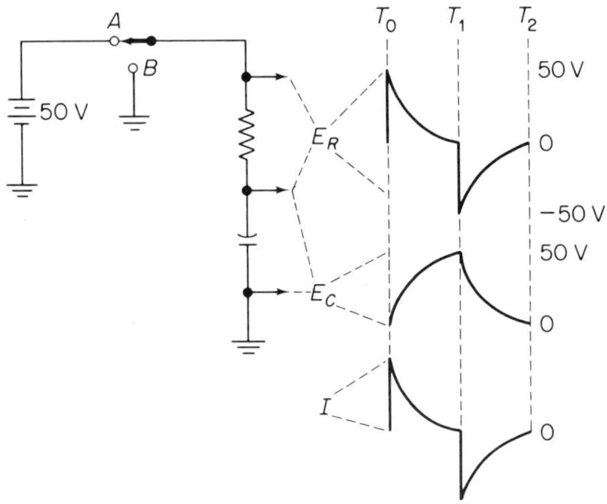

Figure 11-6. Capacitor Charging

The *first instant* after the switch is placed in position *A* the entire applied voltage appears across the resistor. The circuit current jumps from zero to maximum. This is shown at T_0 in the drawing. This current is charging the capacitor. As the charge grows, E_a begins to divide between the capacitor and the resistor. The opposition to current is growing with the charge on the capacitor, and the circuit current is decreasing.

The capacitor charges at an exponential rate and reaches full charge (E_a) at time T_1. At this time, the total applied voltage is across the capacitor; there is no voltage across the resistor; and the circuit current has ceased. The capacitor has blocked the dc. This is the end of the growth cycle, but this cycle does not represent current growth. The growth referred to here is the growth of the charge on the capacitor.

At T_1 we move the switch to position *B*. This removes the battery from the circuit and places ground at both ends of the network. This action initiates the decay cycle.

The capacitor has been charged to E_a with negative polarity at the grounded plate. Current will emanate from this plate, through ground, down through the resistor to the positive plate of the capacitor. The charged capacitor will act as a power supply.

At the first instant, current (in the reversed direction) jumps to the maximum value. This develops the full capacitor voltage ($-E_a$) across the resistor. Current reduces the charge on the capacitor at an exponential rate. As the capacitor charge drops, less voltage appears across the resistor, and the circuit current decreases.

This action continues until time T_2 when the capacitor is fully discharged. This termintes the voltage across both components, stops the current in the circuit, and brings the decay cycle to a halt. Again, the wave shapes can be plotted from mathematical calculations and can also be observed on an oscilloscope.

RC time constants

An *RC time constant* is the time required for the capacitor to charge to 63 percent of E_a on the growth cycle or lose 63 percent of the charge on the decay cycle. Both of the cycles (growth and decay) require five time constants. A time constant (in seconds) is equivalent to the product of resistance and capacitance: $TC = RC$, where TC is one time constant in seconds, R is resistance in ohms, and C is capacitance in farads.

Example 1 (Figure 11-7): One time constant for this circuit takes how many seconds?

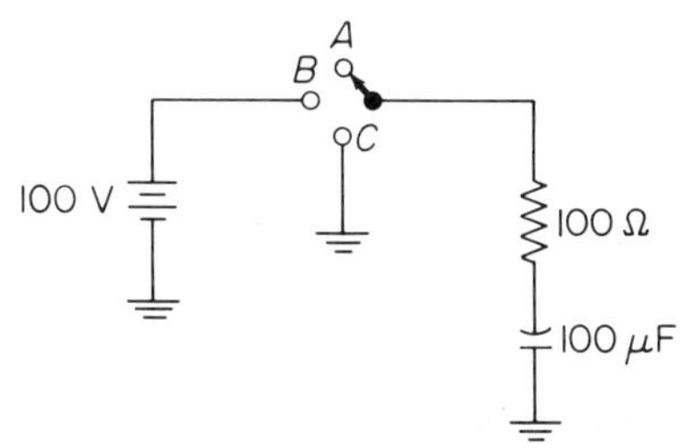

Figure 11-7. Calculate RC Time Constant

$$TC = RC = (100\Omega)(100 \times 10^{-6}) = 0.01\text{s}$$

Example 2 (Figure 11-7): After the switch is moved to position *B*, how many milliseconds are required for the capacitor to charge to 100V?

This requires five time constants. One *TC* is 10ms. Therefore, the capacitor reaches full charge (E_a) in 50ms.

The same chart that we used to estimate the values of current and voltages in the *RL* transient circuits can be used for *RC* transients. The only requirement is that we change the designations on the lines. This chart is shown in Figure 11-8.

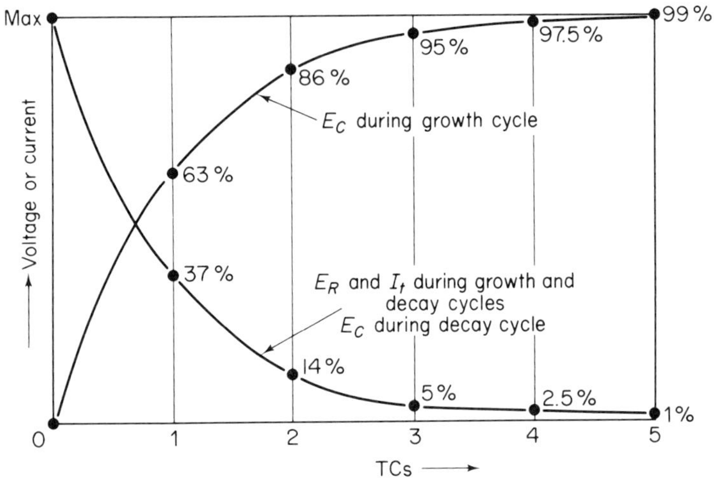

Figure 11-8. RC Time Constant Chart

Again, the percentages are only approximations. Precision work might require mathematical calculations. The formula used for current growth in *RL* circuits may now be adapted to capacitor growth voltage. The current decay formula used previously can be adapted to capacitor decay voltage. The current decay formula may be used for current in this circuit during either the growth or the decay cycle.

timing and wave shaping

RC transient circuits are also useful in timing and wave shaping. The actions are similar to those discussed with *RL* transients, but there are some differences to be watched for. One of these is the component that accomplishes the differentiation. In the *RC* circuit the differentiation occurs *across the resistor.* This is shown by the wave shapes in Figure 11-9.

The character of the time constant (i.e., whether it is long, short, and medium) determines the degree of alternation on the input wave shape. The wave shapes shown in Figure 11-9 have come from a circuit with a short time constant. The three types of *RC* time constants have definitions identical to those described for *RL* time

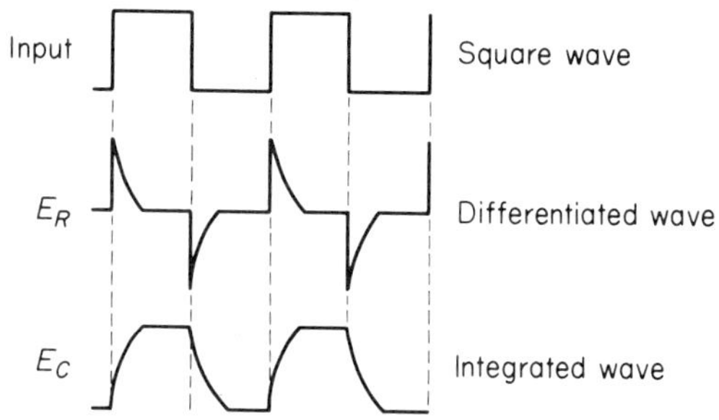

Figure 11-9. Input and Outputs of an RC Time Constant Circuit

constants. For the best *differentiation,* a *short time constant* should be used, and the output should be taken *across the resistor.* For the best *integration,* a *long time constant* should be used, and the output should be taken *across the capacitor.*

It was previously stated that a wave form could be altered to nearly any desired shape. Figure 11-10 illustrates some changes

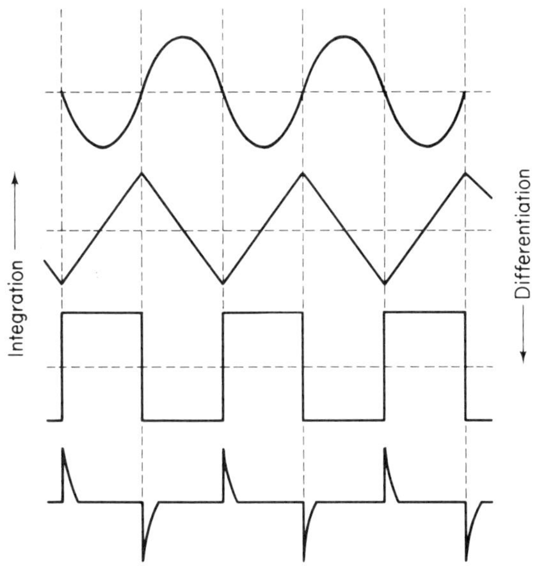

Figure 11-10. Effects of Integration and Differentiation

that may be accomplished. Any of the wave shapes in this drawing may be considered to be either input or output. Moving from bottom to top on the chart, when any wave shape is used as an input to a circuit, the *next higher* wave shape is the output across the *integrating* component. Moving from top to bottom, when any wave shape is used as an input, the *next lower* wave shape is the output across the *differentiation* component. These two conditions apply to both *RL* and *RC* transient circuits. The line across the center of each wave shape represents the zero reference line. Notice that the phase of the wave forms, as well as the shape, is changed.

Thus far, we have seen many combinations of inductors and resistors as well as combinations of capacitors and resistors. Now it is time to mix all three components in the same circuit. This will cause some interesting developments to arise.

SERIES *RLC* CIRCUITS

Circuits encountered previously should have taught us that there is a specific relationship between the current through a component and the voltage across it. This relationship is peculiar to each of the three components, but you should have noticed that for all components of any one of the types the relationship is a fixed factor. These relationships are:

1. For a Resistor: current and voltage are in phase.
2. For an Inductor: current lags voltage by 90°.
3. For a Capacitor: current leads voltage by 90°.

These facts seem to indicate that the relationships of current to the inductor and the capacitor are diametrically opposed to each other. They are in many respects, but when we put both components in the same series circuit, we know that they must at least share the same current. Let's analyze the circuit in Figure 11-11.

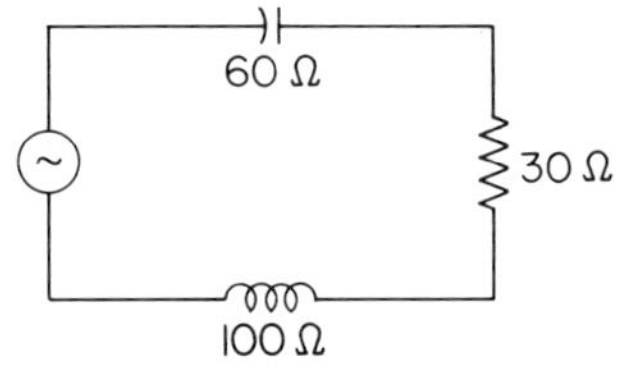

Figure 11-11. Series Circuit

impedance

The vectors for X_L are plotted at 90°; for X_C, they are plotted at —90°. The *R* vector will continue to appear at zero degrees. The vectors for the circuit of Figure 11-11 are illustrated in Figure 11-12.

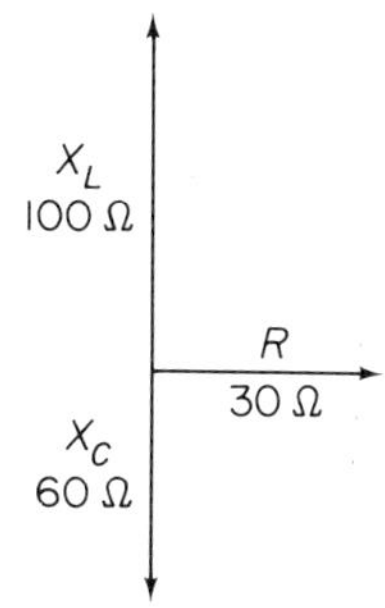

Figure 11-12. Opposition Vectors

A power supply cannot possibly recognize a circuit with two oppositions 180° apart. In effect, the smaller reactance is *swallowed* up by the larger reactance. In the process of cancellation, a portion of the larger reactance is also cancelled. The effective reactance in such a circuit is of the same type as the larger reactance and of a magnitude equivalent to the difference between the two reactances.

Consider the vectors in Figure 11-12. X_L is the larger reactance. The effective reactance in the circuit of Figure 11-11 is determined in this manner:

$$X = X_L - X_C = 100\Omega - 60\Omega = 40\Omega \text{ of } X_L$$

As far as the power supply is concerned, this is a circuit with a 40-Ω inductor and a 30-Ω resistor.

Working with the effective circuit components, we can plot a triangle as in Figure 11-13, and solve it for circuit impedance.

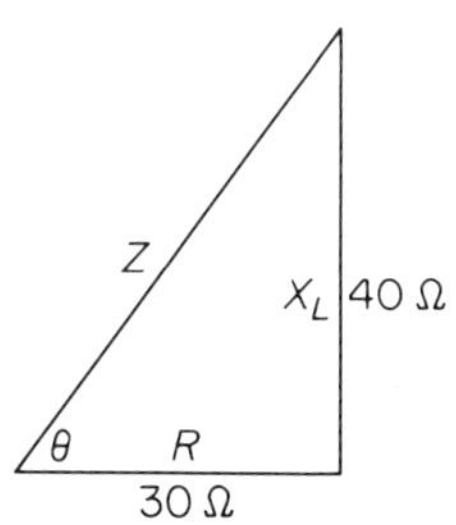

Figure 11-13. Impedance Triangle for Figure 11-11

Solving the triangle is accomplished in the conventional manner, and the result is the circuit impedance at a positive angle θ. In this case, $Z = 50\Omega \angle 53.1°$.

This means that the circuit is predominantly inductive, but that must already have been decided in order for us to have constructed the triangle. The circuit current has a negative phase angle equivalent to the positive angle θ.

When X_C is larger than X_L, the entire picture reverses. The *effective reactance* becomes $X_C - X_L$. The circuit is *predominantly* capacitive; the current vector falls in quadrant one; and the impedance vector appears in quadrant four.

The solution of right triangles has been adequately discussed in previous chapters. The discussion in this chapter will be limited (for the most part) to determining what the triangle should be.

voltages

The voltages in an *RLC* circuit react in a manner similar to the oppositions. Voltage leads current by 90° across a coil and lags current by 90° across a capacitor. Therefore, the voltages across these two components are 180° out of phase. Figure 11-14 shows an *RLC* circuit with its voltage vectors.

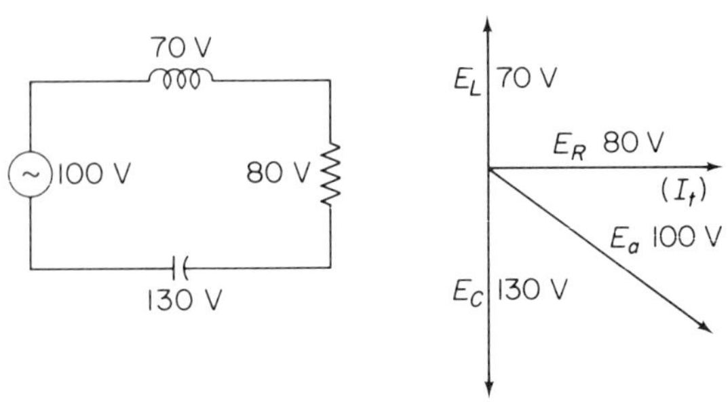

Figure 11-14. Voltage Distribution in a Series RLC Circuit

When showing the phase relation of all four voltages, we must take the E_R vector as our reference because I_t and E_R are in phase. This places I_t in the proper position with respect to both E_L and E_C.

At first glance, it would appear that the individual voltage drops around the circuit would greatly exceed the applied voltage, but this is not the case. These voltages represent vectorial quantities. The 70 volts of E_L cancel 70 of the 130 volts of E_C. This leaves an *effective* E_C of 60 volts and an E_R of 80 volts. These voltages are 90° out of phase. If they are taken as sides of a triangle, the hypotenuse will be 100 volts, which is the applied voltage.

Actually, the circuit current is not at 0° as indicated by the vectors. Since the circuit is effectively capacitive, the current vector is in the first quadrant, and the impedance vector in the fourth quadrant.

With more complex series circuits, like components should be added to reduce the circuit to an equivalent with one resistor and one reactance. We have had practice in this operation, and there should be no need for repetition.

PARALLEL *RLC* CIRCUITS

The situation becomes a bit more complicated with parallel *RLC* circuits, but this complexity will not be enough to cause apprehension. There are two practical approaches to these circuits: (1) the impedance formula solution, (2) and the total current solution. Both of these will be discussed, but you have already had practice with both.

characteristics

Figure 11-15 shows a simple parallel *RLC* circuit.

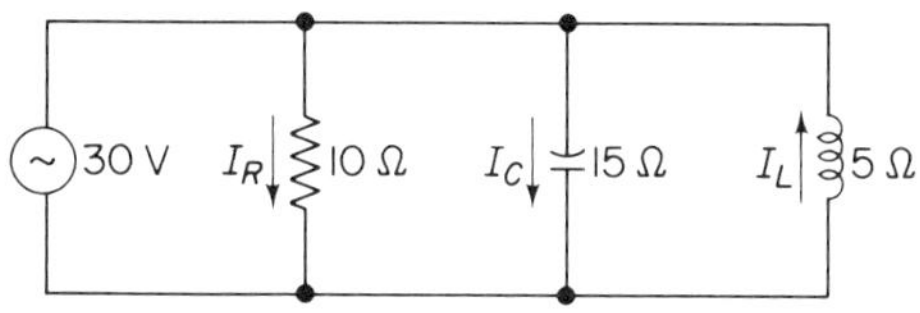

Figure 11-15. Simple Parallel Circuit

The three branch currents begin in the direction indicated for one alternation of the input voltage, then all currents reverse for the next alternation. The current I_C and I_L are always 180° out of phase.

The branch with the smallest reactance has the largest current, and it dominates the reactance of the circuit. In Figure 11-15, there is more current through the inductor than through the capacitor. The current is therefore inductive, and the total circuit current lags the applied voltage.

The applied voltage is impressed across each branch, and Ohm's law can be used for determining the amplitude of each branch current. Applying the formula produces the following currents: $I_C = 2A$; $I_L = 6A$; and $I_R = 3A$.

total current solution

Figure 11-16 shows current vectors which apply specifically to Figure 11-15, but their directions are characteristic of vectors of any parallel *RLC* circuit. Only the I_t vector will change position.

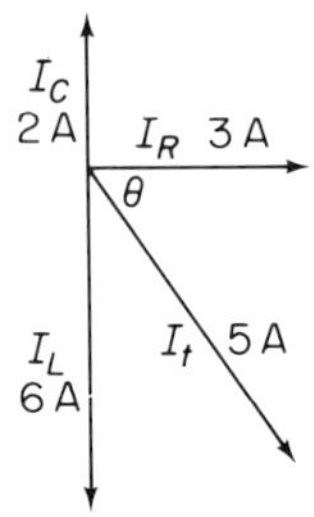

Figure 11-16. Representative Current Vectors

As previously mentioned, I_C and I_L are 180° out of phase. The larger reactive current *cancels* the smaller, and in the process cancels an equal portion of itself. In this case, two amperes of I_L cancel I_C and leave the *effective reactance* current as four amperes of I_L. This *effective current* can be diagrammatically represented as the side of a triangle; I_R will be another side. The hypotenuse of that triangle will be I_t. The triangle for this circuit is shown in Figure 11-17.

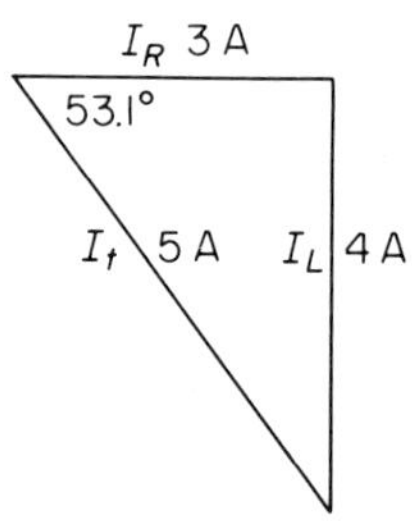

Figure 11-17. Current Triangle for Figure 11-15

Circuit impedance can be calculated by:

$$Z = \frac{E_a \angle 0°}{I_t \angle \theta}$$

In the circuit under discussion, the impedance is $6\Omega \angle 53.1°$. Check this conclusion with the above formula.

This total current approach is recommended in most cases. It involves fewer calculations and fewer opportunities for error than most methods.

Suppose that you are solving a circuit which has the ohmic value of the components given, but the applied voltage is unknown. Remember our assumed voltage method of solving for impedance. It works beautifully here. One need only mentally apply a voltage of convenient amplitude and proceed as if it were the actual voltage. Voltage and currents will be arbitrary, but the impedance will be real and correct.

The foregoing discussion was centered around a circuit that is predominantly inductive. You should reason that, when the circuit is capacitive in nature, the same principles will held true, but the triangles will be in a different quadrant; that is to say, current will have a positive angle, and the angle for impedance will be negative.

impedance approach

Even though you may have a simple parallel circuit (only one type of component per branch), it is possible that you might not desire to use the current solution. If that is the case, you may rely on our standard product over the sum formula. Remember to use two branches, then take that result with the next branch, etc., until all branches have been considered.

Substituting values from Figure 11-15, the equivalent impedance of the two reactive branches is:

$$Z_e = \frac{(15\Omega \angle -90°)(5\Omega \angle 90°)}{(15\Omega \angle -90°) + (5\Omega \angle 90°)} = 7.5\Omega \angle 90°$$

Reconstructing the formula to include the resistor and the results of this calculation, we have:

$$Z_t = \frac{(10\Omega \angle 0°)(7.5\Omega \angle 90°)}{(10\Omega \angle 0°) + (7.5\Omega \angle 90°)} = 6\Omega \angle 53.1°$$

complex branches

More often than not circuits are formed from more or less complex arrangements of components. In a parallel circuit one, several, or all branches may be composed of more than one component. Such an arrangement is illustrated in Figure 11-18. When you encounter such a situation, the proper procedure is to consider each branch to be a separate series circuit (which in fact each is). Apply

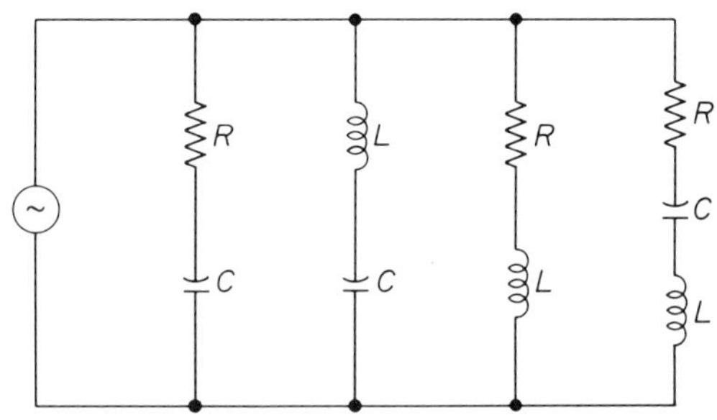

Figure 11-18. Parallel Circuit with Complex Branches

the series circuit rules, and reduce each branch to a single impedance. Use the results of these calculations to solve for total impedance. After finding impedance of each branch, you have a choice of using either the current or impedance approach.

SERIES-PARALLEL *RLC* CIRCUITS

These are the circuit arrangements which frighten most people away. But if you have followed the discussion thus far, they will be easy for you—even the most complicated machine is constructed from an assortment of simple parts. The SP *RLC* circuit appears complicated, but it is composed of two simple parts: a *series circuit* and a *parallel circuit.* You have been thoroughly drilled on how to handle each of the parts. Consider these parts rather than the whole circuit.

the parallel section

Figure 11-19 illustrates a SP *RLC* circuit. Even when viewed as

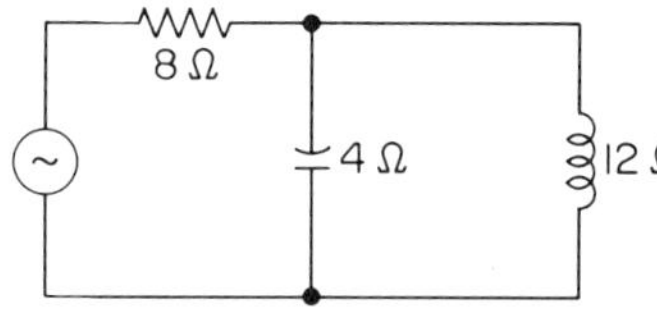

Figure 11-19. Basic *SP* Circuit

a whole, this circuit is not very formidable. But let's consider only the parallel section. Separate it from the circuit (as shown in Figure 11-20). Now apply either the current or impedance approach to

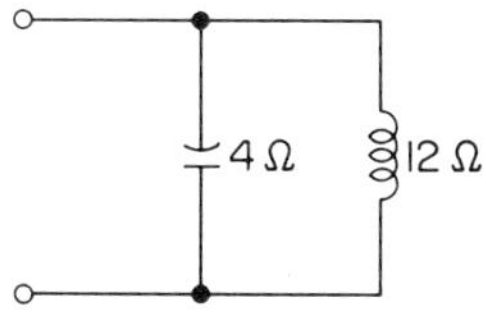

Figure 11-20. Parallel Section of Figure 11-19

determine the equivalent impedance. Is there no applied voltage?—no matter, assume some convenient value of voltage.

If you had fixed the equivalent impedance as six ohms of capacitive reactance, you would have been right. The circuit should be redrawn with this equivalent capacitor replacing the parallel portion of the circuit. The result would be the circuit in Figure 11-21.

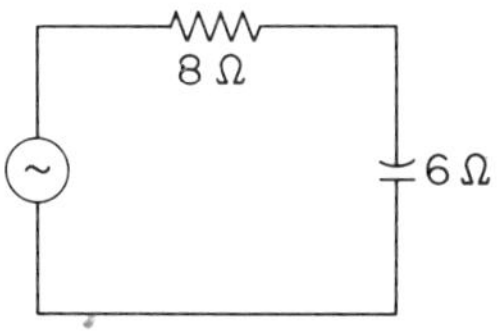

Figure 11-21. Equivalent Circuit for Figure 11-19

the series section

Once you have resolved the circuit to a series circuit, proceed with the rules of a series circuit. That is to say, add like components, subtract unlike reactances, then add the results vectorially. If these procedures are followed carefully, the final equivalent circuit will never be any more difficult than the one in Figure 11-21.

From this point, a single triangle solution yields both the magnitude and the direction of total impedance. In this case, the circuit is *capacitive.* This means that impedance falls in the fourth quadrant, and current has a leading phase angle. This is illustrated in Figure 11-22.

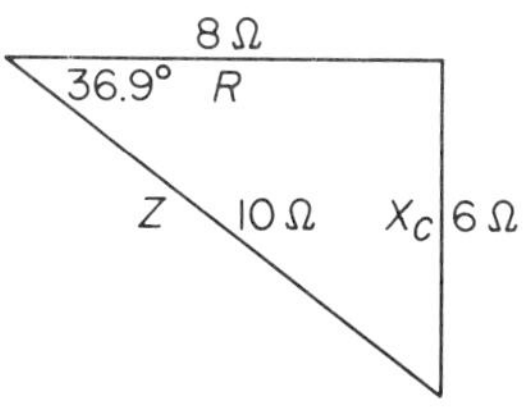

Figure 11-22. Impedance Triangle for Figures 11-19 and 11-21

This completes Chapter 11. This type of information tends to get rather involved, and it is hard for some people to remember. The following review exercises should help in this respect.

REVIEW EXERCISES

(Items 1 through 10 concern *RL* transients.)

1. The current during inductor charge is called ______________ current.
2. During the first instant after voltage is applied, all the voltage appears across the ______________.
3. When the inductor reaches full charge, all the voltage appears across the ______________.
4. The inductor gradually loses its charge during the ______________ cycle.
5. In one *RL* time constant resistor voltage and circuit current will change by approximately ______________ percent.
6. How many time constants are required for the growth cycle? How many for the decay cycle?
7. An *RL* time constant is equivalent to the ______________ divided by the ______________.
8. A circuit has a long time constant when one time constant is at least ______________ times as long as one input ______________.
9. For differentiation, the circuit should have a ______________ *TC*, and the output should be taken across the ______________.
10. For integration, the output should be taken across the ______________ of a ______________ *TC* circuit.

(Items 11 through 18 concern *RC* transients.)

11. The capacitor ______________ during the growth cycle and ______________ during the decay cycle.
12. The first instant after voltage is applied, all the voltage appears across the ______________.
13. When the capacitor reaches full charge, all the voltage appears across the ______________.
14. The capacitor charges at an ______________ rate and reaches full charge in ______________ time constants.

15. An RC time constant is the product of ____________ and ____________.
16. The capacitor changes its charge by ____________ percent during ____________ TC.
17. A sine wave may be obtained by ____________ a triangular wave.
18. A train of pulses is produced when a ____________ wave is differentiated.

(Items 19 through 25 concern nonresonant, series RLC circuits.)

19. The opposition vectors would show X_L at ____________ degrees, X_C at ____________ degrees, and R at ____________ degrees.
20. A circuit with 50Ω of X_L and 30Ω of X_C has an effective reactance of ____________ Ω of ____________.
21. A circuit with an impedance of 100Ω $\angle -36.9°$ must have an 80-Ω resistor and an effective reactance of ____________ ohms of ____________.
22. When X_L is larger than X_C, the circuit is ____________, and the current ____________ the applied voltage.
23. When X_C is larger than X_L, the circuit is ____________, and the current ____________ the applied voltage.
24. The voltage vectors show E_L to be at ____________ degrees, E_C to be at ____________ degrees, and E_R to be at zero degrees.
25. A circuit that has 100V across the inductor, 60V across the capacitor, and 30V across the resistor will have an applied voltage of ____________ volts.

(Items 26 through 30 concern nonresonant parallel RLC circuits.)

26. The current through the capacitive branch is ____________ degrees out of phase with the current through the ____________ branch.
27. The branch with the ____________ current dominates the circuit reactance.
28. When X_L is larger than X_C, the current is ____________, and the circuit current ____________ E_a.
29. A circuit that has an I_t of 10A $\angle$ 36.9° and an I_L of 6A must have an I_R of ____________ amperes and an I_C of ____________ amperes.

30. When complex branches are encountered, each branch should be treated as a separate ______ ______.

(Items 31 through 33 concern series-parallel *RLC* circuits.)

31. The most complex SP circuit is composed of two simple parts. These are the ______ section and the ______ section.
32. The complex SP circuit should be reduced by drawing ______ circuits.
33. The final equivalent circuit of an SP circuit should contain one ______ and one ______.

12

control networks

This chapter concerns impedance networks, which *control* frequency, phase, and time. We will examine both the characteristics and the uses of these relatively simple networks. They are simple in design but are very important because they are used extensively.

There is an interesting circuit condition known as *resonance* which serves a variety of purposes. Resonance may be used in a simple series circuit or in any portion of a more complex circuit. Since you are sure to encounter it many times, you need to learn something about the resonant condition.

SERIES RESONANCE

In the formulas for inductive and capacitive reactance, it was indicated that X_L is directly proportional to frequency, and X_C is inversely proportional to frequency. This means that in a given *RLC* circuit, when frequency increases, X_L increases and X_C decreases.

Take any series *RLC* combination and connect it to a power supply with a variable frequency. At very low frequencies, X_C is high and X_L is low. At very high frequencies, X_L is high and X_C is low. There is a frequency somewhere between these extremes that will cause a condition in which $X_L = X_C$. That is the *resonant frequency* for this particular circuit. When the resonant frequency is applied to a circuit, the circuit is said to be *at resonance* and is referred to as a *resonant circuit.* These points are illustrated on the graph in Figure 12-1.

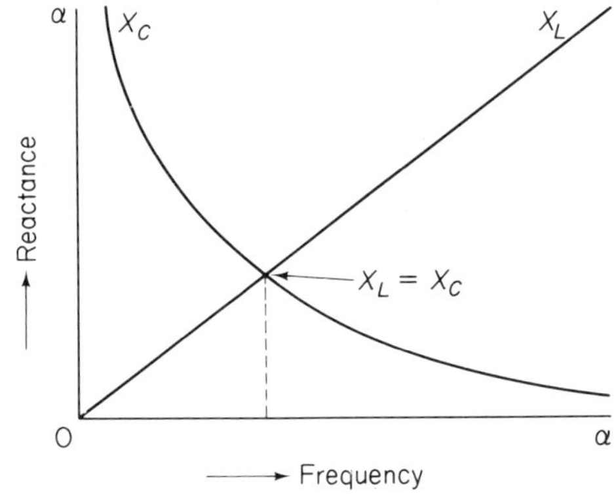

Figure 12-1. Relationship of Reactance and Frequency

The graph shows that when frequency is zero (dc), X_L is zero, and X_C is infinity (open). When frequency is very high, X_C is zero and X_L is infinity. The dotted line marks the frequency that yields X_C and X_L of the same ohmic value (the resonant frequency for this circuit).

resonant frequency

Returning to the reactance formulas, we have:

$$X_L = 2\pi fL \quad \text{and} \quad X_C = \frac{1}{2\pi fC}$$

In a state of resonance, the two formulas are the same.

$$2\pi fL = \frac{1}{2\pi fC}$$

$$f^2 = \frac{1}{4\pi^2 LC}$$

$$f_r = \frac{1}{2\pi\sqrt{LC}}$$

This is the *resonant frequency formula,* and it is applicable to any type of resonant circuit. We have designated the resonant frequency as f_r to distinguish it from all other frequencies.

Some unique things happen in a circuit that is in a state of resonance. Let's examine the impedance first.

impedance

Starting with a frequency *below the resonant frequency,* our circuit will be *capacitive.* The impedance will be high, and the current will have a large leading phase angle.

Increasing the frequency causes X_C to decrease reciprocally as X_L increases linearly. The circuit becomes less capacitive in nature because the increasing X_L cancels more and more of X_C. The phase angle between current and voltage decreases. Resistance is not dependent upon frequency, thus it holds constant. As the reactances cancel each other, impedance decreases and total circuit current increases.

When the increasing frequency reaches resonance, X_L and X_C are *equal and opposite.* All reactance is cancelled from the circuit. The resistive component becomes the total circuit opposition ($Z = R$). The entire applied voltage is dropped across the resistor. Circuit opposition to current is an absolute minimum. Current is maximum and in phase with the applied voltage.

As frequency is increased *above the point of resonance,* X_L continues to increase, and X_C continues to decrease. The circuit becomes *inductive* in nature and grows more inductive with each increase of frequency. The impedance is also increasing, which causes a decrease in current. Since the circuit is inductive, the current begins to lag the applied voltage. This negative angle grows larger as the frequency increases.

Much information has been condensed into the three preceding paragraphs. The impedance graph in Figure 12-2 will serve to emphasize the key points.

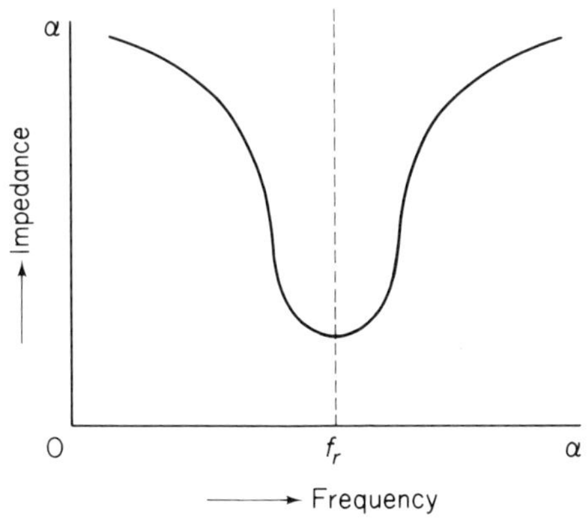

Figure 12-2. Frequency vs. Impedance

Current vs. frequency may also be plotted on a graph. If we had done so, the curve would be the reciprocal of the impedance curve in Figure 12-2.

Let's digest some of these facts by applying them to specific circuits.

solving problems

Example 1 (Figure 12-3): What value of frequency will cause resonance in this circuit?

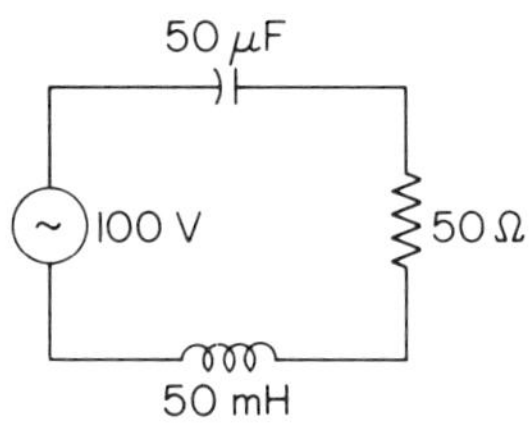

Figure 12-3. Solve for f_r

Solving for f_r:

$$f_r = \frac{1}{2\pi\sqrt{LC}} = \frac{1}{(6.28)\sqrt{(50 \times 10^{-6})(50 \times 10^{-3})}}$$

$$= \frac{1 \times 10^4}{(6.28)(15.8)} = 101\text{Hz}$$

Example 2 (Figure 12-3): How much current is in the circuit?

All of the applied voltage appears across the resistor.

$$I_t = \frac{100\text{V}}{50\Omega} = 2\text{A}$$

Example 3 (Figure 12-4): Calculate the voltage across each component.

Since $X_L = X_C$, the circuit must be in resonance.

This means that:

$$E_a = E_R = IR = 2 \times 50 = 100\text{V}$$

$$E_L = E_C = 2 \times 500 = 1000\text{V}$$

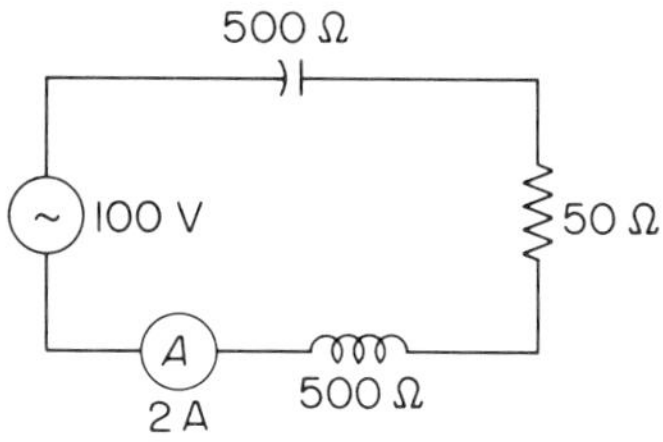

Figure 12-4. Calculate Voltages

voltage

The full applied voltage appears across the resistor, and yet each of the other components has a voltage ten times the amplitude of E_a. Is this possible? It is, because the reactive components are not loading the circuit. Their voltages are 180° out of phase so that one is charging while the other is discharging. Such a circuit provides a strong voltage gain across either of the reactive components.

Circuits of this type can be built with either very low or very high reactances. Once they have been energized, the voltage source can be completely removed and replaced with a short. The current will continue to circulate within the circuit for a prolonged period of time after the external power is removed. The reactances alternately act as power sources to keep the current going. This action is responsible for the term *"tank circuit"* being applied to a resonant circuit.

The *merit* number (or *quality*) of a resonant circuit is a ratio of E_L (or E_C) to E_R. Quality is symbolized by Q, and $Q = E_L/E_R$. Q is also a ratio of X_L to R, $Q = X_L/R$. Both of these formulas are true only when the circuit has the resonant frequency applied.

The term "The Q of the coil" is frequently encountered. The resistive components in our circuit diagrams represent a bunching of all the circuit resistance. Most of this resistance in a resonant circuit is physically located in the windings of the inductor. Thus, the Q of the circuit is the same (or nearly the same) as the quality of the inductor when no physical resistor is used. If the Q is *ten or more,* the circuit or coil is said to be *"high Q."*

The high Q inductors permit higher current with little power loss. In the circuit of Figure 12-4, $Q = 500/50 = 10$. This is classed as a high Q circuit, but it just barely qualifies for that category. It will still have a fairly high loss of power (200W). If the resistance could be reduced to five ohms by using a higher quality inductor, the same level of current would be obtained with only 10 volts applied to the circuit. This would provide the same 1000 volts across each reactance, but the power loss would drop to 20 watts.

Making use of the voltage from one of the reactive components is a tricky business. A resonant circuit is very *sensitive* to any type of loading. In fact, the coil in an ammeter changes the *Q* of the circuit to a noticeable degree. The coupling between the resonant circuit and the circuit it supplies voltage to is a critical item.

use as a filter

When we studied the principles of the dc generator, we encountered a ripple voltage. The frequency of this ripple was two pulses for each revolution of the loop. In many circuits, such a ripple is intolerable. A resonant circuit across the generator output could be used to short out and eliminate this ripple frequency. The circuit would be tuned to resonate at the undesired frequency, and it would be connected as shown in Figure 12-5. Our series resonant circuit becomes a special branch in a more complex parallel circuit.

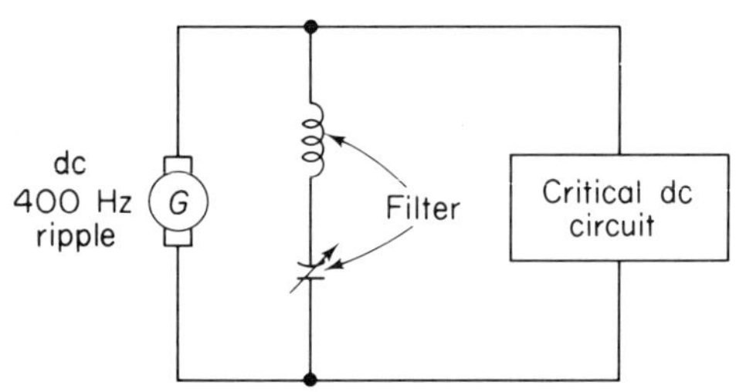

Figure 12-5. Rejection Filter

In this case the undesired ripple on the dc is at a frequency of 400Hz. The filter circuit is tuned so that its resonant frequency is 400Hz. The filter is an *open circuit* to the dc; but to the 400-Hz ripple, the filter is a direct short except for the small dc resistance of the inductor. The 400-Hz ripple is eliminated, leaving a pure dc to operate the circuit.

In this case, the resonant circuit was designed to *reject* an undesirable frequency. It may also be used to select a desired frequency from a source of many frequencies.

frequency response

The use of a resonant circuit to *select* a particular frequency or to *reject* a particular frequency should lead us to consider the degree of selection or rejection. For instance, in Figure 12-5 the filter was designed to reject a frequency of 400Hz. Does this mean that frequencies of 399Hz and 401Hz will pass without trouble? This does not appear likely. Then just where can we draw a line and state which frequencies will be passed through and which others rejected? There is no sharp point of demarcation, but there is a point that has been selected as a practical dividing line.

The dividing line falls at the *half-power points.* Peak power, peak current, and minimum impedance all occur at resonance. Also at resonance, the circuit current has a zero phase angle with the applied voltage, which gives the circuit a power factor of unity. As the frequency moves away from the resonant frequency in either direction, impedance increases sharply with a corresponding decrease in current and power. When current drops to 0.707 of its peak value (at resonance), the power will be down to 50 per cent of its value (at resonance). The points on either side of the resonant frequency where these reactions occur are called the half-power points. The half-power point on the low side of the resonant frequency is called the *low half-power point;* the half-power point on the high side of the resonant frequency is called the *high half-power point.*

The resonant circuit adequately *passes* all frequencies that fall between the low and high half-power points. Of course, its greatest response is to the resonant frequency. All signals with frequencies outside the half-power points are sufficiently attenuated so that we can assume that they are rejected. These points of half-power may be shown on both the current-frequency and the impedance-frequency graphs. On the current-frequency graph they would mark the points on the current curve where current is 0.707 of its maximum value. On the impedance-frequency graph, they would mark the points on the impedance curve where impedance is 1.414 times its minimum value. On either graph, a vertical line dropped to the frequency line will indicate the frequency where the half-power points occur. The half-power points are illustrated in Figure 12-6.

The band of frequencies that falls between the half-power points will pass through a series resonant circuit. For instance, if resonant frequency were 1400kHz, a band of frequencies from 1390kHz to 1410kHz would probably pass. Most circuits are less selective than this, but it should serve as an example. In general, the higher the quality of the circuit, the better the selectivity.

Terms frequently encountered with resonant circuits are *band width* and *band pass.* Band pass has just been described: it is that band between the half-power points. In the example the band pass lies between 1390 and 1410kHz. The band width refers to the num-

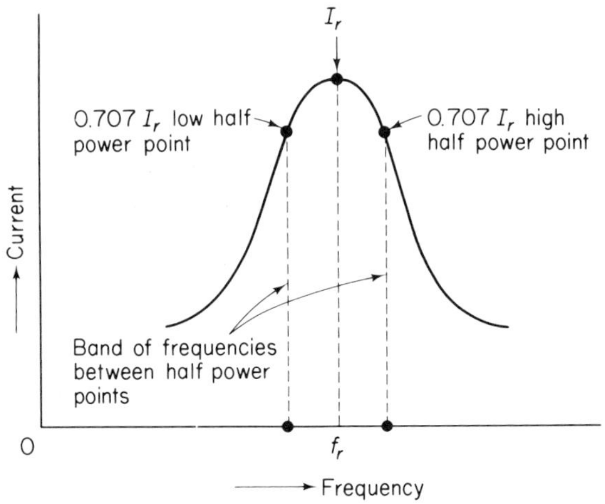

Figure 12-6. Current-Frequency Response

ber of hertz in the band pass. With the band pass described here (1390 to 1410kHz), the band width is 20kHz. This is the difference between the low half-power frequency and the high half-power frequency.

Since the current has dropped to 0.707 of the resonant current at the half-power points, the impedance must have increased to 1.414 times its minimum value. Since the minimum impedance is equivalent to the resistance, the impedance at each half-power point must be 1.414 times resistance. Since resistance is constant for all frequencies, the change in impedance is due to the change in reactance.

At the low half-power point the circuit is capacitive, and the difference between capacitive reactance and inductive reactance is equal to the resistance. The phase angle between the circuit current and the applied voltage is 45° at this time, which gives the circuit a power factor of 0.707. Assuming that the resistance is 10 ohms, we can represent the impedance by the diagram in Figure 12-7.

At the high half-power point, the circuit is inductive, and the current lags the applied voltage by 45°. Assuming the same values, we have drawn Figure 12-8 to represent the impedance at the high half-power point.

These values have been selected because they are easily adapted (by ratio) to fit any resonant circuit. For instance, a circuit with a resistance of 100 ohms would have a triangle for each half-

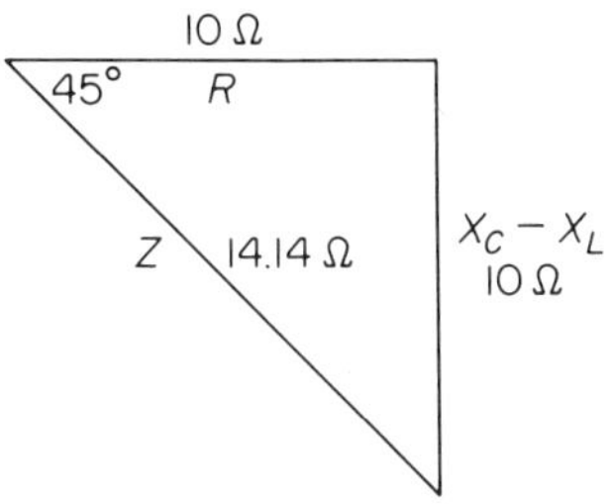

Figure 12-7. Impedance Triangle at the Low Half Power Point

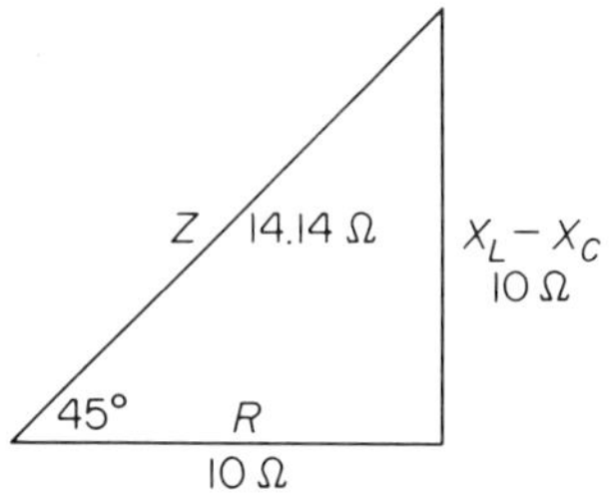

Figure 12-8. Impedance Triangle at the High Half Power Point

power point exactly like that shown in the figure except that the values of the three sides would be ten times as great.

It was mentioned that the parallel portion of more complex circuits could also form a resonant network. Let's investigate this arrangement.

PARALLEL RESONANCE

For any parallel combination of capacitance and inductance there is a frequency that will cause the circuit to become resonant. There are many similarities between series resonance and parallel resonance, but there are a great many characteristics that are totally different. Comparisons of the two types of resonant circuits will be made from time to time. First, let's consider the action in an ideal circuit.

analysis of an ideal circuit

In this circuit we will use a pure capacitor in parallel with a pure inductor, and we will have a circuit that is completely free of resistance. We will energize it momentarily from a dc source and then remove the external power completely. This circuit is illustrated in Figure 12-9.

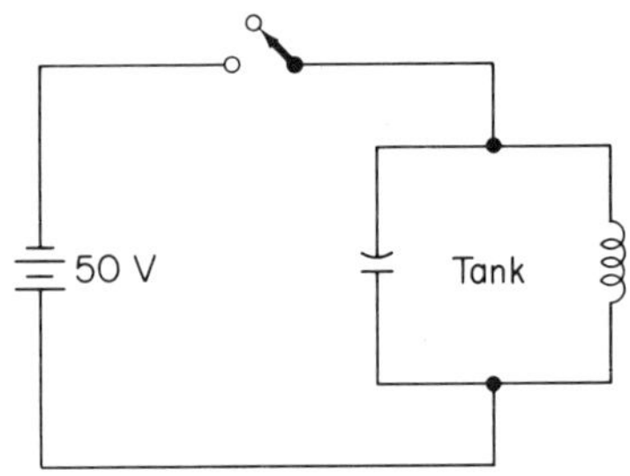

Figure 12-9. The Ideal Parallel Resonant Circuit

We have made no mention of the value of the components because any value will do as well as any other. The switch will be closed for an instant, then opened and left open. Without resistance to slow the action, the capacitor will be instantly charged to the full value of the applied voltage (50V). The polarity of this charge will be negative on the top plate and positive on the bottom. During the instant that the switch was closed, there was no current through the inductor.

When the switch is opened, the capacitor becomes the power supply and impresses its full 50-volt charge across the inductor. The CEMF of the inductor allows only a small current as the magnetic field around the inductor begins to expand. The charge on the capacitor decreases as the field about the inductor builds up.

After a time, the capacitor will be fully discharged (0V). The entire 50-volt charge will have been transferred to the inductor. As far as the capacitor is concerned, the current would have ceased at this instant. But the inductor reacts to keep current moving in the same direction. This current begins to pile electrons onto the bottom plate of the capacitor, and it begins to charge in the opposite direction. The field about the inductor is decaying back toward its center.

After a time, the magnetic field will be completely collapsed. At that instant, the capacitor will have reached a full 50-volt charge with the negative charge on the bottom plate. The capacitor will now take over again to reverse the current and recharge the inductor.

The current will continue to circulate in this manner with the two components alternately assuming the role of power supply.

How long will this continue? Since this is a perfect circuit, it has no losses. Therefore, it can be expected to continue forever.

The circulation of current within the tank circuit is called *oscillation,* and this action is referred to as the *fly wheel effect.* One complete change of events is one *period of the oscillating frequency.* For instance, a chain of events starts when the capacitor is fully charged and ends when the capacitor is fully charged again in the same direction. The oscillating frequency will be the natural resonant frequency of this tank circuit.

The ideal circuit serves a useful purpose in explaining the circulating current in a tank circuit, but unfortunately such a circuit is impossible to construct. Let's move to a more practical circuit.

the practical circuit

Since there are no pure components, our resonant circuit must contain all three elements of opposition. We can, however, group the tank resistance into one resistor in the inductive branch, which we will call R_L. The remaining circuit resistance can be practically grouped into a series resistor, which we will call R_g. The resulting circuit will appear somewhat like the one in Figure 12-10.

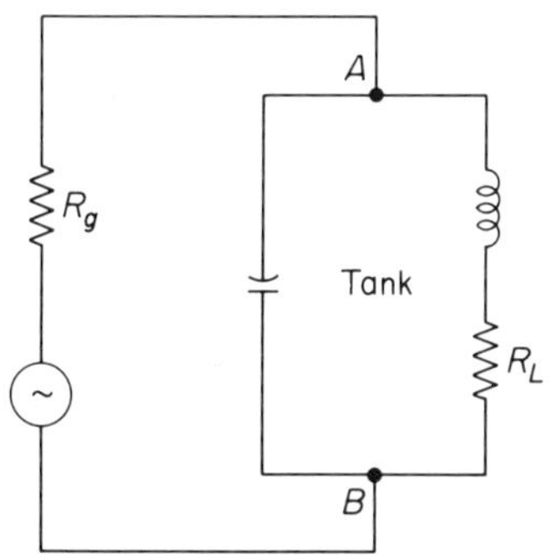

Figure 12-10. The Practical Parallel Resonant Circuit

When dealing with the parallel resonant circuit, it is necessary to differentiate between the current that circulates in the tank and the current in the line leading to and from the tank. That portion of the circuit between *A* and *B* is the *tank* (*resonant*) *circuit.* The remainder of the circuit (i.e., that portion to and from the generator) is the *line.*

In series resonance, we described three conditions of resonance. They were:

1. $X_L = X_C$.
2. Current with a zero phase angle.
3. Current at maximum value.

We may retain these three conditions, but we need to specify which currents we are referring to. The conditions can be restated as follows:

1. $X_L = X_C$.
2. Line current is in phase with E_a.
3. Tank current is maximum.

On a test bench it might be discovered that these three conditions occur at slightly different frequencies. This would be more likely with a low quality circuit. With a resonably high Q circuit, the conditions occur so close together that the difference can be ignored.

When the resonant frequency is applied to a circuit of this type, the impedance of the tank becomes:

$$Z_t = \frac{X_L X_C}{R_L}$$

Since the two reactances are equal, the formula may be restated as:

$$Z_t = \frac{(X_L)^2}{R_L} \quad \text{or} \quad Z_t = \frac{(X_C)^2}{R_L}$$

In a high Q circuit, R_L will be the dc resistance of the inductor and will have a low value. This means that, at resonance, the impedance of the tank circuit is maximum. This is the impedance offered to the power supply. This high impedance reduces the line current to almost zero. For all practical purposes the generator is looking into an open circuit. The only line current will be just enough to replace the losses that occur across R_L.

Nearly all of the applied voltage is dropped across the high impedance of the tank. A very small voltage ($I_t R_g$) is dropped across R_g. This voltage across R_g will be in phase with the applied voltage and with the line current through R_g. This line current is the total circuit current as far as the power supply is concerned, and it is designated as I_t.

Moving inside the tank circuit, we encounter an altogether different set of circumstances. Now we view the tank as a series circuit consisting of the inductor, the resistor, and the capacitor. The reactances are equal and opposite; thus they cancel each other. This leaves R_L as the total impedance in the path of the tank current. The tank is practically self sustaining. The tank current circulates back and forth at the resonant frequency rate. This current is of a high value. The magnitude of this current may be determined by dividing the applied voltage by the value of R_L. How did E_a come into play in the tank circuit?—the full value of E_a is transferred to the capacitor each time the capacitor charges.

For practical purposes, the resonant frequency formula for all resonant circuits is the same: $f_r = 1/2\pi\sqrt{LC}$. (Actually, this is not strictly true for very low Q circuits.)

In a parallel circuit the reactive branch that has the largest current determines the overall reactive nature of the circuit. At frequencies below resonance, the capacitive reactance is greater than the inductive reactance. This means that the inductive branch has the largest current. The generator detects an inductive circuit, and the circuit current will lag the applied voltage. This effect is exactly opposite to that described with the series resonant circuit.

At frequencies above the resonant frequency, the inductive reactance will be greater than the capacitive reactance; the current through the capacitive branch will be the larger current—the circuit is now capacitive. The line current will have a leading phase angle. Again, the effect is opposite to that in a series resonant circuit.

When dealing with parallel resonant circuits, any reference to Q applies strictly to the tank circuit. Since nearly all tank resistance is found within the windings of the inductor, the Q of the inductor is identical to the Q of the tank unless a physical resistor is connected into the tank circuit.

Both band width and band pass have the same meaning and function as previously described for series resonance. However, the half-power points are a bit more elusive. A half-power point occurs when the line current rises to 1.414 times its value at resonance. This is not much help because at resonance the line current is almost too low to permit accurate measurement. A half-power point is also the point where the tank current has dropped to 0.707 of its resonant value. This is even less help because we have no way of measuring the tank circuit current. The most reliable way of detecting the half-power points is to use an oscilloscope and observe the voltage across R_g. When E_g is in phase with E_a, the circuit is at resonance. When E_g is 45° out of phase with E_a, the circuit is at one of the half-power points. At the low half-power point, E_g lags E_a by 45° because the circuit is inductive. At the high half-power point, E_g leads E_a by 45° because the circuit is capacitive. The power factor of the entire circuit is unity at resonance and 0.707 at both half-power points.

use as filters

The parallel resonant circuit finds its greatest task in filtering. It can be employed to select and pass a specified band of frequencies or to reject a band of frequencies. When used as a selection filter (band pass), the output will be taken across the tank as illustrated in Figure 12-11.

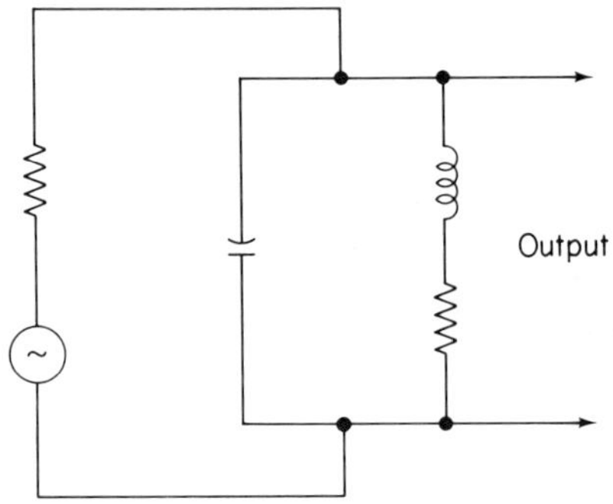

Figure 12-11. Selection Filter

The impedance of the tank reaches a peak value when the resonant frequency is applied. The magnitude of impedance drops off sharply with frequencies either above or below resonance. The band of frequencies between the half-power points will be well developed across the tank and passed on to the next circuit. Don't be misled—the output can be no greater than the input minus the losses in the line. We are not getting a voltage gain. What we do get is almost a one-to-one ratio between input and output with the desired frequencies. All other frequencies will find the tank circuit to be of a fairly low impedance, and they will be poorly developed. Thus, the undesirable frequencies are effectively eliminated while the desired band is retained with very little loss in amplitude.

There are many situations which call for a certain band of frequencies to be rejected. The parallel resonant circuit can do this job very effectively when connected as shown in Figure 12-12.

In such an arrangement a resistor is added in series with the tank circuit. The output will be developed across this resistor. The quality of the tank circuit is not affected by this external resistor.

The band of frequencies between the half-power points is blocked by the high impedance of the tank circuit. Frequencies in this band cause little or no change of current through the output resistor. Thus, they are effectively eliminated from the output.

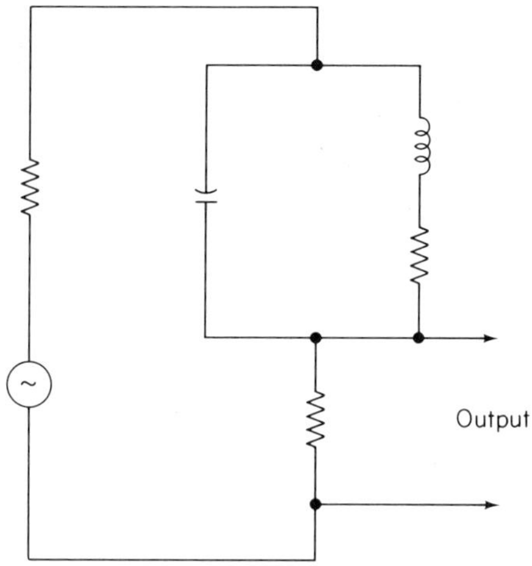

Figure 12-12. Rejection Filter

All frequencies except those between the half-power points will find little opposition from the tank circuit. They will be developed across the output resistor and passed on to the next circuit. Again there is no voltage gain, but the undesirable band of frequencies has been removed from the circuit. When the parallel resonant circuit is used in this fashion it is frequently referred to as a *wave trap.* This expression derives from the fact that the circuit traps the unwanted signals and allows others to pass freely.

solving problems

Example 1 (Figure 12-13): What is the resonant frequency?

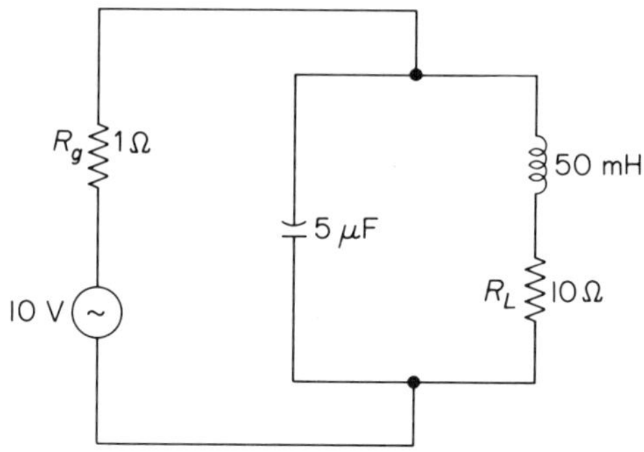

Figure 12-13. Calculate f_r

Calculate f_r:

$$f_r = \frac{1}{2\pi\sqrt{LC}} = \frac{1}{6.28\sqrt{(5 \times 10^{-6})(50 \times 10^{-3})}} = \frac{1 \times 10^4}{31.4} \doteq 318\text{Hz}$$

Example 2 (Figure 12-13): What is the impedance of the tank when the resonant frequency is applied?

Calculate either X_C or X_L, square the result, and divide by R_L. Let's try X_L:

$$X_L = 2\pi fL = (6.28)(318)(50 \times 10^{-3}) = 100\Omega$$

$$(X_L)^2 = 100 \times 100 = 10{,}000$$

$$Z_{\text{tank}} = \frac{10{,}000}{10} = 1000\Omega$$

Example 3 (Figure 12-13): How much current is circulating in the tank when the resonant frequency is applied?

Apply some logic to define the problem. Nearly all of the 10V appear across the tank. The capacitor has charged to 10V. These 10V across R_L determine the tank current; therefore:

$$I_{\text{tank}} = \frac{E_a}{R_L} = \frac{10\text{V}}{10\Omega} = 1\text{A}$$

The line current at resonance is practically zero. This very small current causes a minute voltage drop across the one-ohm resistor (R_g). The remainder of E_a appears across the tank circuit.

PHASE SHIFTING NETWORKS

In many electric circuits, the phase of the input signal is a critical factor. It may not be sufficient to have an input signal of a specified frequency, shape, and amplitude, but it may be necessary to start the input at some specified point along its time axis. If the input were to arrive in just the right phase, it would be purely accidental. *Phase shifting networks* are used to rearrange the phase of a signal to suit the demands of the circuit.

RL *circuits*

These circuits are based on the fact that the current lags the voltage in an *RL* circuit. With a proper arrangement of resistors and inductors, the output can be made to assume any desired phase angle in respect to the input. An *RL* phase shifting network is shown in Figure 12-14.

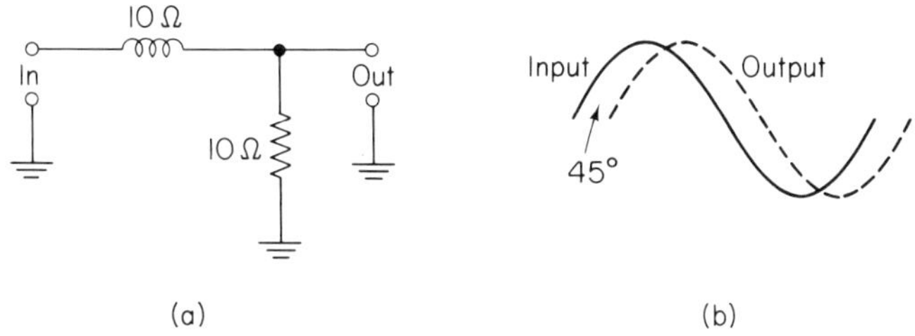

Figure 12-14. Negative 45° Phase Shift

Assuming that the input is a sine wave, the output will also be a sine wave, but it will lag the input by 45°. The input signal represents the applied voltage. The inductor opposes the change of current. The output is developed across the resistor as a result of the circuit current. With components of equal ohmic value, the phase shift is exactly 45°. The equal size components also provide an output with an amplitude of about seven tenths of the input amplitude.

One resistor and one inductor together, as shown in Figure 12-14, may be a section of a more complex phase shifting circuit. A network of three such sections is illustrated in Figure 12-15.

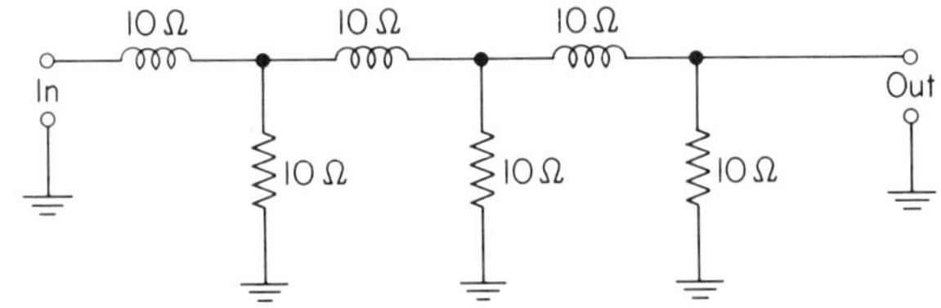

Figure 12-15. Negative 135° Phase Shift

Each of these sections provides a 45° phase shift and reduces the input signal by about 30 percent. A 10-V signal at the input produces an output of about three volts. The output lags the input by 135°.

All *RL* phase shifting circuits need not produce exact multiples of 45°. For a given section to have a greater phase shift, the section must have a higher X_L *to R* ratio. For instance, if the ratio of X_L to *R* were 4 to 3, the phase shift would be 53.1°. Reversing this ratio ($X_L = 3$, $R = 4$) would reduce the phase shift to 36.9°.

In the *RL* phase shifting circuit, the phase is shifted clockwise with respect to the phase of the input signal. The term, "output lags input," is equivalent to assigning a negative phase angle to the output. But suppose that it is desirable to shift the phase forward. In that case, capacitors would be used instead of inductors.

RC *circuits*

Figure 12-16 illustrates an *RC* phase shifting circuit.

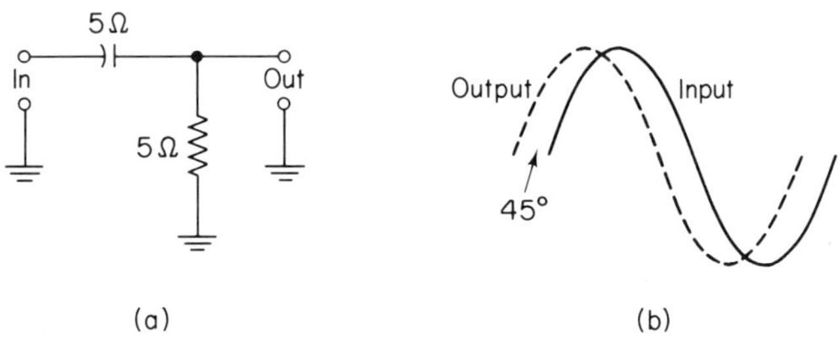

Figure 12-16. Positive 45° Phase Shift

Again, like ohmic values have been chosen in order to produce a 45° phase shift. Now the output is leading the input by 45° because the phase is shifting counterclockwise.

One capacitor and one resistor, as shown here, may be a section of a more complex phase shifting network. Each such section provides an additional 45° phase shift. Different size components may be used to provide shifts that are either less than or more than 45°. The shift is proportional to the ratio, X_C to *R*.

DELAY LINES

Another critical factor in many circuits is the time of arrival of the input signal. Suppose that we have generated a signal, molded it into the proper shape, and shifted it to the proper phase, and then find that it occurs 10μs before the circuit is ready to accept it. What can be done about timing? Can we shift time? No, we cannot manipulate time, but we can delay the arrival time of the signal.

delaying a signal

A *delay line* is constructed by connecting inductors and capacitors as shown in Figure 12-17.

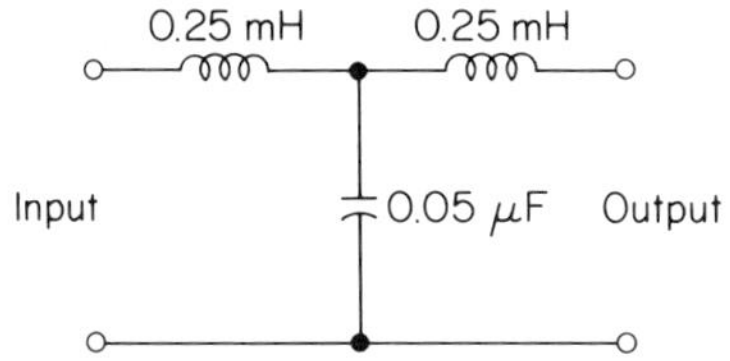

Figure 12-17. *LC* Delay Line

The time relationship between the input signal and the output signal is determined by the time required to charge the capacitor. The inductors determine the rate of capacitor charge. Time is determined by: $T = \sqrt{LC}$, where T is seconds of delay between input and output, L is total inductance in henries, and C is capacitance in farads.

The delay line shown in Figure 12-17 will produce an output 5μs after the arrival of the input. This was determined by $T = \sqrt{LC}$.

$$T = \sqrt{(0.5 \times 10^{-3})(0.05 \times 10^{-6})} = \sqrt{25 \times 10^{-12}}$$
$$= 5 \times 10^{-6} = 5\mu s$$

extending the delay

The delay line used in Figure 12-17 may be a section of a more complex delay line. The line may be extended by adding more identical sections. Each additional section would increase the delay by another 5μs. This is illustrated in Figure 12-18.

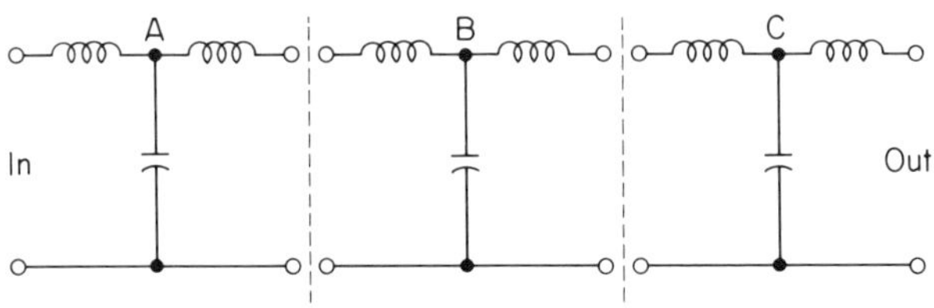

Figure 12-18. Joining Sections of a Delay Line

Assuming that each inductor is 0.5mH and each capacitor is 0.05μF, the delay of each section has already been calculated from Figure 12-17 to be 5μs. When section *B* is joined to section *A*, the total delay will be 10μs. Section *C* may then be joined to *B*, this will provide another 5μs and a total delay of 15μs.

We may continue to add sections in this manner to provide any reasonable amount of delay. Furthermore, taps may be placed on the line where any two sections join. This enables us to take several signals from the same line at various delay intervals. Figure 12-19 illustrates input and output for a delay line of this type.

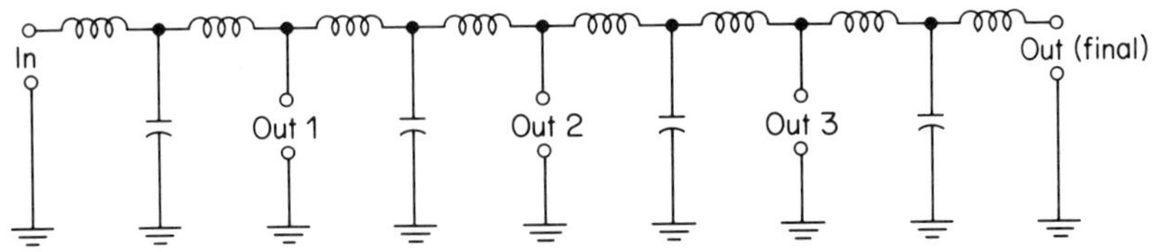

Figure 12-19. Tapped Delay Line

Assuming that the values of inductance and capacitance are the same as those used previously, this delay line provides four outputs at intervals of 5μs with respect to the leading edge of the input signal. The time relationship is illustrated on the graph in Figure 12-20.

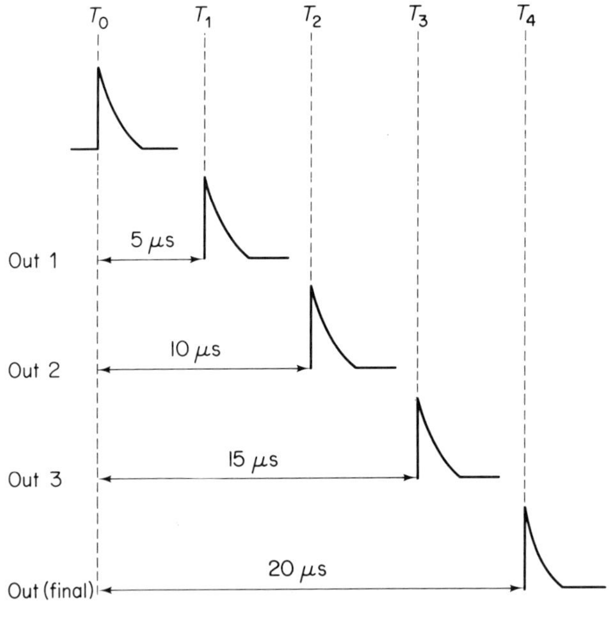

Figure 12-20. Time Relation of Multiple Outputs

Of course, there is some attenuation of the signal as it moves down the delay line. However, delay lines are fairly efficient devices. Much greater delays than these may be accomplished without appreciable attenuation. Tapped delay lines, similar to these described here, are often used to trigger a chain of events in a definite time sequence.

This concludes Chapter 12. The following review exercises will take you back through the chapter and reemphasize some of the key points.

REVIEW EXERCISES

(Items 1 through 14 concern series resonance.)

1. For any series combination of *RLC* components there is a ____________ that will cause resonance.
2. At resonance, $X_C =$ ____________.
3. The formula for calculating resonant frequency is ______ ______.
4. At frequencies below resonance, the circuit is ____________; at resonance, it is ____________; and above resonance, it is ____________.
5. Below resonance, I_t ____________ E_a; at resonance, I_t is in ____________ with E_a; and above resonance, I_t ______ ______ E_a.
6. At resonance, the circuit impedance is ____________, and it ____________ rapidly for frequencies above and below this point.
7. What is the resonant frequency of a circuit with a 0.05μF capacitor and a 0.5 mH inductor?
8. A resonant circuit has 10mA of current, and there are 500V across the inductor. What is the ohmic value of X_C?
9. Dividing X_L by R is one means of determining the ______ ______ of the circuit.
10. At the half-power points, the power factor is ____________.
11. The frequencies between the half-power points represent the ____________ ____________.
12. Subtracting the frequency at the low half-power point from the frequency at the high half-power point will reveal the ____________ ____________.
13. At the low half-power point, the current leads the ______ ______ ____________ by ____________ degrees.

14. At the high half-power point the current __________ E_a by __________ degrees.

(Items 15 through 26 concern parallel resonance.)

15. The fly wheel effect refers to __________ of current in the tank circuit.
16. When the resonant frequency is applied, tank current is __________, X_L is equal to __________, line current is __________ and in phase with E_a.
17. The formula for tank impedance at resonance is __________.
18. At frequencies below resonance, the circuit is __________.
19. At frequencies above resonance, the circuit is __________.
20. When no physical resistor is used, the Q of the inductor is identical to the Q of the __________.
21. Line current lags E_a by __________ degrees at the __________ half-power point and leads E_a by __________ degrees at the __________ half-power point.
22. When the output is taken across the tank, the circuit is a __________ filter.
23. When the output is taken across a resistor which is in series with the tank, the resonant circuit is a __________ filter.
24. The selection filter passes the band of __________ between the __________ points.
25. The rejection filter eliminates the band of __________ between the __________ points.
26. The tank current at resonance is determined by dividing __________ by __________.
27. An RL circuit provides a phase shift which is proportional to the ratio of __________ to __________.
28. A tapped delay line may supply several outputs with varying __________ intervals.

appendix a
bibliography

DE FRANCE, JOSEPH J., *Electrical Fundamentals,* Prentice-Hall, Inc., Englewood Cliffs, N. J., 1969.

JACKSON, HERBERT W., *Introduction to Electric Circuits, 3rd ed.,* Prentice-Hall, Inc., Englewood Cliffs, N. J., 1970.

MANDL, MATTHEW, *Fundamentals of Electronics, 2nd ed.,* Prentice-Hall, Inc., Englewood Cliffs, Inc., N. J., 1965.

MARCUS, ABRAHAM, *Basic Electricity,* Prentice-Hall, Inc., Englewood Cliffs, N. J., 1968.

MATTSON, RAY H., *Electronics,* John Wiley and Sons, Inc., New York, N. Y., 1966.

ROBINSON, VESTER, *Electrical Concepts,* Reston Publishing Company, Reston, Va., 1972.

appendix b

answers to review exercises

chapter 1

(1) Small atomic particles. (2) Electrical. (3) Matter. (4) Original substance. (5) a. Small negative particle of electricity. b. In orbit about the nucleus of an atom. (6) The electrons are much lighter than the protons. (7) Element. (8) Compound. (9) Protons. (10) An atom with a charge. (11) Deficiency. (12) Like charges repel and unlike charges attract. (13) Charge; charges. (14) Electrostatic field. (15) The same as the direction a positive charge would take when placed in the field. (16) Repel. (17) The greatest charge is at the point of greatest curvature. (18) Neutralizing. (19) Battery. (20) It is from the negative terminal to the positive terminal of the voltage source externally and from the positive to the negative terminal internally. (21) Voltage, electrical pressure, difference in potential, and electromotive force. (22) Volts; *E*. (23) Electrons; amperes; *I*. (24) 6.24×10^{18}. (25) An ampere is one coulomb passing a point in one second. (26) Current; ohms; *R*. (27) Current is directly proportional to voltage and inversely proportional to resistance. (28) Ohm's; current; voltage. (29) Silver and copper are good conductors. Oil and mica are good insulators.

(30) 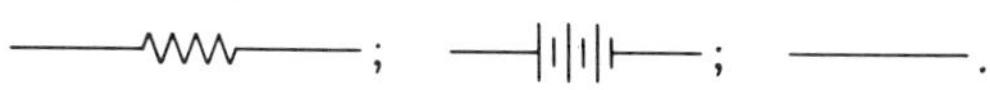

(31) 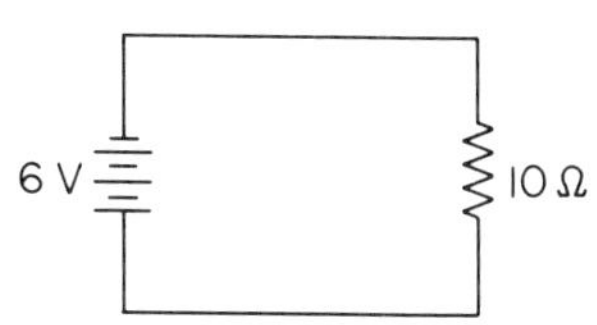

(32) 135V. (33) 0.02A.

(34) a. Current is reduced by half. b. Current doubles. (35) 2Ω. (36) 24.4Ω. (37) 0.9A. (38) 12V.

chapter 2

(1) a. Directly proportional. b. Inversely proportional. (2) a. —(A)—. b. —(V)—. c. —(Ω)—. (3) Reciprocals. (4) 0.1Ω. (5) Directly; inversely; directly. (6) They are inversely proportional. (7) Heat; light; mechanical; chemical. (8) Generator. (9) Cells and batteries. (10) Zinc case, carbon rod, and electrolyte. (11) 4.5; greater. (12) Spongy lead, lead peroxide, and electrolyte. (13) Liquid electrolyte. (14) Capacity; batteries. (15) Discharged. (16) Other water may contain harmful chemicals and minerals. (17) Carbon; wire. (18) Physical size is outward dimensions. Ohmic value is electrical resistance. They have no relation to one another. (19) Heat. (20) Fixed, adjustable, and variable. (21) Adjustable. (22) Current; voltage. (23) A resistor value found in common stock. (24) Ohmic value, tolerance, and percentage of failures. (25) 2,900,000,000Ω, 3% tolerance, 1% failure per 1000 hours. (26) 12Ω, 20% tolerance. (27) a. Switch. b. Rheostat. c. Terminals. (28) 2×10^6. (29) 7.2×10^{19}. (30) a. 0.003A; 3,000μA. b. 4,000kΩ; 4,000,000Ω. c. 6000V; 6,000,000mV.

chapter 3

(1) A circuit in which current is in only one direction. (2) 11.8V. (3) Current in the wrong direction may harm the meter. (4) 0.6A. (5) Ammeter. (6) Ohmmeter. (7) The scale is nonlinear, and the left half is poorly calibrated. (8) To compensate for a weak voltage cell in the meter. (9) Path. (10) $E_a = E_1 + E_2 + E_3$; $I_t = I_1 = I_2 = I_3$.

(11)

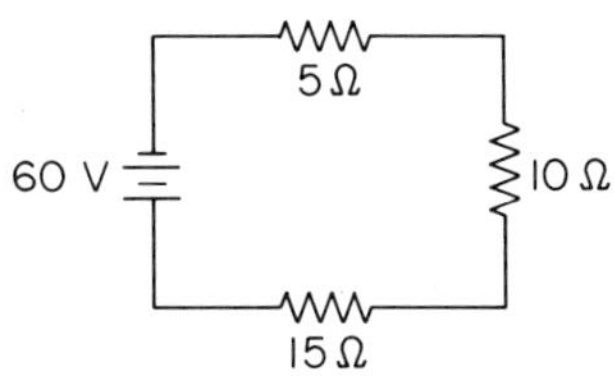

(12) $I_t = 2A$; $R_t = 30\Omega$.

(13)

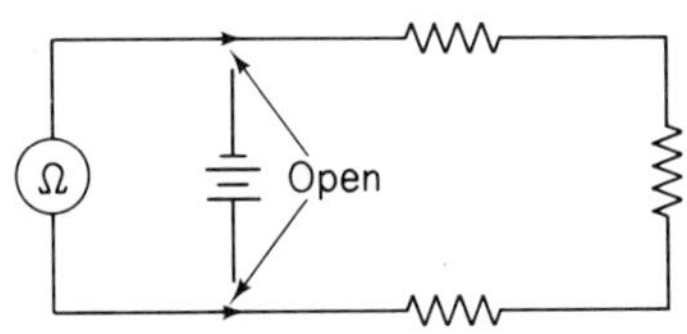

(14) $5\Omega = 10V$; $10\Omega = 20V$; $20\Omega = 40V$; $40\Omega = 80V$. (15) $6M\Omega$. (16) They must be changed to a common power of ten; unlike powers represent unlike quantities. (17) E_a is distributed in proportion to the number and size of the resistors. (18) $20k\Omega$. (19) Time rate. (20) 746. (21) $P = EI; P = I^2R; P = E^2/R$. (22) Summation. (23) Reference; common. (24) Battery. (25) Negative. (26) Fuse. (27) Restored. (28) Shorts; opens. (29) E_a; zero. (30) $\propto$; zero. (31) Ohmmeter; continuous. (32) a. 4A; b. 12.5Ω. (33) a. 4.16Ω; b. 66.67W. (34) 100V.

chapter 4

(1) Current; current. (2) $E_a = E_1 = E_2 = E_3$; $I_t = I_1 + I_2 + I_3$.

(3) $R_t = \dfrac{R_1R_2}{R_1 + R_2}$ $\qquad R_t = \dfrac{1}{\dfrac{1}{R_1} + \dfrac{1}{R_2} + \dfrac{1}{R_3}}$.

(4) Conductance. (5) $10k\Omega = 7.5mA$; $30k\Omega = 2.5mA$; $P_t = 750mW$. (6) $19k\Omega$. (7) Increases; decreases. (8) Adding. (9) A short blows the fuse removing the voltage. An open gives the same voltage reading as a good branch. (10) Otherwise, the meter indicates the equivalent resistance of all good branches. (11) Series; parallel.

(12) $R_t = R_s + \dfrac{1}{\dfrac{1}{R_1} + \dfrac{1}{R_2} + \dfrac{1}{R_3} + \dfrac{1}{R_4}}$.

(13) Series; parallel. (14) $E_a = 130V$; $R_t = 9.28k\Omega$; $P_t = 1.82W$; $I_t = 14mA$. (15) Increase; decrease; decrease; increase, decrease. (16) Voltage. (17) Voltage. (18) Branch; voltage. (19) Load. (20) Variable. (21) Rheostat; series. (22) Resistance. (23) Unbalanced. (24) Equation; loop. (25) $3I_1 - 2I_2 = 10$. (26) 2A. (27) Equivalent.

(28) $R_a = \dfrac{R_1R_3}{\Sigma R\Delta}$; $R_b = \dfrac{R_1R_2}{\Sigma R\Delta}$; $R_c = \dfrac{R_2R_3}{\Sigma R\Delta}$.

(29) $R_1 = (R_aR_b + R_bR_c + R_aR_c)/R_c$
$R_2 = (R_aR_b + R_bR_c + R_aR_c)/R_a$
$R_3 = (R_aR_b + R_bR_c + R_aR_c)/R_b$.

(30) $R_a = 5\Omega$, $R_b = 3.33\Omega$, $R_c = 10\Omega$.

(31) $R_1 = 36.67\Omega$, $R_2 = 110\Omega$, $R_3 = 55\Omega$.

chapter 5

(1) Electricity. (2) Nonmagnetic. (3) Natural, artificial. (4) Temporary and permanent. (5) Retentivity. (6) The magnetism remaining

in a material after the magnetizing influence is removed. (7) Magnetic field. (8) Flux; ends. (9) Poles; north pole. (10) Like poles repel and unlike poles attract. (11) North. (12) Cross; north; south. (13) $F = m_1m_2/d^2$. (14) Ease. (15) 20 dynes. (16) Pole. (17) $H = F/m$; oersteds. (18) The field strength that will exert one dyne of force on a unit pole. (19) Flux density; maxwell. (20) 2×10^5 gauss. (21) A theory is a logical idea used to explain some phenomenon. It becomes a law when it has been thoroughly proven. (22) Weber's. (23) Magnet. (24) Spin. (25) Half. (26) 10^{14}. (27) Diamagnetic, paramagnetic, and ferromagnetic. (28) Diamagnetic. (29) They have a high permeability, a high positive susceptibility, and are strongly attracted to magnets. (30) Current; voltage, resistance, conductance. (31) $\Phi = \text{mmf}/R$. (32) Φ = maxwell, mmf = gilberts, R = rels. (33) It concentrates the flux into a smaller path. (34) Reluctance flux. (35) Flux leakage. (36) Enclose it in a case of low reluctance (high permeability) metal. (37) Flux; current. (38) Flux. (39) Counterclockwise. (40) Repel; attract. (41) a. Attraction. b. Repulsion. (42) Upward. (43) $H = 2\pi I/10r$. (44) A (or left). (45) Solenoid (or coil). (46) Gilberts. (47) Induced. (48) The generator principle. (49) The transformer principle. (50) The polarity of induced EMF will be such as to oppose the force that induced it. (51) $e = N\Phi/t \times 10^8$. (52) Come toward.

chapter 6

(1) Inductance. (2) Inductance; henries. (3) $M = K\sqrt{L_1L_2}$. (4) All flux lines of L_1 would be linking all loops of L_2. (5) Retentivity. (6) Coercive; residual. (7) Magnetic. (8) Coil, core, armature, contacts, and spring. (9) Coil; permanent. (10) Two; one. (11) Voltage times current. (12) Mechanical and electrical. (13) Electrical. (14) Millivolts for a full scale deflection. (15) 0.25Ω. (16) 27.5mA. (17) Series; extend. (18) 500V. (19) 1kΩ/V. (20) Internal. (21) External. (22) Meter movement.

chapter 7

(1) Direction; amplitude. (2) Constantly; periodically. (3) The size. A dynamo produces large quantities of power while the magneto produces very small quantities of power. (4) The generator principle. (5) Nothing; current continues in the same direction. (6) Reverse either (but not both) the direction of the field or the direction of the motion. (7) Periodic. (8) Sine wave. (9) Vector length. (10) Period; sine. (11) They connect a rotating loop to an external circuit. (12) At

0° and 180°. (13) 90 and 270. (14) $e_i = e_m \sin \theta$. (15) 3.5 V. (16) Alternating. (17) Commutator; slip rings. (18) Insulated; commutator. (19) Pulsating; varies; polarity. (20) Conductor; magnetic field. (21) Motor. (22) Rotate; shaft. (23) Torque. (24) 90; 270. (25) The inertia carries the loop through the area of zero torque. (26) Period. (27) dc; ac. (28) Changes dc to ac. (29) Dead center; torque. (30) This happens at the dead center positions when there is no torque being produced.

chapter 8

(1) Magnitude, direction; velocity. (2) Counterclockwise; clockwise. (3) 43,380°. (4) The sexagesimal system. (5) Radian. (6) 2π radians. (7) π. (8) Frequency. (9) One period. (10) Velocity; radians. (11) $\theta = \omega t$. (12) 5×10^6 meters. (13) The sine wave starts at zero amplitude at zero time. The cosine wave starts at maximum amplitude at zero time. (14) They are 90° out of phase. (15) 180°. (16) Acute or complementary; 90. (17) 45°. (18) Opposite; tangent. (19) Cosine. (20) First. (21) III. (22) Pythagorean. (23) 6.75. (24) 45°; large. (25) 40; 61. (26) Magnitude; direction. (27) Complex; vector. (28) $9.2 - j12.8$. (29) 180°. (30) $5 \angle 53.1°$. (31) Rectangular; polar. (32) Right triangle. (33) The ac meter has a rectifier. (34) Effective. (35) Oscilloscope. (36) Time; frequency; phase. (37) Phosphorescent; electrons. (38) Electron; screen. (39) Positive, negative. (40) Amplitude; time. (41) Sawtooth; horizontal. (42) Push-pull. (43) Lissajous. (44) Same; 90. (45) Effective. (46) 28.2. (47) 2206mW.

chapter 9

(1) Resistance; reactance. (2) Inductive and capacitive. (3) Coil; CEMF. (4) Inductive reactance; X_L. (5) Frequency; inductance. (6) 5.6kΩ. (7) Lags; 90.

(8) $L_t = \dfrac{L_1 L_2}{L_1 + L_2}$; $L_t = \dfrac{1}{\dfrac{1}{L_1} + \dfrac{1}{L_2} + \dfrac{1}{L_3}}$.

(9) Divides. (10) Current lags voltage by 90°. (11) Removes; replaces. (12) Magnetic field. (13) Resistor. (14) Mutual. (15) Capacitor. (16) Dielectric field. (17) Area of the plates, distance between the plates, and the dielectric constant of the material between the plates. (18) Dielectric. (19) Normal. (20) Open, short. (21) Less. (22) Capacitive reactance. (23) Decreases. (24) Leads; 90. (25) Block; pass. (26) 15kHz. (27) Increase.

chapter 10

(1) Reactance; direction. (2) Voltage. (3) Zero; three. (4) Apparent power; true power. (5) Cosine; applied voltage. (6) 45; lags. (7) 5A $\angle -53.1°$. (8) Leads. (9)

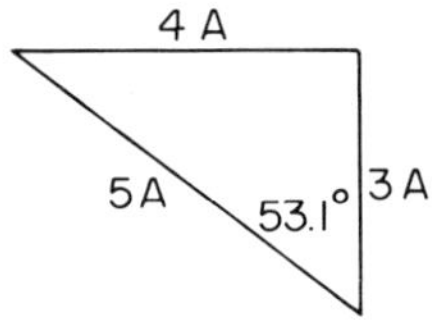

(10) 5A $\angle -53.1°$. (11) 4.8Ω $\angle$ 53.1°. (12) 25Ω $\angle$ 36.9°. (13) 7.07Ω $\angle -45°$. (14) 14.14V.

(15)

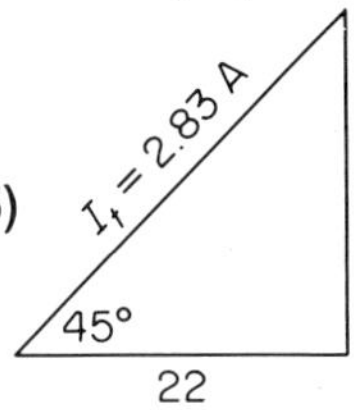

(16) 20W. (17) 0.707.

(18)

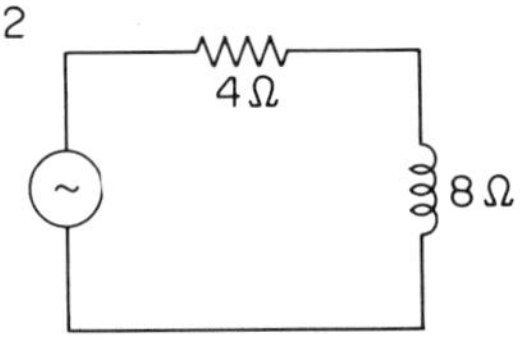

(19) 9.25Ω. (20) a. 4A. b. 64W.

(21)

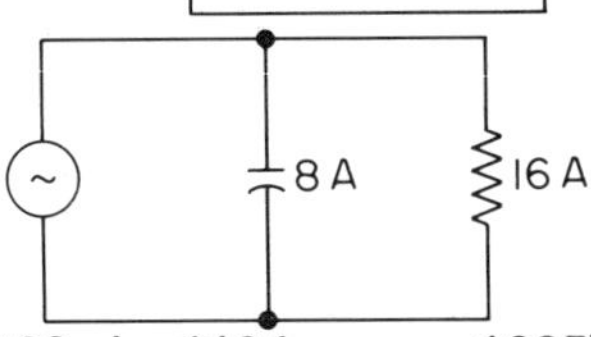

. (22) a. 4Ω. b. 8Ω. (23) a. 0.866. b. 1184v-a. c. 1025W. d. 3.46Ω.

chapter 11

(1) Growth. (2) Inductor. (3) Resistor. (4) Decay. (5) 63. (6) Five for growth and five for decay. (7) Inductance; resistance. (8) 10; alternation. (9) Short; inductor. (10) Resistor; long. (11) Charges; discharges. (12) Resistor. (13) Capacitor. (14) Exponential; 5. (15) Resistance; capacitance. (16) 63; 1. (17) Integrating. (18) Square. (19) 90; —90; 0. (20) 20; X_L. (21) 60; X_L. (22) Inductive; lags. (23) Capacitive; leads. (24) 90; —90. (25) 50. (26) 180; inductive. (27) Largest. (28) Capacitive; leads. (29) 8; 12. (30) Series circuit. (31) Series; parallel. (32) Equivalent. (33) Resistance; reactance.

chapter 12

(1) Frequency. (2) X_L. (3) $f_r = \frac{1}{2}\pi\sqrt{LC}$. (4) Capacitive; resistive; inductive. (5) Leads; phase; lags. (6) Minimum; increases. (7) 31,847 Hz. (8) 50kΩ. (9) Quality. (10) 0.707. (11) Band pass. (12) Band width. (13) Applied voltage; 45. (14) Lags; 45. (15) Circulation. (16) Maximum; X_C; minimum. (17) $Z_t = X_L^2/R_L$. (18) Inductive. (19) Capacitive. (20) Tank. (21) 45; low; 45; high. (22) Selection. (23) Rejection. (24) Frequencies; half-power. (25) Frequencies; half-power. (26) E_a; R_L. (27) X_L; R. (28) Delay.

appendix c
glossary

A. An abbreviation for ampere.

absolute value. A numerical quantity without regard to polarity, direction, or arithmetic sign.

ac. Abbreviation for alternating current.

alloy. A mixture of two or more metals.

alnico. An alloy formed from iron, aluminum, nickel, and cobalt.

alternating current. Electric current which changes amplitude constantly and direction periodically.

ammeter. An instrument for measuring current.

ampere. Unit of measure for current. One ampere = one coulomb past a point in one second.

ampere-turn. Unit of magnetomotive force equivalent to amperes times turns.

aquadag. A suspension of graphite in water used to coat the inside of cathode ray tubes.

armature. Moving part of a magnetic circuit.

atom. Smallest particle of matter that can be identified as an element.

audio frequency. Any frequency within the audible range which extends from approximately 20Hz to 20kHz.

average value. An average of all the instantaneous values in one alternation of a sine wave. Equivalent to 0.636 times peak value.

B+. The positive terminal of a power supply.

B—. The negative terminal of a power supply.

back EMF. A voltage generated in a coil due to a change in current. Always opposes the applied voltage.

band width. The frequencies between low and high half-power points of a tuned circuit.

battery. Two or more voltage cells connected together for a dc power source.

calibration. Method of adjusting an instrument to cause it to conform to a specified standard.

capacitance. The ability of a capacitor to store charge. Measured in farads and symbolized by *C*.

capacitive reactance. The opposition which a capacitor offers to ac. Measured in ohms and symbolized by X_C.

cathode ray tube. An electronic display tube which uses a ray of electrons to trace patterns on a screen.

charge. A displacement of electrons. A positive charge is a deficiency of electrons. A negative charge is a surplus of electrons.

coil. *See inductor.*

color code. A system of colors to indicate resistor values, capacitor characteristics, and circuit connections.

commutator. A segmented conductor used as a collector on the armature of a dc motor or generator.

complex number. A number which expresses both quantity and direction.

compound. A substance formed by chemically combining two or more elements.

conductor. A material which offers little resistance to current.

continuity. A continuous electrical connection between two points.

core. The material which forms the center of an inductor.

coulomb. A quantity of electrical charge equivalent to 6.24×10^{18} electrons.

current. Electron flow. Measured in amperes and symbolized by *I*.

D'Arsonval movement. A common meter movement using a movable coil in a fixed magnetic field.

dc. Abbreviation for direct current.

delay line. A circuit which produces a time interval between input and output.

dielectric. The material separating the plates of a capacitor.

dielectric constant. A ratio comparing the insulative quality of a material to that of air.

direct current. A current which maintains a constant direction.

discharge. The act of neutralizing a charge.

dry cell. A primary voltage cell with an electrolyte of paste.

dual gun oscilloscope. An oscilloscope which features two electron beams and multiple displays.

dynamic electricity. Electricity in motion. A term usually applied to practical electricity which is capable of sustaining a useful current.

dynamo. A large capacity generator.

E. Symbol for voltage.

effective value. The square root of the sum of the square of all the instantaneous values in a sine wave. Equivalent to 0.707 $\times$ peak value.

electricity. Atomic energy capable of producing a flow of electrons.

electrodynamometer. A meter movement characterized by three coils: two fixed and one movable.

electrolyte. The chemical ingredient between the plates of a voltage cell.

electromagnet. A magnet created by current through the loops of an inductor.

electromotive force. A difference in potential between two points. Also, voltage and electrical pressure.

electron. The negative particles of energy orbiting the nucleus of an atom.

electron gun. A device for emitting and controlling a beam of electrons.

electroscope. An instrument which indicates the presence of small charges.

electrostatics. The science of static electricity (electricity at rest).

electrostatic charge. A difference in potential caused by a displacement of electrons.

electrostatic field. The area around an electrostatic charge where its influence can be felt.

element. A primary substance which cannot be broken down by chemical means.

energy. The ability to do work. May be converted into different forms but cannot be created nor destroyed.

epsilon. (Greek letter) ϵ. $\epsilon = 2.71828\ldots =$ base of natural logarithms.

F. Symbol for force or farad.

f. Symbol for frequency.

farad. The unit of capacitance. Equivalent to one coulomb per volt.
ferromagnetic. A substance which is strongly attracted to magnets. Characterized by high permeability.
filter. A circuit which either shorts or attenuates specific ac signals.
flux. The collective lines of force in a region.
flywheel effect. The oscillations of a resonant circuit.
free electron. An electron which is not attached to any atom.
frequency. A measure of the number of periods of ac per unit time. Usually expressed in hertz (Hz) which means periods per second.
fuse. A device which self destructs in order to protect other circuit components from an excessive current.

gauss. Unit of flux density. One gauss is equivalent to one maxwell per square cm.
galvanometer. A sensitive dc meter which registers current in either direction.
generator. An electromagnetic machine which converts mechanical energy to electrical energy.
gilbert. Unit of magnetomotive force. Equivalent to flux times reluctance.
ground. A point of reference and/or a common return path for current.

H. Symbol for henry.
henry. Unit of inductance. The amount of inductance which produces a back EMF of one volt for a current change of one ampere per second.
hertz. Unit of frequency meaning periods per second. Symbolized by Hz.
horsepower. A unit of electrical power equivalent to 746W or mechanical power of 550 foot-pounds per second.
hysteresis. A lag between flux density and magnetizing force.
hysteresis loop. A chart of the hysteresis of a material.
Hz. Symbol for hertz.

I. Symbol for current.
impedance. A combination of reactance and resistance. Symbolized by *Z* and measured in ohms.
inductance. The property of a coil which opposes any change of current in the windings. Measured in henries and symbolized by *L*.

inductive reactance. The opposition which a coil offers to ac. Measured in ohms and symbolized by X_L.
inductor. Several turns of wire wound on a form.
instantaneous value. The amplitude of a sine wave at any selected instant.
insulation. A material characterized by very few free electrons and high resistance. Used to confine current to the desired path.
ion. A charged atom.

j. Symbol for $\sqrt{-1}$. Used to indicate a 90° vectorial rotation.
jack. A plug-in spring terminal. Widely used to facilitate connections between internal and external circuits.
junction. A point in a circuit where two or more conductors are connected.

k. Symbol for kilo.
kilo. A prefix meaning 1000.

L. Symbol for inductance.
l. Symbol for length.
lambda. Greek letter λ. Symbolizes wave length.
lead. (Pronounced leed.) A connecting conductor.
lead acid cell. A voltage cell composed of lead plates with a diluted sulfuric acid electrolyte.
lissajous figures. Patterns on an oscilloscope showing the phase and frequency relation of two ac signals.
lodestone. A magnetite iron ore. Chiefly Fe_3O_4.
loss. Dissipation of energy without accomplishing work.

M. Symbol for mega.
m. Symbol for milli.
magnet. A substance which exhibits an attraction for iron and steel.
magnetic circuit. A magnet with a closed path for flux.
magnetic field. The area about a magnet where its influence can be felt.
magnetomotive force. Magnetic potential difference. Measured in gilberts and symbolized by mmf.
mega. Prefix meaning 1,000,000.
maxwell. One line of magnetic force.
mho. The unit of conductance. Symbolized by ℧.
micro. Prefix meaning 1/1,000,000.
milli. A prefix meaning 1/1000 or 10^{-3}.

molecule. The smallest particle of matter which contains all the characteristics of the original substance.

motor. An electromechanical machine which converts electrical energy to mechanical energy.

mu. Greek letter μ. Symbolizes micro or permeability.

multiplier. A resistor in series with the coil of a voltmeter to increase the range.

mutual inductance. The flux linkage in one coil resulting from a current in another coil.

n. Symbol for nano.

nano. Prefix meaning 10^{-9}.

north pole. The pole of a magnet which seeks the magnetic north pole of the earth.

nucleus. The center of an atom. Consists of protons, neutrons, and other subatomic particles.

oersted. Unit of field strength. One oersted is equivalent to one dyne of force on a unit pole.

ohm. Unit of resistance. Symbolized by Ω.

omega. Greek letter Ω, symbolizes ohms. ω symbolizes angular velocity.

open. A break in the current path.

oscilloscope. An instrument which displays patterns on the screen of a cathode ray tube for the purpose of testing and analyzing.

P. Symbol for power.

p. Symbol for pico.

peak value. The maximum amplitude of a sine wave.

period. The time for one complete set of events in a periodic function. Two consecutive alternations of a sine wave.

permeability. The ease with which a material passes magnetic flux. Symbolized by μ.

phase. The position of a wave shape at any given instant during its period.

phase relation. Comparison of the phase of two signals at a given instant.

phi. Greek letter Φ. Symbolizes magnetic flux and a complementary angle.

phosphor. A substance which emits light when bombarded by electrons.

photoelectric cell. A device which converts light energy to electrical energy.

pi. Greek letter π. Equivalent to 3.1416 . . . which is the ratio of a circle's circumference to its diameter.

pico. Prefix for 10^{-12}. (Also known as micromicro.)

piezoelectric. Pressure electricity. The property of a crystal which causes it to produce a voltage when placed under mechanical stress.

polarity. Direction with respect to an axis.

pole. An end of a magnet.

pole piece. A piece of high permeability material used to reduce the air gap between magnetic poles.

potentiometer. A variable resistor designed for voltage control. Characterized by three contacts; two fixed and one variable.

power. Time rate of doing work. Expressed in watts. Symbolized by *P*.

pulsating dc. A current with a constant direction and a varying amplitude.

pulse. A sharp surge of current.

Q. Symbol for charge. Also for quality of a resonant circuit.

R. Symbol for resistance.

r. Symbol for radius.

radian. The portion of a circle represented by an arc of the radius along the circumference. Equivalent to approximately 57.3°.

reactance. Opposition to a change in current. Symbolized by *X* and measured in ohms.

rectifier. A device for converting ac to dc.

relay. An electromagnetic switch.

reluctance. Opposition to magnetic flux. Measured in rels. Symbolized by $\mathcal{R}$.

residual magnetism. Magnetism which remains in a material after the magnetizing force is removed.

resistance. Opposition to current. Measured in ohms and symbolized by *R*.

resistor. A circuit component designed to limit the current.

resonance. An ac circuit condition where $X_L = X_C$.

resonant frequency. The frequency which produces resonance.

retentivity. The ability of a material to hold magnetism.

rheostat. A variable resistor designed for current control. Characterized by two contacts; one variable and one fixed.

sawtooth. A triangular wave shape.

short. A low resistance path through or around a circuit component.

shunt resistor. A precision resistor parallel to the coil of an ammeter.
signal. A wave form which serves a particular purpose.
sine wave. A time amplitude graph showing the sine function of all angles from 0° to 360°.
slip ring. A collector ring on the armature of an ac motor or generator.
south pole. The pole of a magnet which seeks the south magnetic pole of the earth.
square wave. A wave which switches abruptly back and forth between two stable levels of voltage.
surge. A sudden variation in current or voltage. A pulse.
sweep voltage. A sawtooth wave used to generate the time base on an oscilloscope.

T. Symbol for tera.
tap. A contact in the body of a resistor or inductor.
taut band. A meter movement with moving parts suspended on a steel band.
tera. A prefix meaning 10^{12}.
thermal agitation. Random movement activated by heat.
thermocouple. A device which converts heat energy to electrical energy. Constructed by joining two dissimilar metals. Activated by heating the junction.
tolerance. Permissible variation from a specified value.
trigger. A sharp pulse signal.
true power. The power used by the resistive component of a circuit. Equivalent to $EI \times$ power factor.

unit pole. A standard unit of magnetic pole strength. A force of one dyne is exerted between two unit poles separated by one cm of air.

vector. A quantity having both magnitude and direction.
volt. Unit of electrical pressure.
voltage. Difference in potential between two points. Measured in volts and symbolized by *E*.
voltage divider. A tapped resistor or a multiresistor circuit used to divide the applied voltage among the various loads.
volt-ampere. Unit of apparent power. *EI* without regard to phase.

W. An abbreviation for watt.
watt. Unit of power.

watt-hour. One watt used continuously for one hour.
wave. A periodic variation in voltage or current.
wave form (waveshape). A graph of a wave.
wave length. The length of one period of a wave in meters. Symbolized by λ. $\lambda = 300 \times 10^6/f$.

Z. Symbol for impedance.
zero potential. The reference level for a wave shape. Usually ground.

appendix d
table of trigonometric functions

Natural Trigonometric Functions

deg	function	0.0°	0.1°	0.2°	0.3°	0.4°	0.5°	0.6°	0.7°	0.8°	0.9°
0	sin	0.0000	0.0017	0.0035	0.0052	0.0070	0.0087	0.0105	0.0122	0.0140	0.0157
	cos	1.0000	1.0000	1.0000	1.0000	1.0000	1.0000	0.9999	0.9999	0.9999	0.9999
	tan	0.0000	0.0017	0.0035	0.0052	0.0070	0.0087	0.0105	0.0122	0.0140	0.0157
1	sin	0.0175	0.0192	0.0209	0.0227	0.0244	0.0262	0.0279	0.0297	0.0314	0.0332
	cos	0.9998	0.9998	0.9998	0.9997	0.9997	0.9997	0.9996	0.9996	0.9995	0.9995
	tan	0.0175	0.0192	0.0209	0.0227	0.0244	0.0262	0.0279	0.0297	0.0314	0.0332
2	sin	0.0349	0.0366	0.0384	0.0401	0.0419	0.0436	0.0454	0.0471	0.0488	0.0506
	cos	0.9994	0.9993	0.9993	0.9992	0.9991	0.9990	0.9990	0.9989	0.9988	0.9987
	tan	0.0349	0.0367	0.0384	0.0402	0.0419	0.0437	0.0454	0.0472	0.0489	0.0507
3	sin	0.0523	0.0541	0.0558	0.0576	0.0593	0.0610	0.0628	0.0645	0.0663	0.0680
	cos	0.9986	0.9985	0.9984	0.9983	0.9982	0.9981	0.9980	0.9979	0.9978	0.9977
	tan	0.0524	0.0542	0.0559	0.0577	0.0594	0.0612	0.0629	0.0647	0.0664	0.0682
4	sin	0.0698	0.0715	0.0732	0.0750	0.0767	0.0785	0.0802	0.0819	0.0837	0.0854
	cos	0.9976	0.9974	0.9973	0.9972	0.9971	0.9969	0.9968	0.9966	0.9965	0.9963
	tan	0.0699	0.0717	0.0734	0.0752	0.0769	0.0787	0.0805	0.0822	0.0840	0.0857
5	sin	0.0872	0.0889	0.0906	0.0924	0.0941	0.0958	0.0976	0.0993	0.1011	0.1028
	cos	0.9962	0.9960	0.9959	0.9957	0.9956	0.9954	0.9952	0.9951	0.9949	0.9947
	tan	0.0875	0.0892	0.0910	0.0928	0.0945	0.0963	0.0981	0.0998	0.1016	0.1033
6	sin	0.1045	0.1063	0.1080	0.1097	0.1115	0.1132	0.1149	0.1167	0.1184	0.1201
	cos	0.9945	0.9943	0.9942	0.9940	0.9938	0.9936	0.9934	0.9932	0.9930	0.9928
	tan	0.1051	0.1069	0.1086	0.1104	0.1122	0.1139	0.1157	0.1175	0.1192	0.1210
7	sin	0.1219	0.1236	0.1253	0.1271	0.1288	0.1305	0.1323	0.1340	0.1357	0.1374
	cos	0.9925	0.9923	0.9921	0.9919	0.9917	0.9914	0.9912	0.9910	0.9907	0.9905
	tan	0.1228	0.1246	0.1263	0.1281	0.1299	0.1317	0.1334	0.1352	0.1370	0.1388
8	sin	0.1392	0.1409	0.1426	0.1444	0.1461	0.1478	0.1495	0.1513	0.1530	0.1547
	cos	0.9903	0.9900	0.9898	0.9895	0.9893	0.9890	0.9888	0.9885	0.9882	0.9880
	tan	0.1405	0.1423	0.1441	0.1459	0.1477	0.1495	0.1512	0.1530	0.1548	0.1566
9	sin	0.1564	0.1582	0.1599	0.1616	0.1633	0.1650	0.1668	0.1685	0.1702	0.1719
	cos	0.9877	0.9874	0.9871	0.9869	0.9866	0.9863	0.9860	0.9857	0.9854	0.9851
	tan	0.1584	0.1602	0.1620	0.1638	0.1655	0.1673	0.1691	0.1709	0.1727	0.1745
10	sin	0.1736	0.1754	0.1771	0.1788	0.1805	0.1822	0.1840	0.1857	0.1874	0.1891
	cos	0.9848	0.9845	0.9842	0.9839	0.9836	0.9833	0.9829	0.9826	0.9823	0.9820
	tan	0.1763	0.1781	0.1799	0.1817	0.1835	0.1853	0.1871	0.1890	0.1908	0.1926
deg	function	0′	6′	12′	18′	24′	30′	36′	42′	48′	54′

Natural Trigonometric Functions (cont.)

deg	function	0.0°	0.1°	0.2°	0.3°	0.4°	0.5°	0.6°	0.7°	0.8°	0.9°
11	sin	0.1908	0.1925	0.1942	0.1959	0.1977	0.1994	0.2011	0.2028	0.2045	0.2062
	cos	0.9816	0.9813	0.9810	0.9806	0.9803	0.9799	0.9796	0.9792	0.9789	0.9785
	tan	0.1944	0.1962	0.1980	0.1998	0.2016	0.2035	0.2053	0.2071	0.2089	0.2107
12	sin	0.2079	0.2096	0.2113	0.2130	0.2147	0.2164	0.2181	0.2198	0.2215	0.2232
	cos	0.9781	0.9778	0.9774	0.9770	0.9767	0.9763	0.9759	0.9755	0.9751	0.9748
	tan	0.2126	0.2144	0.2162	0.2180	0.2199	0.2217	0.2235	0.2254	0.2272	0.2290
13	sin	0.2250	0.2267	0.2284	0.2300	0.2318	0.2334	0.2351	0.2368	0.2385	0.2402
	cos	0.9744	0.9740	0.9736	0.9732	0.9728	0.9724	0.9720	0.9715	0.9711	0.9707
	tan	0.2309	0.2327	0.2345	0.2364	0.2382	0.2401	0.2419	0.2438	0.2456	0.2475
14	sin	0.2419	0.2436	0.2453	0.2470	0.2487	0.2504	0.2521	0.2538	0.2554	0.2571
	cos	0.9703	0.9699	0.9694	0.9690	0.9686	0.9681	0.9677	0.9673	0.9668	0.9664
	tan	0.2493	0.2512	0.2530	0.2549	0.2568	0.2586	0.2605	0.2623	0.2642	0.2661
15	sin	0.2588	0.2605	0.2622	0.2639	0.2656	0.2672	0.2689	0.2706	0.2723	0.2740
	cos	0.9569	0.9655	0.9650	0.9646	0.9641	0.9636	0.9632	0.9627	0.9622	0.9617
	tan	0.2679	0.2698	0.2717	0.2736	0.2754	0.2773	0.2792	0.2811	0.2830	0.2849
16	sin	0.2756	0.2773	0.2790	0.2807	0.2823	0.2840	0.2857	0.2874	0.2890	0.2907
	cos	0.9613	0.9608	0.9603	0.9598	0.9593	0.9588	0.9583	0.9578	0.9573	0.9568
	tan	0.2867	0.2886	0.2905	0.2924	0.2943	0.2962	0.2981	0.3000	0.3019	0.3038
17	sin	0.2924	0.2940	0.2957	0.2974	0.2990	0.3007	0.3024	0.3040	0.3057	0.3074
	cos	0.9563	0.9558	0.9553	0.9548	0.9542	0.9537	0.9532	0.9527	0.9521	0.9516
	tan	0.3057	0.3076	0.3096	0.3115	0.3134	0.3153	0.3172	0.3191	0.3211	0.3230
18	sin	0.3090	0.3107	0.3123	0.3140	0.3156	0.3173	0.3190	0.3206	0.3223	0.3239
	cos	0.9511	0.9505	0.9500	0.9494	0.9489	0.9483	0.9478	0.9472	0.9466	0.9461
	tan	0.3249	0.3269	0.3288	0.3307	0.3327	0.3346	0.3365	0.3385	0.3404	0.3424
19	sin	0.3256	0.3272	0.3289	0.3305	0.3322	0.3338	0.3355	0.3371	0.3387	0.3404
	cos	0.9455	0.9449	0.9444	0.9438	0.9432	0.9426	0.9421	0.9415	0.9409	0.9403
	tan	0.3443	0.3463	0.3482	0.3502	0.3522	0.3541	0.3561	0.3581	0.3600	0.3620
20	sin	0.3420	0.3437	0.3453	0.3469	0.3486	0.3502	0.3518	0.3535	0.3551	0.3567
	cos	0.9397	0.9391	0.9385	0.9379	0.9373	0.9367	0.9361	0.9354	0.9348	0.9342
	tan	0.3640	0.3659	0.3679	0.3699	0.3719	0.3739	0.3759	0.3779	0.3799	0.3819
deg	function	0′	6′	12′	18′	24′	30′	36′	42′	48′	54′

Natural Trigonometric Functions (cont.)

deg	function	0.0°	0.1°	0.2°	0.3°	0.4°	0.5°	0.6°	0.7°	0.8°	0.9°
21	sin	0.3584	0.3600	0.3616	0.3633	0.3649	0.3665	0.3681	0.3697	0.3714	0.3730
	cos	0.9336	0.9330	0.9323	0.9317	0.9311	0.9304	0.9298	0.9291	0.9285	0.9278
	tan	0.3839	0.3859	0.3879	0.3899	0.3919	0.3939	0.3959	0.3979	0.4000	0.4020
22	sin	0.3746	0.3762	0.3778	0.3795	0.3811	0.3827	0.3843	0.3859	0.3875	0.3891
	cos	0.9272	0.9265	0.9259	0.9252	0.9245	0.9239	0.9232	0.9225	0.9219	0.9212
	tan	0.4040	0.4061	0.4081	0.4101	0.4122	0.4142	0.4163	0.4183	0.4204	0.4224
23	sin	0.3907	0.3923	0.3939	0.3955	0.3971	0.3987	0.4003	0.4019	0.4035	0.4051
	cos	0.9205	0.9198	0.9191	0.9184	0.9178	0.9171	0.9164	0.9157	0.9150	0.9143
	tan	0.4245	0.4265	0.4286	0.4307	0.4327	0.4348	0.4369	0.4390	0.4411	0.4431
24	sin	0.4067	0.4083	0.4099	0.4115	0.4131	0.4147	0.4163	0.4179	0.4195	0.4210
	cos	0.9135	0.9128	0.9121	0.9114	0.9107	0.9100	0.9092	0.9085	0.9078	0.9070
	tan	0.4452	0.4473	0.4494	0.4515	0.4536	0.4557	0.4578	0.4599	0.4621	0.4642
25	sin	0.4226	0.4242	0.4258	0.4274	0.4289	0.4305	0.4321	0.4337	0.4352	0.4368
	cos	0.9063	0.9056	0.9048	0.9041	0.9033	0.9026	0.9018	0.9011	0.9003	0.8996
	tan	0.4663	0.4684	0.4706	0.4727	0.4748	0.4770	0.4791	0.4813	0.4834	0.4856
26	sin	0.4384	0.4399	0.4415	0.4431	0.4446	0.4462	0.4478	0.4493	0.4509	0.4524
	cos	0.8988	0.8980	0.8973	0.8965	0.8957	0.8949	0.8942	0.8934	0.8926	0.8918
	tan	0.4877	0.4899	0.4921	0.4942	0.4964	0.4986	0.5008	0.5029	0.5051	0.5073
27	sin	0.4540	0.4555	0.4571	0.4586	0.4602	0.4617	0.4633	0.4648	0.4664	0.4679
	cos	0.8910	0.8902	0.8894	0.8886	0.8878	0.8870	0.8862	0.8854	0.8846	0.8838
	tan	0.5095	0.5117	0.5139	0.5161	0.5184	0.5206	0.5228	0.5250	0.5272	0.5295
28	sin	0.4695	0.4710	0.4726	0.4741	0.4756	0.4772	0.4787	0.4802	0.4818	0.4833
	cos	0.8829	0.8821	0.8813	0.8805	0.8796	0.8788	0.8780	0.8771	0.8763	0.8755
	tan	0.5317	0.5340	0.5362	0.5384	0.5407	0.5430	0.5452	0.5475	0.5498	0.5520
29	sin	0.4848	0.4863	0.4879	0.4894	0.4909	0.4924	0.4939	0.4955	0.4970	0.4985
	cos	0.8746	0.8738	0.8729	0.8721	0.8712	0.8704	0.8695	0.8636	0.8678	0.8669
	tan	0.5543	0.5566	0.5589	0.5612	0.5635	0.5658	0.5681	0.5704	0.5727	0.5750
30	sin	0.5000	0.5015	0.5030	0.5045	0.5060	0.5075	0.5090	0.5105	0.5120	0.5135
	cos	0.8660	0.8652	0.8643	0.8634	0.8625	0.8616	0.8607	0.8599	0.8590	0.8581
	tan	0.5774	0.5797	0.5820	0.5844	0.5867	0.5890	0.5914	0.5938	0.5961	0.5985
deg	function	0′	6′	12′	18′	24′	30′	36′	42′	48′	54′

Natural Trigonometric Functions (cont.)

deg	function	0.0°	0.1°	0.2°	0.3°	0.4°	0.5°	0.6°	0.7°	0.8°	0.9°
31	sin	0.5150	0.5165	0.5180	0.5195	0.5210	0.5225	0.5240	0.5255	0.5270	0.5284
	cos	0.8572	0.8563	0.8554	0.8545	0.8536	0.8526	0.8517	0.8508	0.8499	0.8490
	tan	0.6009	0.6032	0.6056	0.6080	0.6104	0.6128	0.6152	0.6176	0.6200	0.6224
32	sin	0.5299	0.5314	0.5329	0.5344	0.5358	0.5373	0.5388	0.5402	0.5417	0.5432
	cos	0.8480	0.8471	0.8462	0.8453	0.8443	0.8434	0.8425	0.8415	0.8406	0.8396
	tan	0.6249	0.6273	0.6297	0.6322	0.6346	0.6371	0.6395	0.6420	0.6445	0.6469
33	sin	0.5446	0.5461	0.5476	0.5490	0.5505	0.5519	0.5534	0.5548	0.5563	0.5577
	cos	0.8387	0.8377	0.8368	0.8358	0.8348	0.8339	0.8329	0.8320	0.8310	0.8300
	tan	0.6494	0.6519	0.6544	0.6569	0.6594	0.6619	0.6644	0.6669	0.6694	0.6720
34	sin	0.5592	0.5606	0.5621	0.5635	0.5650	0.5664	0.5678	0.5693	0.5707	0.5721
	cos	0.8290	0.8281	0.8271	0.8261	0.8251	0.8241	0.8231	0.8221	0.8211	0.8202
	tan	0.6745	0.6771	0.6796	0.6822	0.6847	0.6873	0.6899	0.6924	0.6950	0.6976
35	sin	0.5736	0.5750	0.5764	0.5779	0.5793	0.5807	0.5821	0.5835	0.5850	0.5864
	cos	0.8192	0.8181	0.8171	0.8161	0.8151	0.8141	0.8131	0.8121	0.8111	0.8100
	tan	0.7002	0.7028	0.7054	0.7080	0.7107	0.7133	0.7159	0.7186	0.7212	0.7239
36	sin	0.5878	0.5892	0.5906	0.5920	0.5934	0.5948	0.5962	0.5976	0.5990	0.6004
	cos	0.8090	0.8080	0.8070	0.8059	0.8049	0.8039	0.8028	0.8018	0.8007	0.7997
	tan	0.7265	0.7292	0.7319	0.7346	0.7373	0.7400	0.7427	0.7454	0.7481	0.7508
37	sin	0.6018	0.6032	0.6046	0.6060	0.6074	0.6088	0.6101	0.6115	0.6129	0.6143
	cos	0.7986	0.7976	0.7965	0.7955	0.7944	0.7934	0.7923	0.7912	0.7902	0.7891
	tan	0.7536	0.7563	0.7590	0.7618	0.7646	0.7673	0.7701	0.7729	0.7757	0.7785
38	sin	0.6157	0.6170	0.6184	0.6198	0.6211	0.6225	0.6239	0.6252	0.6266	0.6280
	cos	0.7880	0.7869	0.7859	0.7848	0.7837	0.7826	0.7815	0.7804	0.7793	0.7782
	tan	0.7813	0.7841	0.7869	0.7898	0.7926	0.7954	0.7983	0.8012	0.8040	0.8069
39	sin	0.6293	0.6307	0.6320	0.6334	0.6347	0.6361	0.6374	0.6388	0.6401	0.6414
	cos	0.7771	0.7760	0.7749	0.7738	0.7727	0.7716	0.7705	0.7694	0.7683	0.7672
	tan	0.8098	0.8127	0.8156	0.8185	0.8214	0.8243	0.8273	0.8302	0.8332	0.8361
40	sin	0.6428	0.6441	0.6455	0.6468	0.6481	0.6494	0.6508	0.6521	0.6534	0.6547
	cos	0.7660	0.7649	0.7638	0.7627	0.7615	0.7604	0.7593	0.7581	0.7570	0.7559
	tan	0.8391	0.8421	0.8451	0.8481	0.8511	0.8541	0.8571	0.8601	0.8632	0.8662
deg	function	0′	6′	12′	18′	24′	30′	36′	42′	48′	54′

Natural Trigonometric Functions (cont.)

deg	function	0.0°	0.1°	0.2°	0.3°	0.4°	0.5°	0.6°	0.7°	0.8°	0.9°
41	sin	0.6561	0.6574	0.6587	0.6600	0.6613	0.6626	0.6639	0.6652	0.6665	0.6678
	cos	0.7547	0.7536	0.7524	0.7513	0.7501	0.7490	0.7478	0.7466	0.7455	0.7443
	tan	0.8693	0.8724	0.8754	0.8785	0.8816	0.8847	0.8878	0.8910	0.8941	0.8972
42	sin	0.6691	0.6704	0.6717	0.6730	0.6743	0.6756	0.6769	0.6782	0.6794	0.6807
	cos	0.7431	0.7420	0.7408	0.7396	0.7385	0.7373	0.7361	0.7349	0.7337	0.7325
	tan	0.9004	0.9036	0.9067	0.9099	0.9131	0.9163	0.9195	0.9228	0.9260	0.9293
43	sin	0.6820	0.6833	0.6845	0.6858	0.6871	0.6884	0.6896	0.6909	0.6921	0.6934
	cos	0.7314	0.7302	0.7290	0.7278	0.7266	0.7254	0.7242	0.7230	0.7218	0.7206
	tan	0.9325	0.9358	0.9391	0.9424	0.9457	0.9490	0.9523	0.9556	0.9590	0.9623
44	sin	0.6947	0.6959	0.6972	0.6984	0.6997	0.7009	0.7022	0.7034	0.7046	0.7059
	cos	0.7193	0.7181	0.7169	0.7157	0.7145	0.7133	0.7120	0.7108	0.7096	0.7083
	tan	0.9657	0.9691	0.9725	0.9759	0.9793	0.9827	0.9861	0.9896	0.9930	0.9965
45	sin	0.7071	0.7083	0.7096	0.7108	0.7120	0.7133	0.7145	0.7157	0.7169	0.7181
	cos	0.7071	0.7059	0.7046	0.7034	0.7022	0.7009	0.6997	0.6984	0.6972	0.6959
	tan	1.0000	1.0035	1.0070	1.0105	1.0141	1.0176	1.0212	1.0247	1.0283	1.0319
46	sin	0.7193	0.7206	0.7218	0.7230	0.7242	0.7254	0.7266	0.7278	0.7290	0.7302
	cos	0.6947	0.6934	0.6921	0.6909	0.6896	0.6884	0.6871	0.6858	0.6845	0.6833
	tan	1.0355	1.0392	1.0428	1.0464	1.0501	1.0538	1.0575	1.0612	1.0649	1.0686
47	sin	0.7314	0.7325	0.7337	0.7349	0.7361	0.7373	0.7385	0.7396	0.7408	0.7420
	cos	0.6820	0.6807	0.6794	0.6782	0.6769	0.6756	0.6743	0.6730	0.6717	0.6704
	tan	1.0724	1.0761	1.0799	1.0837	1.0875	1.0913	1.0951	1.0990	1.1028	1.1067
48	sin	0.7431	0.7443	0.7455	0.7466	0.7478	0.7490	0.7501	0.7513	0.7524	0.7536
	cos	0.6691	0.6678	0.6665	0.6652	0.6639	0.6626	0.6613	0.6600	0.6587	0.6574
	tan	1.1106	1.1145	1.1184	1.1224	1.1263	1.1303	1.1343	1.1383	1.1423	1.1463
49	sin	0.7547	0.7559	0.7570	0.7581	0.7593	0.7604	0.7615	0.7627	0.7638	0.7649
	cos	0.6561	0.6547	0.6534	0.6521	0.6508	0.6494	0.6481	0.6468	0.6455	0.6441
	tan	1.1504	1.1544	1.1585	1.1626	1.1667	1.1708	1.1750	1.1792	1.1833	1.1875
50	sin	0.7660	0.7672	0.7683	0.7694	0.7705	0.7716	0.7727	0.7738	0.7749	0.7760
	cos	0.6428	0.6414	0.6401	0.6388	0.6374	0.6361	0.6347	0.6334	0.6320	0.6307
	tan	1.1918	1.1960	1.2002	1.2045	1.2088	1.2131	1.2174	1.2218	1.2261	1.2305
deg	function	0′	6′	12′	18′	24′	30′	36′	42′	48′	54′

Natural Trigonometric Functions (cont.)

deg	function	0.0°	0.1°	0.2°	0.3°	0.4°	0.5°	0.6°	0.7°	0.8°	0.9°
51	sin	0.7771	0.7782	0.7793	0.7804	0.7815	0.7826	0.7837	0.7848	0.7859	0.7869
	cos	0.6293	0.6280	0.6266	0.6252	0.6239	0.6225	0.6211	0.6198	0.6184	0.6170
	tan	1.2349	1.2393	1.2437	1.2482	1.2527	1.2572	1.2617	1.2662	1.2708	1.2753
52	sin	0.7880	0.7891	0.7902	0.7912	0.7923	0.7934	0.7944	0.7955	0.7965	0.7976
	cos	0.6157	0.6143	0.6129	0.6115	0.6101	0.6088	0.6074	0.6060	0.6046	0.6032
	tan	1.2799	1.2846	1.2892	1.2938	1.2985	1.3032	1.3079	1.3127	1.3175	1.3222
53	sin	0.7986	0.7997	0.8007	0.8018	0.8028	0.8039	0.8049	0.8059	0.8070	0.8080
	cos	0.6018	0.6004	0.5990	0.5976	0.5962	0.5948	0.5934	0.5920	0.5906	0.5892
	tan	1.3270	1.3319	1.3367	1.3416	1.3465	1.3514	1.3564	1.3613	1.3663	1.3713
54	sin	0.8090	0.8100	0.8111	0.8121	0.8131	0.8141	0.8151	0.8161	0.8171	0.8181
	cos	0.5878	0.5864	0.5850	0.5835	0.5821	0.5807	0.5793	0.5779	0.5764	0.5750
	tan	1.3764	1.3814	1.3865	1.3916	1.3968	1.4019	1.4071	1.4124	1.4176	1.4229
55	sin	0.8192	0.8202	0.8211	0.8221	0.8231	0.8241	0.8251	0.8261	0.8271	0.8281
	cos	0.5736	0.5721	0.5707	0.5693	0.5678	0.5664	0.5650	0.5635	0.5621	0.5606
	tan	1.4281	1.4335	1.4388	1.4442	1.4496	1.4550	1.4605	1.4659	1.4715	1.4770
56	sin	0.8290	0.8300	0.8310	0.8320	0.8329	0.8339	0.8348	0.8358	0.8368	0.8377
	cos	0.5592	0.5577	0.5563	0.5548	0.5534	0.5519	0.5505	0.5490	0.5476	0.5461
	tan	1.4826	1.4882	1.4938	1.4994	1.5051	1.5108	1.5166	1.5224	1.5282	1.5340
57	sin	0.8387	0.8396	0.8406	0.8415	0.8425	0.8434	0.8443	0.8453	0.8462	0.8471
	cos	0.5446	0.5432	0.5417	0.5402	0.5388	0.5373	0.5358	0.5344	0.5329	0.5314
	tan	1.5399	1.5458	1.5517	1.5577	1.5637	1.5697	1.5757	1.5818	1.5880	1.5941
58	sin	0.8480	0.8490	0.8499	0.8508	0.8517	0.8526	0.8536	0.8545	0.8554	0.8563
	cos	0.5299	0.5284	0.5270	0.5255	0.5240	0.5225	0.5210	0.5195	0.5180	0.5165
	tan	1.6003	1.6066	1.6128	1.6191	1.6255	1.6319	1.6383	1.6447	1.6512	1.6577
59	sin	0.8572	0.8581	0.8590	0.8599	0.8607	0.8616	0.8625	0.8634	0.8643	0.8652
	cos	0.5150	0.5135	0.5120	0.5105	0.5090	0.5075	0.5060	0.5045	0.5030	0.5015
	tan	1.6643	1.6709	1.6775	1.6842	1.6909	1.6977	1.7045	1.7113	1.7182	1.7251
60	sin	0.8660	0.8669	0.8678	0.8686	0.8695	0.8704	0.8712	0.8721	0.8729	0.8738
	cos	0.5000	0.4985	0.4970	0.4955	0.4939	0.4924	0.4909	0.4894	0.4879	0.4863
	tan	1.7321	1.7391	1.7461	1.7532	1.7603	1.7675	1.7747	1.7820	1.7893	1.7966
deg	function	0′	6′	12′	18′	24′	30′	36′	42′	48′	54′

Natural Trigonometric Functions (cont.)

deg	function	0.0°	0.1°	0.2°	0.3°	0.4°	0.5°	0.6°	0.7°	0.8°	0.9°
61	sin	0.8746	0.8755	0.8763	0.8771	0.8780	0.8788	0.8796	0.8805	0.8813	0.8821
	cos	0.4848	0.4833	0.4818	0.4802	0.4787	0.4772	0.4756	0.4741	0.4726	0.4710
	tan	1.8040	1.8115	1.8190	1.8265	1.8341	1.8418	1.8495	1.8572	1.8650	1.8728
62	sin	0.8829	0.8838	0.8846	0.8854	0.8862	0.8870	0.8878	0.8886	0.8894	0.8902
	cos	0.4695	0.4679	0.4664	0.4648	0.4633	0.4617	0.4602	0.4586	0.4571	0.4555
	tan	1.8807	1.8887	1.8967	1.9047	1.9128	1.9210	1.9292	1.9375	1.9458	1.9542
63	sin	0.8910	0.8918	0.8926	0.8934	0.8942	0.8949	0.8957	0.8965	0.8973	0.8980
	cos	0.4540	0.4524	0.4509	0.4493	0.4478	0.4462	0.4446	0.4431	0.4415	0.4399
	tan	1.9626	1.9711	1.9797	1.9883	1.9970	2.0057	2.0145	2.0233	2.0323	2.0413
64	sin	0.8988	0.8996	0.9003	0.9011	0.9018	0.9026	0.9033	0.9041	0.9048	0.9056
	cos	0.4384	0.4368	0.4352	0.4337	0.4321	0.4305	0.4289	0.4274	0.4258	0.4242
	tan	2.0503	2.0594	2.0686	2.0778	2.0872	2.0965	2.1060	2.1155	2.1251	2.1348
65	sin	0.9063	0.9070	0.9078	0.9085	0.9092	0.9100	0.9107	0.9114	0.9121	0.9128
	cos	0.4226	0.4210	0.4195	0.4179	0.4163	0.4147	0.4131	0.4115	0.4099	0.4083
	tan	2.1445	2.1543	2.1642	2.1742	2.1842	2.1943	2.2045	2.2148	2.2251	2.2355
66	sin	0.9135	0.9143	0.9150	0.9157	0.9164	0.9171	0.9178	0.9184	0.9191	0.9198
	cos	0.4067	0.4051	0.4035	0.4019	0.4003	0.3987	0.3971	0.3955	0.3939	0.3923
	tan	2.2460	2.2566	2.2673	2.2781	2.2889	2.2998	2.3109	2.3220	2.3332	2.3445
67	sin	0.9205	0.9212	0.9219	0.9225	0.9232	0.9239	0.9245	0.9252	0.9259	0.9265
	cos	0.3907	0.3891	0.3875	0.3859	0.3843	0.3827	0.3811	0.3795	0.3778	0.3762
	tan	2.3559	2.3673	2.3789	2.3906	2.4023	2.4142	2.4262	2.4383	2.4504	2.4627
68	sin	0.9272	0.9278	0.9285	0.9291	0.9298	0.9304	0.9311	0.9317	0.9323	0.9330
	cos	0.3746	0.3730	0.3714	0.3697	0.3681	0.3665	0.3649	0.3633	0.3616	0.3600
	tan	2.4751	2.4876	2.5002	2.5129	2.5257	2.5386	2.5517	2.5649	2.5782	2.5916
69	sin	0.9336	0.9342	0.9348	0.9354	0.9361	0.9367	0.9373	0.9379	0.9385	0.9391
	cos	0.3584	0.3567	0.3551	0.3535	0.3518	0.3502	0.3486	0.3469	0.3453	0.3437
	tan	2.6051	2.6187	2.6325	2.6464	2.6605	2.6746	2.6889	2.7034	2.7179	2.7326
70	sin	0.9397	0.9403	0.9409	0.9415	0.9421	0.9426	0.9432	0.9438	0.9444	0.9449
	cos	0.3420	0.3404	0.3387	0.3371	0.3355	0.3338	0.3322	0.3305	0.3289	0.3272
	tan	2.7475	2.7625	2.7776	2.7929	2.8083	2.8239	2.8397	2.8556	2.8716	2.8878
deg	function	0′	6′	12′	18′	24′	30′	36′	42′	48′	54′

Natural Trigonometric Functions (cont.)

deg	function	0.0°	0.1°	0.2°	0.3°	0.4°	0.5°	0.6°	0.7°	0.8°	0.9°
71	sin	0.9455	0.9461	0.9466	0.9472	0.9478	0.9483	0.9489	0.9494	0.9500	0.9505
	cos	0.3256	0.3239	0.3223	0.3206	0.3190	0.3173	0.3156	0.3140	0.3123	0.3107
	tan	2.9042	2.9206	2.9375	2.9544	2.9714	2.9887	3.0061	3.0237	3.0415	3.0595
72	sin	0.9511	0.9516	0.9521	0.9527	0.9532	0.9537	0.9542	0.9548	0.9553	0.9558
	cos	0.3090	0.3074	0.3057	0.3040	0.3024	0.3007	0.2990	0.2974	0.2957	0.2940
	tan	3.0777	3.0961	3.1146	3.1334	3.1524	3.1716	3.1910	3.2106	3.2305	3.2506
73	sin	0.9563	0.9568	0.9573	0.9578	0.9583	0.9588	0.9593	0.9598	0.9603	0.9608
	cos	0.2924	0.2907	0.2890	0.2874	0.2857	0.2840	0.2823	0.2807	0.2790	0.2773
	tan	3.2709	3.2914	3.3122	3.3332	3.3544	3.3759	3.3977	3.4197	3.4420	3.4646
74	sin	0.9613	0.9617	0.9622	0.9627	0.9632	0.9636	0.9641	0.9646	0.9650	0.9655
	cos	0.2756	0.2740	0.2723	0.2706	0.2689	0.2672	0.2656	0.2639	0.2622	0.2605
	tan	3.4874	3.5105	3.5339	3.5576	3.5816	3.6059	3.6305	3.6554	3.6806	3.7062
75	sin	0.9659	0.9664	0.9668	0.9673	0.9677	0.9681	0.9686	0.9690	0.9694	0.9699
	cos	0.2588	0.2571	0.2554	0.2538	0.2521	0.2504	0.2487	0.2470	0.2453	0.2436
	tan	3.7321	3.7583	3.7848	3.8118	3.8391	3.8667	3.8947	3.9232	3.9520	3.9812
76	sin	0.9703	0.9707	0.9711	0.9715	0.9720	0.9724	0.9728	0.9732	0.9736	0.9740
	cos	0.2419	0.2402	0.2385	0.2368	0.2351	0.2334	0.2317	0.2300	0.2284	0.2267
	tan	4.0108	4.0408	4.0713	4.1022	4.1335	4.1653	4.1976	4.2303	4.2635	4.2972
77	sin	0.9744	0.9743	0.9751	0.9755	0.9759	0.9763	0.9767	0.9770	0.9774	0.9778
	cos	0.2250	0.2232	0.2215	0.2198	0.2181	0.2164	0.2147	0.2130	0.2113	0.2096
	tan	4.3315	3.4662	4.4015	4.4374	4.4737	4.5107	4.5483	4.5864	4.6252	4.6646
78	sin	0.9781	0.9785	0.9789	0.9792	0.9796	0.9799	0.9803	0.9806	0.9810	0.9813
	cos	0.2079	0.2062	0.2045	0.2028	0.2011	0.1994	0.1977	0.1959	0.1942	0.1925
	tan	4.7046	4.7453	4.7867	4.8288	4.8716	4.9152	4.9594	5.0045	5.0504	5.0970
79	sin	0.9816	0.9820	0.9823	0.9826	0.9829	0.9833	0.9836	0.9839	0.9842	0.9845
	cos	0.1908	0.1891	0.1874	0.1857	0.1840	0.1822	0.1805	0.1788	0.1771	0.1754
	tan	5.1446	5.1929	4.2422	5.2924	5.3435	5.3955	5.4486	5.5026	5.5578	5.6140
80	sin	0.9848	0.9851	0.9854	0.9857	0.9860	0.9863	0.9866	0.9869	0.9871	0.9874
	cos	0.1736	0.1719	0.1702	0.1685	0.1668	0.1650	0.1633	0.1616	0.1599	0.1582
	tan	5.6713	5.7297	5.7894	5.8502	5.9124	5.9758	6.0405	6.1066	6.1742	6.2432
deg	function	0′	6′	12′	18′	24′	30′	36′	42′	48′	54′

Natural Trigonometric Functions (cont.)

deg	function	0.0°	0.1°	0.2°	0.3°	0.4°	0.5°	0.6°	0.7°	0.8°	0.9°
81	sin	0.9877	0.9880	0.9882	0.9885	0.9888	0.9890	0.9893	0.9895	0.9898	0.9900
	cos	0.1564	0.1547	0.1530	0.1513	0.1495	0.1478	0.1461	0.1444	0.1426	0.1409
	tan	6.3138	6.3859	6.4596	6.5350	6.6122	6.6912	6.7720	6.8548	6.9395	7.0264
82	sin	0.9903	0.9905	0.9907	0.9910	0.9912	0.9914	0.9917	0.9919	0.9921	0.9923
	cos	0.1392	0.1374	0.1357	0.1340	0.1323	0.1305	0.1288	0.1271	0.1253	0.1236
	tan	7.1154	7.2066	7.3002	7.3962	7.4947	7.5958	7.6996	7.8062	7.9158	8.0285
83	sin	0.9925	0.9928	0.9930	0.9932	0.9934	0.9936	0.9938	0.9940	0.9942	0.9943
	cos	0.1219	0.1201	0.1184	0.1167	0.1149	0.1132	0.1115	0.1097	0.1080	0.1063
	tan	8.1443	8.2636	8.3863	8.5126	8.6427	8.7769	8.9152	9.0579	9.2052	9.3572
84	sin	0.9945	0.9947	0.9949	0.9951	0.9952	0.9954	0.9956	0.9957	0.9959	0.9960
	cos	0.1045	0.1028	0.1011	0.0993	0.0976	0.0958	0.0941	0.0924	0.0906	0.0889
	tan	9.5144	9.6768	9.8448	10.02	10.20	10.39	10.58	10.78	10.99	11.20
85	sin	0.9962	0.9963	0.9965	0.9966	0.9968	0.9969	0.9971	0.9972	0.9973	0.9974
	cos	0.0872	0.0854	0.0837	0.0819	0.0802	0.0785	0.0767	0.0750	0.0732	0.0715
	tan	11.43	11.66	11.91	12.16	12.43	12.71	13.00	13.30	13.62	13.95
86	sin	0.9976	0.9977	0.9978	0.9979	0.9980	0.9981	0.9982	0.9983	0.9984	0.9985
	cos	0.0698	0.0680	0.0663	0.0645	0.0628	0.0610	0.0593	0.0576	0.0558	0.0541
	tan	14.30	14.67	15.06	15.46	15.89	16.35	16.83	17.34	17.89	18.46
87	sin	0.9986	0.9987	0.9938	0.9989	0.9990	0.9990	0.9991	0.9992	0.9993	0.9993
	cos	0.0523	0.0506	0.0488	0.0471	0.0454	0.0436	0.0419	0.0401	0.0384	0.0366
	tan	19.08	19.75	20.45	21.20	22.02	22.90	23.86	24.90	26.03	27.27
88	sin	0.9994	0.9995	0.9995	0.9996	0.9996	0.9997	0.9997	0.9997	0.9998	0.9998
	cos	0.0349	0.0332	0.0314	0.0297	0.0279	0.0262	0.0244	0.0227	0.0209	0.0192
	tan	28.64	30.14	31.82	33.69	35.80	38.19	40.92	44.07	47.74	52.08
89	sin	0.9998	0.9999	0.9999	0.9999	0.9999	1.000	1.000	1.000	1.000	1.000
	cos	0.0175	0.0157	0.0140	0.0122	0.0105	0.0087	0.0070	0.0052	0.0035	0.0017
	tan	57.29	63.66	71.62	81.85	95.49	114.6	143.2	191.0	286.5	573.0
deg	function	0′	6′	12′	18′	24′	30′	36′	42′	48′	54′

INDEX

O

P

R

W

Y